MEANINGS OF WATER IN EARLY MEDIEVAL ENGLAND

STUDIES IN THE EARLY MIDDLE AGES

VOLUME 47

General Editor
Tom Pickles, University of Chester

Editorial Board
Charles West, University of Sheffield
Caroline Goodson, University of Cambridge
Gabor Thomas, University of Reading
Catherine Clarke, University of Southampton
Tom O'Donnell, Fordham University

Previously published volumes in this series
are listed at the back of the book.

Meanings of Water in Early Medieval England

Edited by
CAROLYN TWOMEY AND
DANIEL ANLEZARK

BREPOLS

© 2021, Brepols Publishers n. v., Turnhout, Belgium.

All rights reserved. No part of this publication may be reproduced, stored in a retrieval system, or transmitted, in any form or by any means, electronic, mechanical, photocopying, recording, or otherwise without the prior permission of the publisher.

D/2021/0095/207
ISBN 978-2-503-58888-9
eISBN 978-2-503-58889-6
DOI 10.1484/M.SEM-EB.5.119727

ISSN 1377-8099
eISSN 2294-835X

Printed in the EU on acid-free paper.

Table of Contents

List of Illustrations   7

Acknowledgements   9

Abbreviations   11

Introduction: Worlds of Water
Carolyn TWOMEY and Daniel ANLEZARK   13

The Sacred Nature of Rivers, Wells, Springs, and Other Wetlands in Anglo-Saxon England
Della HOOKE   33

Rivers and Rituals: Baptism in the Early English Landscape
Carolyn TWOMEY   59

Swimming in Anglo-Saxon England
Simon TRAFFORD   85

Sounds of Salvation: Nautical Noise in Old English and Anglo-Latin Literature
Rebecca SHORES   109

The Sailors, the Sea Monster, and the Saviour: Depicting Jonah and the *Ketos* in Anglo-Saxon England
Elizabeth A. ALEXANDER   127

Pearls before Paradise: Considering the Material Associations of Heavenly Waters, Precious Stones, and Liminality in the Art of the Medieval West
Meg BOULTON   145

'Streams of Wholesome Learning': The Waters of Genesis in Early Anglo-Saxon Exegesis
John J. GALLAGHER   167

Aquas ab Aquis*: Aqueous Creation in *Andreas
Michael BINTLEY   191

Water, Wisdom, and Worldliness in the Anglo-Saxon Prose Lives of Guthlac
Helen APPLETON   211

Drawing Alfredian Waters: The Old English *Metrical Epilogue* to the *Pastoral Care*, Boethian *Metre* 20, and *Solomon and Saturn II*
Daniel ANLEZARK   241

***Modor is monigra mærra wihta*: Watering the World in Exeter Book Riddle 84**
Jill FREDERICK   267

Index   283

List of Illustrations

Della Hooke

Figure 1.1.	Lindisfarne and the Farne Islands, Northumberland. Island and coastal sites of monastic communities and early churches.	37
Figure 1.2.	North-west Wales, Anglesey and the Lleyn peninsula. Select island and coastal sites of monastic communities and early churches.	39
Figure 1.3.	Glastonbury monastic centre (also showing churches and areas liable to flood). The dotted line following the foot of the upland to the west of the tor represents the dug canal which has now produced evidence of an Anglo-Saxon date.	42
Figure 1.4.	Glastonbury Tor rising above the surrounding lowlands.	44
Figure 1.5.	River Avon minsters in Worcestershire with an inset photograph of Evesham Abbey today.	47
Figure 1.6.	Kempsey church beside the River Severn in Worcestershire. An early minster site on raised ground in the river meadows.	48
Figure 1.7a.	Withington holy well in a boundary clause (S 1556) contained in an eleventh-century manuscript.	50
Figure 1.7b.	Withington holy well today.	50
Figure 1.8.	The site of Abingdon Abbey, its early water channels and canal.	52

Carolyn Twomey

Figure 2.1.	'Baptism of Christ', Benedictional of St Æthelwold, London, British Library, Add MS 49598, fol. 25^r. AD 963–84.	60

Simon Trafford

Figure 3.1.	Detail of a swimming figure from Psalm 66 in the Bury Psalter. Vatican City, Biblioteca Apostolica Vaticana, MS Reg. lat. 12. fol. 71^v. Second quarter of the eleventh century.	98

Elizabeth A. Alexander

Figure 5.1.	'Jonah Lowered from the Boat', Antwerp, Museum Plantin-Moretus, MS M. 17. 4, fol. 9^v. Early ninth century.	133
Figure 5.2.	'Jonah Regurgitated', Antwerp, Museum Plantin-Moretus, MS M. 17. 4, fol. 10^r. Early ninth century.	134
Figure 5.3.	'Illustration of Psalm 104 (103)', London, British Library, MS Harley 603, fol. 51^v. Early eleventh century.	138

Meg Boulton

Figure 6.1.	Detail of the apse mosaic at San Vitale, Ravenna, consecrated 547.	147
Figure 6.2.	Oysters in Bestiary, Paris, Bibliothèque nationale de France, lat. 14429, fol. 117^v, twelfth century.	150
Figure 6.3.	Detail of the apse mosaic at Sant'Apollinare in Classe, consecrated 549.	153
Figure 6.4a.	The Easby Cross (N. Yorks), Victoria & Albert Museum, c. 800.	158
Figure 6.4b.	Detail of the Easby Cross (N. Yorks), Victoria & Albert Museum, c. 800.	159

Acknowledgements

This volume originated from the generous interdisciplinary collaboration of scholars pursuing the many intersecting identities of water in the early medieval world. The editors are grateful to the Institute of Historical Research of the School of Advanced Study at the University of London and the Boston College Institute for the Liberal Arts for facilitating the initial 2015 colloquium at Senate House that formed the genesis of this book. The editors wish to thank all speakers and participants at the IHR colloquium 'Water in Anglo-Saxon England' and at the ensuing conference sessions at the International Medieval Congress at the University of Leeds for their lively discussions on the topic of water. In particular, Caroline Goodson, Alan Thacker, Michael Bintley, Meg Boulton, and IHR staff members Vanessa Rockel and Gemma Dormer deserve special thanks for their support.

Many thanks are owed to each of the individual contributors to this volume for their hard work and dedication throughout the ebb and flow of the publication process. We especially would like to thank Thomas Pickles, the general editor of the Studies in the Early Middle Ages series, for his guidance and advice, Guy Carney and the editorial team at Brepols, and the anonymous reviewer for their careful consideration of the volume. The editors also wish to acknowledge St Lawrence University and the Australian Research Council for their support of this project.

Carolyn Twomey, Canton, NY, USA

Daniel Anlezark, Sydney, Australia

Abbreviations

ACMRS	Arizona Center for Medieval and Renaissance Studies
BAR	British Archaeological Reports
Bede, HE	Bede, *Bede's Ecclesiastical History of the English People*, ed. and trans. by Bertram Colgrave and R. A. B. Mynors (Oxford: Clarendon Press, 1969)
Bede, VCP	Bede, *Vita S. Cuthberti prosaica*, in *Two Lives of Saint Cuthbert: A Life by an Anonymous Monk of Lindisfarne and Bede's Prose Life*, ed. and trans. by Bertram Colgrave (Cambridge: Cambridge University Press, 1940), pp. 141–307
EETS	Early English Text Society
CCSL	Corpus Christianorum Series Latina
CSEL	Corpus Scriptorum Ecclesiasticorum Latinorum
DOE	*Dictionary of Old English*
MGH	Monumenta Germaniae Historica
VCA	*Vita S. Cuthberti auctore anonymo*, in *Two Lives of Saint Cuthbert: A Life by an Anonymous Monk of Lindisfarne and Bede's Prose Life*, ed. and trans. by Bertram Colgrave (Cambridge: Cambridge University Press, 1940), pp. 59–139

CAROLYN TWOMEY AND DANIEL ANLEZARK

Introduction: Worlds of Water

All water on earth in the early Middle Ages is for a greater part the same physical water that is on the earth now. Most of the earth's water moves through the hydrological cycle through rainfall, running off into rivers, lakes, seas, and oceans relatively quickly.[1] On average, soil retains moisture for one to two months before evaporation, or water passes along rivers for anywhere on average from two to six months. Water that falls as snow, where snow is seasonal, can stay as snow up to six months before it either enters the soil or flows into river systems. Water can evaporate at any moment in this cycle, and spend, on average, nine days in the planet's atmosphere. Water can also be locked in water tables for up to 10,000 years, and some of the frozen water on Antarctica has been there for 800,000 years. Reflecting on the water moving through the hydrological cycle, we might muse that the water molecules that were used, for example, to baptize King Edwin of Northumbria would now have passed through the cycle at least a few thousand times, on each occasion dispersing further and further, mixing in the earth's atmosphere and oceans. It is possible that some of these molecules were in your morning cup of coffee. But water is found not only on Earth. When and how our water got here is not exactly clear; it now covers 71 per cent of the earth's surface and is where life began. It has been discovered that water is everywhere in the universe, as a by-product of star formation: Mars used to have much more water than what remains in the planet's icy polar caps, the matter of comets contains water in high proportion, and water floats in gigantic vapour clouds billions of light years from Earth and our galaxy.[2]

---

1   The literature on the water cycle is vast and diffuse. This overview provided here is summarized from Bengtsson and others, *The Earth's Hydrological Cycle*. See also Green, *The Water Cycle*, and Bronstert and others, *Coupled Models for the Hydrological Cycle*.
2   See Dyches and Choa, 'The Solar System and Beyond Is Awash in Water'.

Carolyn Twomey • (ctwomey@stlawu.edu) is a Visiting Assistant Professor of European History at St Lawrence University.
Daniel Anlezark • (daniel.anlezark@sydney.edu.au) is the McCaughey Professor of Early English Literature and Language at the University of Sydney.

While the water of the early Middle Ages might have been materially the same as our water today, and many of its uses the same — human beings and other animals still need to drink water, plants still need water to grow — understandings of the physical element and its properties have changed. Early medieval science had no concept of the molecule, though the idea of atoms had been around for centuries. In early medieval England there was little interest in exploring this aspect of matter, but there was, among the educated elite, a clear understanding that water was one of the four elements that combined to make up the material universe. Water, like the other elements, was believed not to exist in a pure form anywhere in the world below the heavens, but it was theoretically held that each of the four existed in their ideal Platonic forms above. In the case of water, this explained water's property of rising upwards whenever it is blocked or dammed, seeking out its true home above. And as the Venerable Bede explains in his commentary *On Genesis*, water is also found in solid crystalline form in the celestial spheres.[3] Scientific ideas about water's properties were important in medieval medical theory, inherited from Classical Antiquity, which associated water with a phlegmatic temperament, as well as the brains and the lungs. The apparent general disinterest in scientific theory does not mean early medieval people lacked the associated practice, and the restoration of balanced humours underlies many of their medical remedies, with water — sacred or otherwise — as a key ingredient. There is no doubt now that about 60 per cent of the human body is made up of water, and so the brain is watery still. However, we have many more elements than four, and describe water as made up of two of them — hydrogen and oxygen. We can easily break water molecules into their constituent elements by electrolysis, a process unknown (and indeed unknowable) to the science of the Middle Ages.

This brief discussion of the science of water, then and now, serves to underline a central aspect of the many identities of water found in this volume. When we discuss water in the early Middle Ages, we are discussing an aspect of the material world which has objective realities and physical continuity, but which is also embedded culturally in differing understandings of what water is. Water in the early medieval landscape acted on those who drank, swam, blessed, and traversed it, and was an element imbued with natural and supernatural significances. Meaning encoded in water through human actions continued to form and reform the cultural and social experiences of its inhabitants, each generation stirring their own currents and undercurrents of meaning into their landscapes of water.[4] The essays in this collection explore, from the perspectives of literature, history, art history, archaeology, linguistics, and religious studies, some of the ways in which water was present to the English cultural imagination

---

3 Bede, *Libri quatuor in principium Genesis*, ed. by Jones, pp. 10–11. See also Ferrand, 'The Hydrologic Cycle in Bede's *De Natura Rerum*'.
4 For this approach to the study of the landscape, see Tilley, *A Phenomenology of Landscape* and Walsham, *The Reformation of the Landscape*.

in the early Middle Ages. The natural flexibility and fluidity of water allows the material to take on particular historical qualities, shapes, and signifiers while maintaining its universal identity as a single element in history.[5]

Physical water was obviously present to the senses of those living on the island of Britain. They saw rain falling from the sky and felt it on their skin as it wet their clothes. They moved goods and pilgrims across channels and seas in foul weather and fair. They dug wells to find clean drinking water, tasted clear water from springs, and, more precariously in an age unaware of microbial life, would have drunk water from streams and lakes. In winter, they observed snow blanketing the earth and in colder places would have seen ice forming on water. Various poems surviving in Latin and Old English as well as contemporary visual and material evidence marvel at the transformative power of water and give these natural observations and sensory experiences cultural expression. Such transformations demonstrated the majesty of a Creator who could imbue an earthly element with mysterious power. Water brought both life and death: it is essential for sustaining the life of the body and the agrarian community, yet in the same water human beings can easily drown in its abundance or perish in its absence. Aquatic creatures provided — and still provide — a valuable food source for human beings, but rivers can break their banks in flood, and seas can surge in stormy weather, bringing death and destruction. Most human beings have lived close to bodies of water, and the historical record of the early Middle Ages reveals that communities were far more vulnerable to destruction by water than is the case now, though the danger is ever present.[6] In this lived context it is perhaps unsurprising that Bede's Judgement Day poem idealizes human community with nature in a *locus amoenus* around a clear spring, but that a familiar image associated with the suddenness of the Last Judgement was the rising of the seas to obliterate the land. Water is fundamentally threatening to human beings even beyond the perils of too much and too little of it: metaphysically, water suggests a loss of self.[7] Watery places are often sites of anxiety where categories of order and chaos, individual and community, and sacred and secular distort and collapse, and it is within this tension that the early medieval identities of this book will be explored.[8]

The recent rise of medieval water studies has diversified the approaches with which scholars interpret the many manifestations of this essential element. The field of medieval environmental history has led many scholars

---

5 Strang, *The Meaning of Water*, pp. 49–50; Oestigaard, 'Water', p. 38; Hastrup, 'Water and the Configuration of Social Worlds', p. 61.
6 See, for example, the report of a catastrophic flood that 'covered nearly the whole of Frisia' in 839, reported in the Annals of St-Bertin; see *The Annals of St-Bertin*, trans. by Nelson, p. 42.
7 Strang, *Water*, pp. 58–59; Strang, *The Meaning of Water*, p. 72.
8 Strang, *Water*, pp. 60–61; Strang, 'Common Senses', p. 115. See also Klein, 'Navigating the Anglo-Saxon Seas'.

to consider the role of water in landscapes both literal and conceptual.[9] Water — along with other interdisciplinary topics in the history of the premodern environment, such as forests, cities, and agriculture — provides a new lens through which to examine the relationships between people and the natural world.[10] Identities of men and women in the past were inextricable from their landscapes; the study of any aspect of the environment is the study of its early medieval inhabitants. Scholars have pursued this relationship from the perspective of science and water technology, and investigated how human interactions with the built environment in the form of regulated dams, canals, fishponds, navigational aids, bridges, and ports managed water resources for the economic and agricultural benefit of their communities.[11] Such studies reveal that what scholars may have once idealized as natural, unspoiled wilderness were in fact carefully curated places. In more than physical ways, the human inhabitants of the medieval environment conceptually reworked their relationships to its watery features as essential parts of defining themselves.[12] Seas and oceans receive the most attention from the fields of literature and history as not only maritime regions of industry and trade, but also cultural zones of their own replete with watery agency.[13] Ecocritical scholars have gone further to challenge the actions and attitudes of human beings in past — and present — landscapes as harmful power imbalances of a humanity opposed, rather than equal, to nature.[14] Moving beyond the utilitarian identity of water as resource, ecocriticism decentres the human in order to think about — and with — the medium of water.[15] Other approaches examine how the watery environments of marshland, rivers, and springs provided monastic communities and chroniclers with powerful metaphors with which to negotiate regional

---

9 The recent historiography of the environmental turn in late antique and medieval history is helpfully summarized in Eisenberg and others, 'The Environmental History of the Late Antique West'; and Arnold, 'An Introduction to Medieval Environmental History'.
10 See Bintley and Shapland, *Trees and Timber in the Anglo-Saxon World*; Banham and Faith, *Anglo-Saxon Farms and Farming*; and, most recently on water, Hyer and Hooke, *Water and the Environment*.
11 Blair, *Waterways and Canal-Building in Medieval England*; Squatriti, *Working with Water in Medieval Europe*; Squatriti, *Water and Society in Early Medieval Italy*; Magnusson, *Water Technology in the Middle Ages*.
12 Howe and Wolfe, *Inventing Medieval Landscapes*; see also Arnold, 'An Introduction to Medieval Environmental History'.
13 Hyer and Hooke, *Water and the Environment*; Klein, Schipper, and Lewis-Simpson, *The Maritime World of the Anglo-Saxons*; and wider historical studies of oceans and seas such as Klein and Mackenthun, *Sea Changes*; Gillis, *The Human Shore*; and Armitage, Bashford, and Sivasundaram, *Oceanic Histories*. For the Mediterranean Sea specifically, Horden and Purcell, *The Corrupting Sea*; Harris, *Rethinking the Mediterranean*; and Abulafia, *The Great Sea*.
14 Estes, *Anglo-Saxon Literary Landscapes*; see also Siewers, *Strange Beauty*; and, for water in medieval literature in general, Classen, *Water in Medieval Literature*.
15 Cohen and Duckert, *Elemental Ecocriticism*; Chen, MacLeod, and Neimanis, *Thinking with Water*. See also Smith and Howes, 'Medieval Water Studies'.

and religious identities.[16] Water existed within local sacred landscapes, it occupied a conceptual world of miracles and a divinely ordered cosmos, yet water in its capacity for ecological devastation brought a clear tension between different understandings of water for both clerical authors and lay farmers and sailors.

Navigating the conflicting and often contradictory meanings of water is the heart of this book. This perceived tension between the numerous identities of water in early England prompted a 2015 colloquium at the Institute of Historical Research of the School of Advanced Study at the University of London and subsequent sessions organized at the International Medieval Congress at the University of Leeds. The project of publishing these essays originated from within these cross-disciplinary scholarly conversations and sought to bring a wide array of new and established scholars — separated by oceans in North America, Europe, and Australia — into dialogue with one another. Many essays in this volume build on existing approaches to combine the histories of religion, material culture, art, and literature with the study of water in the environment to speak to the cultural resonances of this essential element in the early medieval world. The essays concentrate their study on early England, though several authors draw from parallel and comparative sources from Rome, Ireland, Scandinavia, and elsewhere in the early medieval West. Eleven chapters proceed gradually from physical to conceptual waters in the landscape while building upon one another and allowing for crucial overlap, as each essay, to different degrees, considers literal bodies of water alongside wider meanings of water at work in the early medieval world view. The waters explored in these pages are vast: the book begins with the natural settings of rivers and seas for monastic community and ritual, where sailors shouted, heroes swam, and sea creatures descended; it proceeds to articulate how the divine waters of heaven were understood and depicted in medieval scientific and biblical traditions; it continues to interpret water as intellectual metaphors for wisdom and creation; and it concludes with what water says about itself in one particular Old English riddle.

Water in this volume is many things yet one thing — always in motion yet constant — a conceptual trap familiar to water studies.[17] All the essays demonstrate the various ways in which water defined and was defined by the early medieval world; however, more than a catalogue of instances, they also collectively speak to the power of water to transform and change. Never a passive resource in this book, water is an agent of its own history — whether a dangerous riddle-creature and watery hell directly challenging human beings, or 'living waters' slaking thirst for holy wisdom and initiating the believer, it destroys as it creates. Unlike other elements, water in early medieval England

---

16 Arnold, *Negotiating the Landscape*; Wickham-Crowley, 'Fens and Frontiers'; Smith, *Water in Medieval Intellectual Culture*. See also Anlezark, *Water and Fire*.
17 Smith and Howes, 'Medieval Water Studies', p. 9.

had this ability to transform boundaries and communities both temporal and eternal. As a created thing itself, water was a medium for understanding the power of God in the medieval imagination, beyond the role of symbol. The divine *was* water and the water *was* divine, physically embodied in all its beauty and terror. The apparent contradictions that we see today in the roles played by water — the very tensions that inspired this project — were part of the fullness of Creation and bore little conflict for people in the early medieval past. Water overlaps our modern categories of sacred and secular, temporal and eternal, material and immaterial, and it floods the quotidian agricultural landscapes of early England at the same time it pours from Christ's side at the Crucifixion. The following essays embrace this broader early medieval meaning of water within their individual pursuits of its diverse manifestations in littoral and literary landscapes.

The seas and rivers of early medieval England provide the physical settings necessary for considering the many meanings of water in this book. Water offered the easiest way of travelling at speed in the Middle Ages, when what roads there were became boggy with mud after rain. Waterways were the missionary and economic highways, transporting Christ as well as the migrations of people and products between the Insular world and the Continent.[18] The reliance on such routes created shared cultures and networks based around coastlines and watersheds rather than what modern maps often delineate.[19] Before the great works of land reclamation of the later Middle Ages, larger parts of the English landscape were either perennially or seasonally marshy than is now the case. Water was omnipresent, essential for the successful functioning of daily life and sought out for spiritual edification. Many of the island retreats favoured by hermit saints now stand as low hillocks. One example is associated with Plegmund (d. 923), appointed by Alfred the Great as archbishop of Canterbury in 890. According to local legend, Plegmund lived on 'the Isle of Chester' at Plemstall in Cheshire, on which now stands St Peter's Church; two hundred metres west of this former island is 'St Plegmund's Well' dressed in medieval stone, some of which may be early in date.[20] More famous in the period was another Mercian, St Guthlac (d. 714), who retreated from the monastic community at Repton along the River Trent at the end of the seventh century to fight demons in the wild fens of Crowland. His isolation on his Neolithic barrow home in the midst of waters was ameliorated by the fact that his religious brothers could visit him by boat. What is more important in Guthlac's act of withdrawal is the symbolic isolation offered by his island, in

---

18 Pelteret, 'The Role of Rivers and Coastlines in Shaping Early English History'; Gardiner, 'Inland Waterways and Coastal Transport'. See also essays by Martin Carver, John Baker and Stewart J. Brooks, Juliet Mullins, David Pelteret, and Gale Owen-Crocker in Klein, Schipper, and Lewis-Simpson, *The Maritime World of the Anglo-Saxons*.
19 Williamson, *Sutton Hoo and its Landscape*; Klein, 'Navigating the Anglo-Saxon Seas', pp. 6–9.
20 See Matthews, *St Plegmund's Well*.

which the surrounding waters mirrored the desert oasis into which the early Egyptian hermit St Anthony retreated.[21]

More famous still among the island saints is the Northumbrian St Cuthbert (d. 687), whose sacred topography is surveyed in the first chapter of this volume by Della Hooke. After serving as a monk and then bishop of Lindisfarne, Cuthbert permanently withdrew to the Farne Islands in the North Sea, off the Northumbrian coast, which had previously been a favoured retreat of St Aidan (d. 651), the founder of Lindisfarne's monastic community. Cuthbert's isolation on his island was real, especially in the frequent stormy weather, yet Inner Farne, where Cuthbert settled, is less than two kilometres east of the great Northumbrian royal centre at Bamburgh. Bede reported that a torch signalled his death to the community at Lindisfarne across the water.[22] Cuthbert's eremitical isolation was as real as it was symbolic, with the waters of the North Sea constituting both physical reality and spiritual metaphor. His penitential exercise of standing neck-deep in the cold waters of the North Sea echoed the watery asceticism inherited in the North of England from teachers like the Irish Aidan.[23]

Hooke brings together a comprehensive geographic and ritual discussion of such watery monastic retreats and further complicates the relationship coastal and island monasteries had with sites of secular power. This chapter introduces the long history of water as a liminal symbol from prehistoric Britain to the establishment of Christianity in England and Wales in the fifth through ninth centuries, and shows how the use of watery sites continued over time despite vast religious and political changes. Irish and Roman missionaries chose such watery places for the locations of early monastic centres, such as the small raised hermitage of Beckery Chapel enclosed in a bend of the River Brue south-west of Glastonbury, where a burial of seven individuals provides the earliest archaeological evidence for monasticism in Britain. Using literary, documentary, and place-name evidence, Hooke examines the breadth of early medieval monastic communities in England and their relationships to the watery sites of islands, rivers, and fenlands, setting up a wide historical overview of the geography of religious foundations and a comprehensive backdrop against which to site the following essays. While the initial desires of early Irish and English hermits were to seek out watery deserts for contemplation in imitation of late antique forebears and to avoid the affairs of the world, Hooke points out that such key locations near bodies of water ironically contributed to their entanglements with local secular authorities as churches began to profit financially from their positions in the landscape.

---

21 Pickles, 'Anglo-Saxon Monasteries as Sacred Places'; Mullins, '*Herimum in mari*'.
22 Bede, *VCP*, ch. 39, p. 284.
23 Herity, 'Early Irish Hermitages in the Light of the *Lives* of Cuthbert', pp. 51–53; Ireland, 'Penance and Prayer in Water'.

In the early Middle Ages, as now, rivers and seas both united and divided human communities as mediums of travel and as borders. For the retreats of hermits, these waters served to link the saints to the outside world and to form a perimeter that defined the bounds of the sacred space within which the saint would combat his demons. Charter bounds often use water courses, wells, and ponds to delimit lands and their ownership, while lands with adjacent water would have been valuable for agriculture or animal husbandry. The River Humber looms large in the political imagination of the island, dividing not only Mercia in the south from Northumbria in the north, but in contemporary nomenclature splitting the whole of early England into two: Northumbria and Southumbria.[24] At the Battle of the Trent in 679, fought in the Kingdom of Lindsey, the Mercians eclipsed Northumbrian power in the region, which they held until the vikings came in the ninth century. For the vikings, skilled navigators and sailors, the coastal waters and river systems of England made possible first their rapid raids and later the conquest of vast territories. The River Trent provided an easy access route into the Mercian heartland in the third quarter of the ninth century, though the vikings failed to follow this up with a full dominance of the Thames and of Wessex. In King Alfred's treaty with the viking King Guthrum (878–90), the Rivers Thames and Lea form the boundary between their territories. Kings would meet at the banks of rivers, places both central and peripheral in the landscape as neutral liminal ground or areas of conflict for the clash of armies.[25] These were also key places targeted by early missionaries for mass baptisms, as explored by Carolyn Twomey in her essay.

Diving deeper within the waterscapes discussed in the previous chapter, Twomey pursues the use of rivers for the Christian rite of baptism in the outdoor landscape. Baptism was the most important ritual use of water in early medieval England. By late antiquity, Christian exegetes had transformed Jewish ritual bathing into the gospel event of Jesus's own baptism in the River Jordan by John the Baptist, an act that forever sanctified all water on Earth in Bede's understanding.[26] The Baptist himself was one of the most revered saints of the first millennium, a liminal character in Christian interpretation — the last Jewish prophet and one of the first Christian saints. In early medieval representations of the Baptism of Christ, John and Jesus are often depicted standing together in the waters of the river, at the birth of the Christian sacrament of initiation. The Trinitarian symbolism found in these images — with the Father's hand above, and the Holy Spirit's descending dove — is expressed in the formula of baptism still used by many Christian denominations today.

---

24 See also the consideration of watery landscapes as borders in Klein, 'Navigating the Anglo-Saxon Seas', pp. 6–7.
25 Barrow, 'Chester's Earliest Regatta?'.
26 Bede, *Homeliarium evangelii libri II*, ed. by Hurst, I. 12, l. 31; Bede, *Homilies*, trans. by Martin and Hurst, I. 12, at p. 114. For how this idea was taken up by Ælfric, see Bedingfield, *The Dramatic Liturgy of Anglo-Saxon England*, pp. 186–87.

Using a well-known tenth-century image of the baptismal scene to anchor her analysis, Twomey considers the literary evidence for the use of rivers for baptism alongside the importance of rivers in the landscape as central places of community identity in order to argue that riverine baptism was not merely convenient but indeed strategically planned by missionaries such as Paulinus and Cuthbert. Rivers and their watersheds defined not only settlement patterns but also cultural zones; patterns of burial, assemblies, and watery depositions indicate that rivers played important roles in locating expressions of identity and community gatherings for the living and the dead. Twomey synthesizes evidence from archaeological, anthropological, and landscape studies to re-examine the brief accounts of river baptisms found in Bede's *Historia Ecclesiastica*, the lives of St Cuthbert, and Adomnán's life of St Columba. Rather than taking architectural form inside baptisteries or baptismal fonts, baptism in early England imitated the natural setting of the Baptism of Christ in the River Jordan and perpetuated the importance of watery places across the coming of Christianity. In this way, Twomey situates the practice of baptism within other contemporary acts of religious inculturation such as the reuse of temples for Christian worship advised by Pope Gregory the Great to Bishop Mellitus to accomplish the successful conversion of England.

Baptism signifies the passing from the death of sin to life in Christ, an immersion into Jesus's victory over death and the Devil. This potent imaginary is inseparable from Cuthbert's ascetical bathing, Plegmund's well, and the encounters of opposing armies facing each other across waters — from the Christian and 'heathen' armies represented in the late Old English poem *The Battle of Maldon*, to the Christianized Israelites celebrating the destruction of the Egyptian army in the poem *Exodus*. In the Christian society that early medieval England became, baptism not only constituted an initiation in the universal community of the Church, but also made the people of early England subjects of a Christian king.[27] About a decade before King Alfred made his treaty with Guthrum, and soon after he had defeated his viking army in battle and divided their kingdoms along the banks of rivers, the English king had stood at the waters of baptism and become Guthrum's godfather.[28] Water could be a potent spiritual and political force in early medieval England.

The next chapter revives the study of another crucial early medieval form of bodily immersion in water: the act of swimming. Simon Trafford addresses the watery landscapes of England from the perspective of swimming and swimmers in early English documentary, literary, and visual culture. Water and passage through water was a transformative act in Old English heroic poetry, though the cultural behaviour of swimming appears in few early

---

27 The term *fideles* indicated both the faithful of Christ and loyal vassals. Phelan, *The Formation of Christian Europe*, p. 1.
28 *The Anglo-Saxon Chronicle* [A], ed. by Bately, s.a. 878, p. 51.

English sources.[29] Trafford begins by considering the most famous swimmer in the literature, Beowulf, and compares his poetic swimming episode and new surviving evidence from England with the more numerous examples from the Continent, Ireland, and Scandinavia. His chapter elevates the discussion of an everyday action in light of recent scholarship on other human interactions with and immersions in water, including overlapping categories of bathing, Christian baptisms, and mass military drownings, the last of which authors used to emphasize the fullest extent of an enemy's defeat. Trafford offers us a tantalizing watery image from an eleventh-century manuscript suggestive of an early medieval swimmer enmeshed in the waters of the page of Psalms with his arms and legs forming the athletic position of the breaststroke. This familiar bodily image of swimming demonstrates how even quotidian interactions with water and watery places were embedded in early medieval culture.

The saints and swimmers inhabiting dynamic watery environments experienced them in all five physical senses in addition to those of scripture and ritual. Continuing focus on the littoral landscapes of monastic communities and introducing key figures from Old English literature, Rebecca Shores examines the sense of sound in several ocean-going scenes from hagiography and poetry. Despite the common topos of watery seclusion, the seas of early medieval England teemed with many animal and human noises — as any modern visitor to Cuthbert's island of Inner Farne during tern nesting season knows. Deploying a sensory approach to water in her essay, Shores pursues how the sounds and swells of the sea created a sonic environment that — like the built environment itself — reflected and redirected human experience and emotion. The sea is an inhabited vocal space; the nautical sounds of human encroachment in *The Seafarer* and the lives of Ceolfrith, Wilfrid, and Oswald permeate the supposedly remote wildernesses of the islands and fens. Shores shows how the ship-bound voices of sailors and monks — chanting the rhythm-keeping shouts of classical *celeuma* — animate these watery settings into soundscapes of holiness that audibly situate saints in their physical and spiritual milieux. The divine singing and communal wailing that accompanied Ceolfrith's ritualized departure from St Peter's Church at Wearmouth is a transformative moment of divine song and silence across the Wear estuary, a river that Shores shows was part of the liturgical architecture of the Northumbrian coastal monastery. This chapter demonstrates the value of approaching watery landscapes through the fullness of early medieval sensory experience to pursue the ways in which early medieval listeners defined and redefined their relationships with water and with themselves.

Under the surface of the sonic sea, the inhabitants of water were essential for the maintenance of everyday diets and foodways of local communities.[30] More than simply aquatic resources, however, these fish and whales also

---

29 Klein, 'Navigating the Anglo-Saxon Seas', pp. 7–8.
30 Reynolds, 'Food from the Water'.

possessed remarkable spiritual identities as metaphors of sacrality and monstrosity in early English literary and visual culture.³¹ Christ fed the five thousand with the sustaining miracle of loaves and fishes, yet wicked beasts of the Bible roamed the deep to symbolize the threat of worldly sin and test holy men and women. The English collected the bones of whales and carved them into special objects, such as the powerful jawbone of the Franks Casket.³² The obedience of water itself as well as its inhabitants testified to the power of God to command the natural created world, from Columba's victory over the monster of the River Ness to Cuthbert's otters drying the salty wet feet of the saint.³³ Creatures of the water, like water itself, operated as both physical and conceptual beings.

The next chapter concentrates on the story of one sailor and his fate in the belly of a creature under the waves: Jonah and the sea monster. Elizabeth A. Alexander approaches the visual culture of the Old Testament figure as a prefiguration of the sacrifice of Christ in the manuscript tradition of early medieval England. Sailors transporting the disobedient Jonah threw the prophet into the ocean in order to save the ship tossed by the Lord's storm, after which he spent three days and three nights in the belly of a *ketos*. The *ketos* was a specific type of sea creature bearing the head of a canine with sharp teeth, pointed ears, front paws, and a long twisting body terminating in a tail. Through a detailed study of the history and iconography of the *ketos*, Alexander argues that those responsible for the surviving ninth- to eleventh-century artistic representations of Jonah understood this creature to belong to the abyss of the sea and the depths of hell. Her essay brings together hints of this beast from early medieval manuscripts such as the Antwerp Sedulius and Harley Psalter with commentaries by Jerome and Bede to articulate how the images of Jonah — with and without depictions of water itself — prefigured the victory of Christ over sin and death for their audiences. The *ketos* dwelled in a watery hell and was also a manifestation of hell itself, its iconography adapted to denote the Leviathan of Revelation, the serpent of the Adam and Eve story, and the literal mouth of hell into which Christ descends to liberate the souls of the damned. More than just a living creature of the ocean, the *ketos* articulated early medieval understandings of water as a place of terror.

The liminality of waters — as a place where life can meet death, friend can meet foe — constitutes a symbolism deeply embedded in most human cultures; in the Christian cultures of early medieval England this combined

---

31 See Esser-Miles, '"King of the Children of Pride"'; Momma, 'Ælfric's Fisherman and the *Hronrad*'; and Riddler, 'The Archaeology of the Anglo-Saxon Whale'.
32 Neuman de Vegvar, '*Hronæs ban*'; see also Paz, *Nonhuman Voices in Anglo-Saxon Literature and Material Culture*, ch. 3.
33 Thomas, 'The "Monster" Episode in Adomnan's *Life* of St Columba'; Borsje, 'The Monster in the River Ness in *Vita Sancti Columbae*'; Ward, 'The Spirituality of St Cuthbert', p. 72. For the hagiographic motif of nature's obedience to the saint, see Gusakova, 'A Saint and the Natural World'.

with other rich metaphors such as baptism in the symbolism of divine water. Water that had been sanctified by contact with a holy person or through the ritual actions of clergy had the power to heal the sick and purify bodies and places in the landscape. The sacred power of early medieval relics could spread to other materials and locations through the medium of water. The water that washed the holy bones of St Oswald transferred a miraculous quality to the soil in which it was poured out, and the damp earth subsequently had the power to cure those possessed by demons.[34] Water made holy in this way or collected from blessed springs became a common trope in hagiographies and histories as divine water of heaven that could cure those who washed with or drank it.[35] English artistic and literary landscapes teemed with such liminal waters.

While we have seen how water could represent hell in early medieval art, the next essay in this collection explores how water was also a force of heaven. Meg Boulton pursues the artistic representations of divine waters, articulating how depictions of water and its transforming properties took material form as visual metaphors in the borders and frames of late antique and early medieval art. The four rivers of Paradise that flowed from the Tree of Life in the heavenly Jerusalem were popular depictions of divine water with multiple biblical parallels in early Christian mosaics. Boulton pursues these and other more unexpected motifs of water in new forms on English monuments and sculpture while also considering the role of the pearl in late antique and early medieval art. Pearls were the physical materializations of heavenly water and often — like other jewels as well as waters — framed the boundaries of heaven in different media to delineate sacred spaces. Decentring the central spaces themselves, Boulton pursues these overlooked watery borders and argues they ought to be understood as cohesive iconographies in their own right. Depictions of biblical waters at Rome and Ravenna were adapted in Insular tradition to serve as potent iconographic devices that negotiated the thresholds between heavenly and earthly realms for the viewer. Water in Christian iconography was never just water; it was deployed to reify heaven within the sacred spaces and places of the church building.

Water flooded the early medieval imagination with biblical resonance. As we have seen, a single drop in the Old and New Testaments had the potential to represent key moments in salvation history and to unite past, present, and future in the life of the early medieval Christian. Water began and ended life on Earth from Genesis and the rebirth of Noah's Flood, to the end times of

---

34 Bede, *HE*, III. 11. See Hooke, 'Rivers, Wells and Springs in Anglo-Saxon England', and also Stocker and Everson, 'The Straight and Narrow Way' on how the bones themselves sanctified the entirety of the River Witham. For other examples of the deaths and relics of saints creating miraculous wells, see Bord, *Cures and Curses*, pp. 113–21.

35 Gittos, *Liturgy, Architecture, and Sacred Places*, p. 21 n. 5; Bede, *HE*, III. 2, 9, 11, 13; IV. 3; and V. 18; Bord and Bord, *Sacred Waters*, pp. 20–33 and 96–105.

Revelation, where living water streamed from either side of the Tree of Life.[36] John J. Gallagher and Michael Bintley each expand on different aspects of the cosmological role of divine water in Latin exegesis and Old English poetry.

Gallagher approaches the commentaries on water in the Book of Genesis produced by the late seventh-century Canterbury School. Water fills the world of Genesis: the primordial abyss; the separation of the waters into the firmament, sea, and dry land; the notable absence of rain at Creation; and the purifying destruction of the Flood of Noah. For the exegete, understanding water was integral to understanding the creation of the earth and the Bible in its literal sense. This chapter explores how early medieval exegetes perceived the nuances of supercelestial, primordial, and diluvial waters to reconcile biblical accounts of water in Genesis with the observed behaviour of the element in the natural world. In doing so, the scholars at Canterbury relied on an intellectual network from Ireland to continental Europe to consult manuscripts of commentaries by the Church Fathers and Irish intellectuals. Gallagher demonstrates how the Canterbury Commentator developed late antique traditions of Latin exegesis by bringing together the Bible and contemporary scientific understandings of the hydrological cycle, in particular, the history of rain in the meteorology of Genesis. Biblical water and everyday water were one and the same in early England.

Bintley further pursues the relationship between medieval waters in the environment with divine water in the Old English poem *Andreas*. His examination of the disparate meanings of water in the literary landscapes of St Andrew shows how the poet used water as an act of Creation, ordering the waters of the poem as God ordered the waters of Genesis. Andrew's long sea journey to the city of Mermedonia is the ship-bound setting for the poet to expound on the role of the ocean as the uncertain and dangerous waters through which all human beings travel over the course of their lives. However, this space can be a navigable terrain for adept seafarers who have adequate training and preparation against the wages of sin. When Andrew arrives in Mermedonia, he finds a rocky countryside that is bathed in blood — including the saint's own — rather than water. The flood that subsequently swells from a stone pillar and drowns the Mermedonians in divine retribution swallows up the city in a single act that collapses Christian metaphors of the Flood of Noah, Moses crossing the Red Sea, Christ's calming of the storm, the waters of the heavenly Jerusalem, and baptism. The *Andreas* poet demonstrates his role in salvation history by effortlessly adapting the conceptual fluidity of the element both to describe water and to become water himself. Bintley argues that, rather than serving an incoherent role in the poem, these many different linked understandings of water replicate the process of creation found in Genesis. The poet's use of water imagery and metaphors imitates

---

36 For a comprehensive treatment of the waters of the Flood of Genesis, see Anlezark, *Water and Fire*.

the cosmogonic act itself: as God divided the waters at Creation, so does the poet become a conduit for divine water itself in *Andreas*.

Like the poet of *Andreas*, the Latin and Old English authors of the lives of St Guthlac pour out their words in streams of living water as they defined Christian sanctity in the watery world of the fens. Helen Appleton returns us to the island retreats of the saints to explore how the language of physical and metaphorical waters at Crowland promoted saintly wisdom for those seeking salvation within a turbulent and sinful world of water. Water was both nourishing and destructive in the life of Guthlac, as a Gregorian image of knowledge yet also emblematic of the worldly chaos from which the saint fled.[37] The changing treatment of water between Felix's eighth-century Latin hagiography and the later Old English prose version emphasized the translator's intent to articulate a different spiritual understanding of water more suitable to a vernacular audience. Appleton's detailed literary analysis of the saint's lives suggests the Old English version — appearing in the late tenth-century Vercelli and late eleventh-century Vespasian manuscripts — was simplified and streamlined, devoid of the sophisticated watery allusions and complexities of Felix's Latin. References to well-ordered waters as metaphors for wisdom were reduced: Guthlac's vision of hell was no longer a Vergilian abyss filled with watery fire in the Old English version; instead, the translator chose to concentrate on the more familiar physical descriptions of the saint's wet environment. These more literal landscapes emphasized the power of the saint against a far more threatening fenland to the vernacular listener as opposed to Felix's neat envelope patterns that promoted Guthlac's place in a long ecclesiastical history of spiritual waters.

The watery landscapes of early England not only teemed with creatures divine and malign, but water too was a creature (*creatura aquae*) that could both act and be acted upon.[38] Daniel Anlezark and Jill Frederick explore the contradictory natural agencies of water in Old English literature. Anlezark continues to pursue the association between water and wisdom begun in the previous chapter to elucidate a wide range of watery discourses in the epilogue to the late ninth-century Alfredian *Pastoral Care* and the early tenth-century dialogues of *Solomon and Saturn*. In these Old English interrogations of the nature of living, restless water from the court of King Alfred, early medieval authors fused science and religion. Waters were Christian authors and teachers, both their floods of wisdom as well as their words themselves, but they required appropriate conduits to bring divine wisdom to the listener and avoid disastrous dams of idle chatter and drowned fools. In these complex metaphors of flowing water, Anlezark reveals how each of the texts expresses contemporary understandings of the natural movements of water in the landscape for their authors to create appropriately effective metaphors. In

---

37 See also Frederick, 'From Whale's Road to Water under the Earth'.
38 Dendle, 'Demons of the Water', pp. 192–94.

this way, the author of the *Metrical Epilogue* had some knowledge of fluid dynamics and water infrastructure in the early medieval landscape, perhaps based on the urban topography of Gregorian Rome. Similarly, the author of *Solomon and Saturn II* understood the scientific properties of water — as well as water's theological import — enough to discuss the compressing of water as snow and ice. By going beyond the literal, the character of Solomon pushes Saturn to consider both natural and supernatural lessons of living water, although with the ultimate understanding in *Solomon and Saturn I* that the Pater Noster has divine power over all water and its inhabitants. This chapter brings together these complex discussions of actively flowing, living waters to demonstrate the close relationships between key Alfredian texts.

Water speaks of its own identities in the final essay of this collection. Jill Frederick's chapter continues to examine the changeable nature of water's fluid movements and metaphors by focusing on the damaged Old English Riddle 84, solved by consensus as 'water'. Here, personified water directly addresses its human reader/listener, challenging us — as we are challenged in this volume — to identify what it is. This riddle creature of the tenth-century Exeter Book collapses sacred and secular agency, expressing a vicious, violent identity that Frederick shows is also maternal and salvific. The essay investigates the particular lapidary language of the riddle that contrasts water's brightness and beauty with the more threatening language attached to water found in other early English riddles and poetry. Frederick attributes the seemingly contradictory appearance of water in the poem not to the fragmentary nature of the manuscript but rather to the choice of the poet to best describe the agency of water and its capacity for change. The generative qualities of water in the riddle to nurture and create, as a 'modor is monigra mærra wihta' (mother of many greater creatures), lead Frederick to describe the identity of the poem's water-mother subject as distinctly feminine and godlike, with a maternal ferocity to protect her children in the mode of Grendel's mother from *Beowulf*. Water was a dangerous force — and possessed within it dangerous forces — to be feared at the same time that water could be an inclusive channel to eternal life. It is with the voice of water itself that this collection ends its analysis where it began, with the enmeshment of early medieval people within their many worlds of water.

The eleven contributions assembled here approach the meanings of water in the early medieval landscape from different and overlapping disciplinary strands: from the everyday needs of human life and settlement to exegetical arguments about biblical and liturgical water, from artistic depictions of water beasts and water travel to the danger of watery places in Old English and Latin literature. The transformative power of water to both create and destroy life in its physical and spiritual senses imbued the experience of the single element with manifold meanings that found diverse expression in early English culture. Using water as a lens that refracts and focuses understanding, this book reveals interactions between the natural world and the constructed worlds of literature, art, and religion by studying humans as embedded in their watery environments.

The essays assembled here pursue these overlapping conceptual, scientific, and religious worlds that framed experiences of water, beginning with how humans encountered physical watery landscapes and ending with the dynamic agency of water itself. This approach and other spatial and ecocritical perspectives in recent studies of water in the Middle Ages, and in the early medieval Insular world specifically, testify to the continued resonance that water bears in both the medieval past and the changing climate of our present. In all, these essays emphasize the complex intersecting meanings of water in early medieval England.

Works Cited

Primary Sources

The Anglo-Saxon Chronicle: A Collaborative Edition, vol. III: *MS A*, ed. by Janet M. Bately (Cambridge: D. S. Brewer, 1986)
The Annals of St-Bertin: Ninth-Century Histories, vol. 1, trans. by Janet L. Nelson (Manchester: Manchester University Press, 1991)
Bede, *Bede's Ecclesiastical History of the English People*, ed. and trans. by Bertram Colgrave and R. A. B. Mynors (Oxford: Clarendon Press, 1969)
———, *Homeliarium evangelii libri II*, ed. by David Hurst, CCSL, 122 (Turnhout: Brepols, 1955)
———, *Homilies on the Gospels*, trans. by Lawrence T. Martin and David Hurst, 2 vols (Kalamazoo: Cistercian Publications, 1991)
———, *Libri quatuor in principium Genesis usque ad nativitatem Isaac et eiectionem Ismahelis adnotationum*, vol. 1 of *Bedae Venerabilis Opera: Opera exegetica*, ed. by Charles W. Jones, CCSL, 118A (Turnhout: Brepols, 1967)
———, *Vita S. Cuthberti prosaica*, in *Two Lives of Saint Cuthbert: A Life by an Anonymous Monk of Lindisfarne and Bede's Prose Life*, ed. and trans. by Bertram Colgrave (Cambridge: Cambridge University Press, 1940), pp. 141–307

Secondary Works

Abulafia, David, *The Great Sea: A Human History of the Mediterranean* (Oxford: Oxford University Press, 2011)
Anlezark, Daniel, *Water and Fire: The Myth of the Flood in Anglo-Saxon England* (Manchester: Manchester University Press, 2006)
Armitage, David, Alison Bashford, and Sujit Sivasundaram, eds, *Oceanic Histories* (Cambridge: Cambridge University Press, 2018)
Arnold, Ellen F., 'An Introduction to Medieval Environmental History', *History Compass*, 6.3 (2008), 898–916
———, *Negotiating the Landscape: Environment and Monastic Identity in the Medieval Ardennes* (Philadelphia: University of Pennsylvania Press, 2013)
Banham, Debby, and Rosamond Faith, *Anglo-Saxon Farms and Farming* (Oxford: Oxford University Press, 2014)

Barrow, Julia, 'Chester's Earliest Regatta? Edgar's Dee-Rowing Revisited', *Early Medieval Europe*, 10 (2001), 81–93

Bedingfield, M. Bradford, *The Dramatic Liturgy of Anglo-Saxon England* (Woodbridge: Boydell & Brewer, 2002)

Bengtsson, Lennart, R.-M. Bonnet, M. Calisto, G. Destouni, R. Gurney, J. Johannessen, Y. Kerr, W. A. Lahoz, and M. Rast, eds, *The Earth's Hydrological Cycle* (Dordrecht: Springer Netherlands, 2014)

Bintley, Michael D. J., and Michael Shapland, eds, *Trees and Timber in the Anglo-Saxon World* (Oxford: Oxford University Press, 2013)

Blair, John, ed., *Waterways and Canal-Building in Medieval England* (Oxford: Oxford University Press, 2007)

Bord, Janet, *Cures and Curses: Ritual and Cult at Holy Wells* (Loughborough: Heart of Albion, 2006)

Bord, Janet, and Colin Bord, *Sacred Waters: Holy Wells and Water Lore in Britain and Ireland* (London: Granada, 1985)

Borsje, Jacqueline, 'The Monster in the River Ness in *Vita Sancti Columbae*: A Study of a Miracle', *Peritia*, 8 (1994), 27–34

Bronstert, Axel, Jesus Carrera, Pavel Kabat, and Sabine Lütkemeier, eds, *Coupled Models for the Hydrological Cycle* (Berlin: Springer, 2005)

Chen, Cecilia, Janine MacLeod, and Astrida Neimanis, eds, *Thinking with Water* (Montreal: McGill-Queen's University Press, 2013)

Classen, Albrecht, *Water in Medieval Literature: An Ecocritical Reading* (Lanham, MD: Lexington Books, 2018)

Cohen, Jeffrey Jerome, and Lowell Duckert, eds, *Elemental Ecocriticism: Thinking with Earth, Air, Water, and Fire* (Minneapolis: University of Minnesota Press, 2015)

Dendle, Peter, 'Demons of the Water: Anglo-Saxon Responses to the Gerasene Demoniac', in *The Maritime World of the Anglo-Saxons*, ed. by Stacey S. Klein, William Schipper, and Shannon Lewis-Simpson, Essays in Anglo-Saxon Studies, 5 (Tempe: ACMRS, 2014), pp. 187–208

Dyches, Preston, and Felicia Choa, 'The Solar System and Beyond Is Awash in Water', National Aeronautics and Space Administration, 7 April 2015, <https://www.nasa.gov/jpl/the-solar-system-and-beyond-is-awash-in-water> [accessed 16 February 2020]

Eisenberg, Merle, David J. Patterson, Jamie Kreiner, Ellen F. Arnold, and Timothy P. Newfield, 'The Environmental History of the Late Antique West: A Bibliographic Essay', in *Environment and Society in the Long Late Antiquity*, ed. by Adam Izdebski and Michael Mulryan, Late Antique Archaeology, 11 (Leiden: Brill, 2018), pp. 31–50

Esser-Miles, Carolin, '"King of the Children of Pride": Symbolism, Physicality and the Old English Whale', in *The Maritime World of the Anglo-Saxons*, ed. by Stacey S. Klein, William Schipper, and Shannon Lewis-Simpson, Essays in Anglo-Saxon Studies, 5 (Tempe: ACMRS, 2014), pp. 275–302

Estes, Heide, *Anglo-Saxon Literary Landscapes: Ecotheory and the Environmental Imagination* (Amsterdam: Amsterdam University Press, 2017)

Ferrand, Lin, 'The Hydrologic Cycle in Bede's *De Natura Rerum*', in *The Nature and Function of Water, Baths, Bathing and Hygiene from Antiquity through the Renaissance*, ed. by Cynthia Kosso and Anne Scott (Leiden: Brill, 2009), pp. 361–80

Frederick, Jill, 'From Whale's Road to Water under the Earth', in *Water and the Environment in the Anglo-Saxon World*, ed. by Maren Clegg Hyer and Della Hooke, Exeter Studies in Medieval Europe (Liverpool: Liverpool University Press, 2017), pp. 15–32

Gardiner, Mark, 'Inland Waterways and Coastal Transport: Landing Places, Canals and Bridges', in *Water and the Environment in the Anglo-Saxon World*, ed. by Maren Clegg Hyer and Della Hooke, Exeter Studies in Medieval Europe (Liverpool: Liverpool University Press, 2017), pp. 152–66

Gillis, John R., *The Human Shore: Seacoasts in History* (Chicago: University of Chicago Press, 2012)

Gittos, Helen, *Liturgy, Architecture, and Sacred Places in Anglo-Saxon England* (Oxford: Oxford University Press, 2013)

Green, C. C., *The Water Cycle: Concepts and Processes* (Melbourne: Cassell, 1970)

Gusakova, Olga, 'A Saint and the Natural World', in *God's Bounty? The Churches and the Natural World*, ed. by Peter Clarke and Tony Claydon (Woodbridge: Boydell & Brewer, 2010), pp. 42–52

Harris, W. V., ed., *Rethinking the Mediterranean* (Oxford: Oxford University Press, 2005)

Hastrup, Kirsten, 'Water and the Configuration of Social Worlds: An Anthropological Perspective', *Journal of Water Resource and Protection*, 5.4A (2013), 59–66

Herity, Michael, 'Early Irish Hermitages in the Light of the *Lives* of Cuthbert', in *St Cuthbert, his Cult and his Community to AD 1200*, ed. by Gerald Bonner, David Rollason, and Clare Stancliffe (Woodbridge: Boydell, 1989), pp. 45–63

Hooke, Della, 'Rivers, Wells and Springs in Anglo-Saxon England: Water in Sacred and Mystical Contexts', in *Water and the Environment in the Anglo-Saxon World*, ed. by Maren Clegg Hyer and Della Hooke, Exeter Studies in Medieval Europe (Liverpool: Liverpool University Press, 2017), pp. 107–35

Horden, Peregrine, and Nicholas Purcell, *The Corrupting Sea: A Study of Mediterranean History* (Oxford: Wiley-Blackwell, 2000)

Howe, John M., and Michael Wolfe, *Inventing Medieval Landscapes: Senses of Place in Western Europe* (Gainesville: University Press of Florida, 2002)

Hyer, Maren Clegg, and Della Hooke, eds, *Water and the Environment in the Anglo-Saxon World*, Exeter Studies in Medieval Europe (Liverpool: Liverpool University Press, 2017)

Ireland, Colin, 'Penance and Prayer in Water: An Irish Practice in Northumbrian Hagiography', *Cambrian Medieval Celtic Studies*, 34 (1997), 51–66

Klein, Bernhard, and Gesa Mackenthun, eds, *Sea Changes: Historicizing the Ocean* (New York: Routledge, 2004)

Klein, Stacey S., 'Navigating the Anglo-Saxon Seas', in *The Maritime World of the Anglo-Saxons*, ed. by Stacey S. Klein, William Schipper, and Shannon Lewis-Simpson, Essays in Anglo-Saxon Studies, 5 (Tempe: ACMRS, 2014), pp. 1–20

Klein, Stacey S., William Schipper, and Shannon Lewis-Simpson, eds, *The Maritime World of the Anglo-Saxons*, Essays in Anglo-Saxon Studies, 5 (Tempe: ACMRS, 2014)

Magnusson, Roberta J., *Water Technology in the Middle Ages: Cities, Monasteries, and Waterworks after the Roman Empire* (Baltimore: Johns Hopkins University Press, 2002)

Matthews, K. J., *St Plegmund's Well: An Archaeological and Historical Survey*, Archaeological Service Evaluation Report, 37 (Chester: Chester City Council, 1995)

Momma, Haruko, 'Ælfric's Fisherman and the *Hronrad*: A Colloquy on the Occupation', in *The Maritime World of the Anglo-Saxons*, ed. by Stacey S. Klein, William Schipper, and Shannon Lewis-Simpson, Essays in Anglo-Saxon Studies, 5 (Tempe: ACMRS, 2014), pp. 303–22

Mullins, Juliet, '*Herimum in mari*: Anglo-Saxon Attitudes towards *Peregrinatio* and the Ideal of a Desert in the Sea', in *The Maritime World of the Anglo-Saxons*, ed. by Stacey S. Klein, William Schipper, and Shannon Lewis-Simpson, Essays in Anglo-Saxon Studies, 5 (Tempe: ACMRS, 2014), pp. 59–74

Neuman de Vegvar, Carol, '*Hronæs ban*: Exoticism and Prestige in Anglo-Saxon Material Culture', in *The Maritime World of the Anglo-Saxons*, ed. by Stacey S. Klein, William Schipper, and Shannon Lewis-Simpson, Essays in Anglo-Saxon Studies, 5 (Tempe: ACMRS, 2014), pp. 323–36

Oestigaard, Terje, 'Water', in *Handbook of the Archaeology of Ritual and Religion*, ed. by Timothy Insoll (Oxford: Oxford University Press, 2011), pp. 38–50

Paz, James, *Nonhuman Voices in Anglo-Saxon Literature and Material Culture* (Manchester: Manchester University Press, 2017)

Pelteret, David A. E., 'The Role of Rivers and Coastlines in Shaping Early English History', *Haskins Society Journal*, 21 (2009), 21–46

Phelan, Owen M., *The Formation of Christian Europe: The Carolingians, Baptism, and the Imperium Christianum* (Oxford: Oxford University Press, 2014)

Pickles, Thomas, 'Anglo-Saxon Monasteries as Sacred Places: Topography, Exegesis and Vocation', in *Sacred Text, Sacred Space: Architectural, Spiritual and Literary Convergences in England and Wales*, ed. by Joseph Sterrett and Peter Thomas (Leiden: Brill, 2011), pp. 35–56

Reynolds, Rebecca, 'Food from the Water: Fishing', in *Water and the Environment in the Anglo-Saxon World*, ed. by Maren Clegg Hyer and Della Hooke, Exeter Studies in Medieval Europe (Liverpool: Liverpool University Press, 2017), pp. 136–51

Riddler, Ian, 'The Archaeology of the Anglo-Saxon Whale', in *The Maritime World of the Anglo-Saxons*, ed. by Stacey S. Klein, William Schipper, and Shannon Lewis-Simpson, Essays in Anglo-Saxon Studies, 5 (Tempe: ACMRS, 2014), pp. 337–54

Siewers, Alfred K., *Strange Beauty: Ecocritical Approaches to Early Medieval Landscape* (New York: Palgrave Macmillan, 2009)

Smith, James L., *Water in Medieval Intellectual Culture: Case Studies from Twelfth-Century Monasticism* (Turnhout: Brepols, 2017)

Smith, James L., and Hetta Howes, 'Medieval Water Studies: Past, Present and Promise' *Open Library of Humanities*, 5.1 (2019), 1–13, <https://doi.org/10.16995/olh.443>

Squatriti, Paolo, *Water and Society in Early Medieval Italy, AD 400–1000* (Cambridge: Cambridge University Press, 1998)

——, ed., *Working with Water in Medieval Europe: Technology and Resource-Use*, Technology and Change in History, 3 (Leiden: Brill, 2000)

Stocker, David, and Paul Everson, 'The Straight and Narrow Way: Fenland Causeways and the Conversion of the Landscape in the Witham Valley, Lincolnshire', in *The Cross Goes North: Processes of Conversion in Northern Europe, AD 300–1300*, ed. by Martin Carver (Woodbridge: Boydell & Brewer, 2003), pp. 271–88

Strang, Veronica, 'Common Senses: Water, Sensory Experience and the Generation of Meaning', *Journal of Material Culture*, 10 (2005), 92–120

——, *The Meaning of Water* (Oxford: Bloomsbury Academic, 2004)

——, *Water: Nature and Culture* (London: Reaktion Books, 2015)

Tilley, Christopher, *A Phenomenology of Landscape: Places, Paths and Monuments*, Explorations in Anthropology (Oxford: Bloomsbury Academic, 1994)

Thomas, Charles, 'The "Monster" Episode in Adomnan's *Life* of St Columba', *Cryptozoology*, 7 (1988), 38–45

Walsham, Alexandra, *The Reformation of the Landscape: Religion, Identity, and Memory in Early Modern Britain and Ireland* (Oxford: Oxford University Press, 2011)

Ward, Benedicta, 'The Spirituality of St Cuthbert', in *St Cuthbert, his Cult and his Community to AD 1200*, ed. by Gerald Bonner, David Rollason, and Clare Stancliffe (Woodbridge: Boydell, 1989), pp. 65–76

Wickham-Crowley, Kelley M., 'Fens and Frontiers', in *Water and the Environment in the Anglo-Saxon World*, ed. by Maren Clegg Hyer and Della Hooke, Exeter Studies in Medieval Europe (Liverpool: Liverpool University Press, 2017), pp. 68–88

Williamson, Tom, *Sutton Hoo and its Landscape: The Context of Monuments* (Oxford: Windgather Press, 2008)

DELLA HOOKE

The Sacred Nature of Rivers, Wells, Springs, and Other Wetlands in Anglo-Saxon England

Water has had a metaphorical significance since prehistoric times in pre-Christian Britain. Although much of this changed within a Christian milieu, water retained a symbolic value manifest in literary and documentary sources, as a source of purification essential for liturgical practices including baptism. While wetlands might be perceived as dangerous places, they also represented places detached to a degree from the everyday world. Island locations were particularly favoured. 'Holy wells' are also met with in place-names and on some charter boundaries, some of which sites can still be located today. Such associations were to continue, occasionally expressed in later hagiography and folklore. As the Church increased in wealth as a landowner, the riverine and coastal location of many minsters was beneficial to their long-term economic success.

The Role of Water in Religious Belief

Water played a special role in most early religions. Here the human world was impinging on the aqueous domain: such sites were places of liminality, perhaps a realm of gods and ancestors. In Britain, a spring on the edge of Salisbury Plain had the capacity to turn flints dipped into its waters a bright magenta colour, something which seems to have attracted attention in Mesolithic times and may have been a factor in determining the sacred role of this area throughout the subsequent prehistoric period. The sources of rivers had sacred significance — Silbury Hill, an enigmatic monument, also in Wiltshire, began as a small mound around 2450 BC but was continually added to over

> **Della Hooke** • After a career as a university research fellow, lecturer, and subsequent freelance consultant in Archaeology and Historical Landscapes, Della Hooke (della.hooke@blueyonder.co.uk) is a journal editor (*Landscape History* and *The Transactions of the Birmingham and Warwickshire Archaeological Society*) and an Associate Member of the School of Geography, Earth and Environmental Science in the University of Birmingham.

a period of fifty-five and 155 years; it stands above springs at the head of the West Kennet not far distant from the Neolithic Avebury henge, a tributary river which may have been perceived as the source of the Thames.¹ The largest henge in that county, at Marden, stands at the head of the River Avon. Votive offerings were frequently cast into rivers and wetland, a practice which reached a peak in the Bronze Age. Even sacrificed human bodies have been found, dating from as late as the Roman period, in bogs and other wetland locations. Springs retained a special significance, emanating from this other world, and were much revered. In Roman Britain sacred springs continued to play an important role before the widespread adoption of Christianity in the fourth century AD, especially those thought to possess medicinal properties, and were often seen as related to particular goddesses, endowed with powers of prophecy. Thus water seems to have enjoyed a liminal quality, separating the everyday world from the sacred. This attribute appears to have continued in England after the Christian conversion of the Anglo-Saxons,² helping religious communities also to avoid secular involvement.

Christianity regarded wetlands as places of danger in hagiographical literature, frequented by demons who might try the faith of a Christian hermit saint. One example is the fens of Cambridgeshire where Guthlac chose to make his home.³ Guthlac, the son of a Mercian nobleman, had won fame through his military adventures, especially when he had led a war-band fighting on the borders of Wales, but subsequently decided instead to devote himself to the service of God and entered the monastery at Repton. Around 700, however, he sought out the more solitary life he had read about while studying the 'desert fathers', and he departed to 'the foggy marshes, bogs and black waters of the fens, where he lived on Crowland'.⁴ This was a place he was told 'no man could ever live':

> Wid is þes westen,   wræcsetla fela,
> eardas onhæle   earmra gæsta.
> Sindon wærlogan   þe þa wic bugað.⁵
>
>> [This wilderness is wide — [there are] many places of exile and secret dwelling places of wretched spirits. They are devils who dwell in this place.]⁶

However, he chose to scorn 'the temptation of the accursed spirits and was strengthened with heavenly support so that he began to live alone amidst

---

1   Learey and Symonds, 'The Many Faces of Silbury Hill'.
2   These themes are also discussed in Hooke, 'Rivers, Wells and Springs in Anglo-Saxon England'. See also Pickles, 'Anglo-Saxon Monasteries as Sacred Places'.
3   See in this volume Helen Appleton, 'Water, Wisdom, and Worldliness in the Anglo-Saxon Prose Lives of Guthlac'.
4   Hunt, *Warriors, Warlords and Saints*, p. 51.
5   'St Guthlac', in *The Exeter Book*, ed. by Krapp and Dobbie, p. 58.
6   Neville, *Representations of the Natural World*, p. 127 n. 165.

the swampy thickets of the wilderness'.[7] Other early churchmen also chose to become hermits in such watery locations but especially upon islands cut off from the everyday world where they felt that they could commune with God, undistracted by everyday life. Many of these island locations were to become the sites of monastic foundations, as will be shown below, and, though inland, rivers might also provide a similar degree of liminality. In time, such sites even offered the more worldly commercial benefits of accessibility for river and sea transport.

Christians also saw water as a powerful symbol: heathen shrines could be purified by sprinkling on 'holy' water, and many springs and wells were to be linked to Christian saints. In an early wave of enthusiasm following the conversion under Edwin of Northumbria 'tantus autem fertur tunc fuisse feruor fidei ac desiderium lauacri salutaris genti Nordanhymbrorum' (so great is said to have been the fervour of the faith of the Northumbrians and their longing for the washing of salvation) that Bishop Paulinus carried out baptisms from morning till evening, instructing the crowds in the teaching of Christ, 'atque instructam in fluuio Gleni, qui proximus erat, lauacro remissionis abluere' (when they had received instruction he washed them in the waters of regeneration [cleansing waters of baptism] in the river Glen, which was close at hand).[8] Mass baptisms as a means of salvation were also carried out in other rivers: within the kingdom of Deira in the River Swale at Catterick and in the kingdom of Lindsey in the River Trent near Littleborough.[9] Water continued to be an essential part of Christian baptism as a means of purification and blessing. A number of early churches were established at sites where springs were present. There were two wells within Glastonbury Abbey while Wells in Somerset, which was to become a diocesan centre in AD 909, took its name from the freshwater springs which still flow to the south-east of the cathedral. The cathedral of Winchester, too, close beside the River Itchen in Hampshire, had springs actually within the Old Minster.[10]

Blair warns that some minsters may have been so small that they missed most listings. 'Minster', the modern translation of OE *mynster* and its Latin equivalent *monasterium*, can be defined as 'a complex ecclesiastical settlement which is headed by an abbess, abbot or man in priest's orders, which contains nuns, monks, priests, or laity in a variety of possible combinations, and is united to a greater or lesser extent by their liturgy and devotions', especially having a collegiate lifestyle (staffed by communities of clerics).[11]

---

7 'Guthlac', in *Anglo-Saxon Prose*, ed. and trans. by Swanton, p. 44. The meaning of the name is discussed by Watts, *CDEPN*, p. 172, either 'land in a bend', referring to the River Welland, or 'land taken in from marsh'.
8 Bede, *HE*, II. 14, pp. 188–89, with insert from Sherley-Price, p. 132.
9 Bede, *HE*, II. 16, pp. 192–93.
10 Hooke, 'Rivers, Wells and Springs in Anglo-Saxon England', p. 122.
11 Blair, *The Church in Anglo-Saxon Society*, pp. 3–4, 80–83, esp. p. 3.

The Location of Christian Communities

Island Locations

The love of the early saints of the Irish Church for establishing their monasteries upon islands is well known. Inheriting a tradition of the eremitic, ascetic model from the eastern desert fathers, islands offered places suitable for hermits, alongside other places that were 'hidden' or difficult of access such as caves and clefts in mountainsides,[12] or any large area of relatively 'empty' land. Almost every small island off the coasts of Britain attracted early hermits and often, in time, churches and monasteries.

A small rocky island off the coast of south-west Ireland, for instance, Skellig Michael (or Great Skellig, *Sceilig Mhichíl*), was the site of a monastery and hermitage allegedly founded by St Fionán in the sixth century, with stone beehive-shaped huts perched high on a ledge close to the top of one of the two peaks on the island. According to tradition, a small monastery was established in 563 on the small island of Iona, *Ì Chaluim Chille* in Scottish Gaelic, in the Inner Hebrides off the Ross of Mull on the west coast of Scotland, by the monk Columba, an exile from Ireland.[13] This became 'the mothership of Celtic Christianity in Britain'.[14] The succeeding abbey is near the eastern shore of the island, facing the Isle of Mull, and recent excavations have identified the stone foundation of a wooden hut dated by charcoal samples to 540–650 which seems to confirm its use by Columba during his lifetime (563–597).[15] From this base Columba is said to have converted the Picts of present-day Scotland to Christianity in the late sixth century.

From Iona, an Irish monk Aidan was sent to another island base: Lindisfarne off the Northumbrian coast, which became known as Holy Island (Figure 1.1). Its earliest name may have been of Brittonic origin as *insula Medcaut/Metcaut*, perhaps from Latin *medicata* 'drugged, charmed, healed', perhaps even 'holy', changing to Lindisfarne only with its establishment as an Irish Christian missionary centre in 635, noted by Bede as *Lindisfarena ea* or *insula Lindisfarnea*, perhaps derived from 'an Archaic Irish name-expression **Lindis-ferrana*, meaning something like "domain at/of Lindis", [... referring to] the once-tidal streamlet flowing from the pool on the holy island itself'.[16] Here Aidan founded a monastery at a southern tip of the island where he

---

12 Brown, *Pagans and Priests*, pp. 95–100.
13 *Ì Chaluim Chille*, earlier *c*. 700 Latin *Ioua insula* '?yew place': Watson, *The History of the Celtic Place-Names of Scotland*, pp. 87–90.
14 Mayhew-Smith, *Britain's Holiest Places*, p. 463; but see below.
15 Society of Antiquaries of London, 'The Hut that St Columba Built', research and excavations led by Ewan Campbell, University of Glasgow, commissioned by Historic Environment Scotland.
16 Coates, 'Un-English Reflections in Lindisfarne', pp. 241–55. Brittonic is a British-Celtic language which was to split into Welsh, Cumbric, Cornish, and Breton.

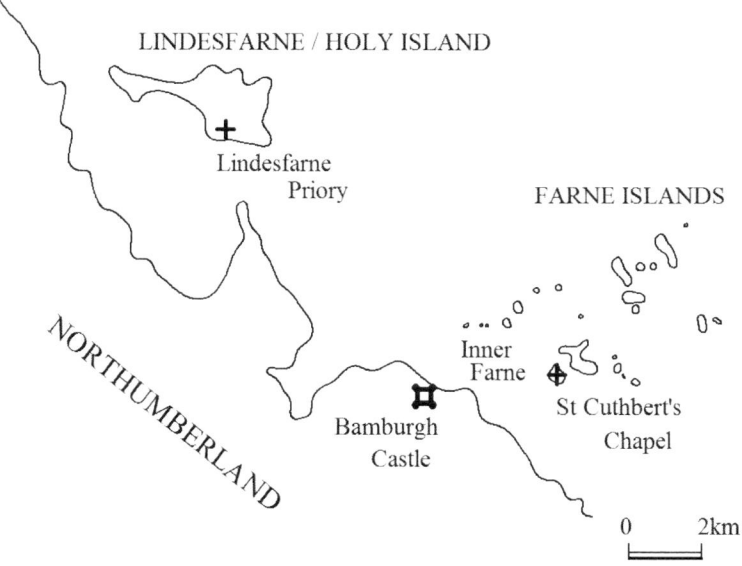

Figure 1.1. Lindisfarne and the Farne Islands, Northumberland. Island and coastal sites of monastic communities and early churches. Map by author.

lived a life of poverty and prayer with his disciples, and this was to become the base for Christian evangelism in the North of England. Aidan may have also been responsible for building the first Christian church on a coastal headland at the royal centre of Bamburgh on the Northumberland mainland, the capital of the Anglo-Saxon kingdom of Bernicia, from which he could see Lindisfarne some six miles to the north. Cuthbert, one of the monks of Melrose, a daughter house of Lindisfarne, had become the prior there, often travelling to remote upland areas to receive the confessions of the people.[17] However, he sought to escape 'worldly glory' and was invited to the monastery on the island of Lindisfarne. Bede's *Life of St Cuthbert* describes this as lying 'undique in medio mari fluctibus circumcinctam' (in the midst of the sea and surrounded on every side by water)[18] or, in his *Historia Ecclesiastica*, 'as the tide ebbs and flows, this place is surrounded by sea twice a day like an island, and twice a day the sand dries and joins it to the mainland'.[19] Still longing for 'the silence and secrecy of the hermit's life', he eventually moved to a small island in inner Farne nine miles from the church of Lindisfarne.[20]

---

17   Bede, *HE*, IV. 27, pp. 431–34.
18   Bede, *VCP*, ch. 17, pp. 214–15.
19   Bede, *HE*, III. 3, trans. by Sherley-Price, p. 147.
20   Farne may have meant 'island of the domain', i.e. 'of Lindisfarne', Coates, 'Un-English Reflections in Lindisfarne', p. 256.

Moreover, this was a place 'aquae prorsus et frugis et arboris inops, sed et spirituum malignorum frequentia humanae habitationi minus accommodus' (utterly lacking in water, corn, and trees; and as it was frequented by evil spirits, it was ill suited for human habitation).[21] Cuthbert confined himself to a cell constructed of stone and turf 'de quibus nisi sursum coelum uidere nihil potuit' (from which he could see nothing except the heavens above).[22] Here he was able to live a life of great austerity.[23] Bishop Aidan retired here in 651, and this continued to be an anchorite cell until 1246. Lindisfarne or 'Holy Island' is today joined to the mainland by a causeway but is still cut off at high tide whereas the Farne Islands are still reachable only by boat, the preserve of seals and migrating seabirds.

Cuthbert was eventually forced to return to the Northumbrian mainland as bishop of Hexham in 684 until his death in 687 when his body was returned to Lindisfarne, as he had wished, and 'atque in aecclesia beati apostoli Petri ad dexteram altaris petrino in sarcophago repositum' (placed in a stone sarcophagus in the church of the blessed apostle Peter on the right side of the altar) at Lindisfarne.[24] His body was removed again when Danes attacked in 793 and 'the heathen miserably devastated God's church in Lindisfarne island by looting and slaughter', an attack portended by 'immense flashes of lightning, and fiery dragons [...] seen flying in the air'. These may have been 'long-tailed comets' which were regarded as portents of disaster.[25] Lindisfarne monastery was in a position to become the hub of the Irish and Roman Christianity of Northumbrian churches, and it housed the diocesan see for a while in the late seventh / early eighth century. But the island site also left it especially vulnerable to Viking attack, and the site eventually had to be abandoned in 875.[26] A Benedictine house was established at Lindisfarne in Norman times in 1093. The church of St Mary the Virgin (restored in the nineteenth century but with the outlines of a Saxon doorway remaining) stands on the original priory site. Recent excavations on the island have revealed new evidence of the early medieval community including several name-stone fragments and two early structures — an early church and signal tower — on the Heugh. A chapel, still standing but heavily renovated in the nineteenth century, was built in the early fourteenth century, perhaps incorporating parts of an earlier building, and was used by the community until the Dissolution. Another small island off Northumberland is Coquet Island, which before 684 was a cell dependent upon the monastery at Tynemouth.

In north Wales, too, early monasteries were established upon several offshore islands, apparently as early as the sixth century. Church sites abound

---

21 Bede, *HE*, IV. 28, pp. 434–37.
22 *VCA*, pp. 96–97.
23 Bede, *HE*, III. 3.
24 Bede, *VCP*, ch. 40, pp. 288–89.
25 *The Anglo-Saxon Chronicle*, ed. and trans. by Swanton, pp. 54–55, and n. 6.
26 See too Bonner, 'St Cuthbert at Chester-le-Street'.

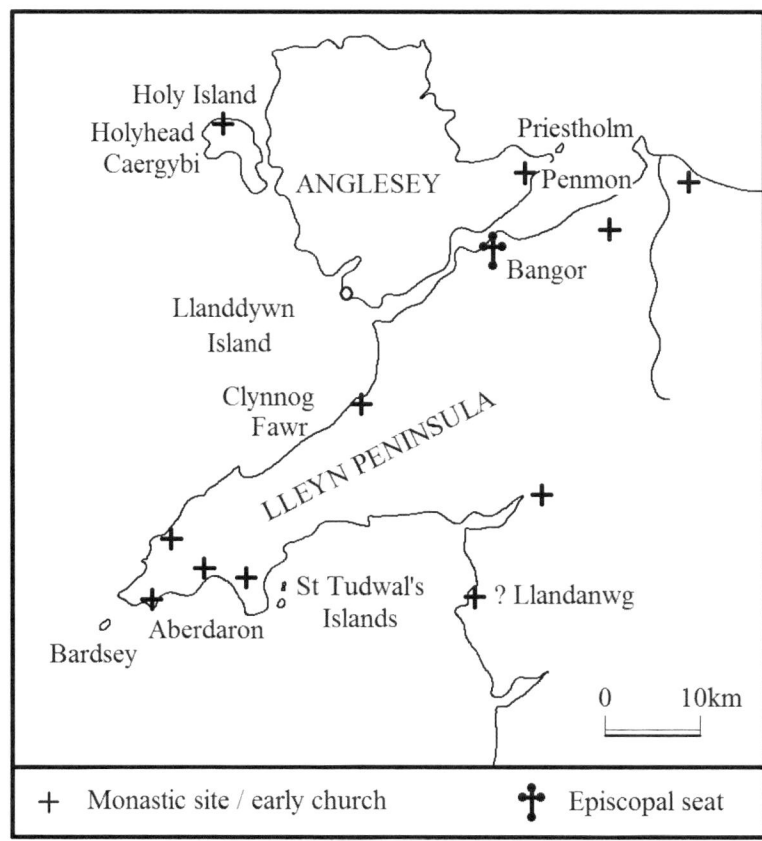

Figure 1.2. North-west Wales, Anglesey, and the Lleyn peninsula. Select island and coastal sites of monastic communities and early churches. Map by author.

along the coast of the Lleyn peninsula, as at Clynnog and Aberdaron, some of which were to give rise to monastic communities (Figure 1.2).[27] Clynnog is the earliest foundation that has been identified archaeologically in Gwynedd (a building possibly dating from the early seventh century when the monastery of St Beuno was established and where it is claimed the saint was buried).[28] The island of Bardsey, *Ynys Enlli* or 'strong current', off the end of the Lleyn peninsula, was said to be the oldest religious house in Wales and became known as 'the burial ground of 20,000 saints', including St Cadfan, a sixth-century saint originally from Brittany, St Dubricius (St Dyfrig), another early saint

---

27 Mayhew-Smith, *Britain's Holiest Places*, pp. 394–400, 423–33.
28 RCHM, Wales and Monmouthshire, *An Inventory of the Ancient Monuments in Caernarvonshire*, II, 36–37.

whose relics were transferred to Llandaff in 1120, and St Deiniol. Bardsey became a major site of pilgrimage in the medieval period, when it was used as a monastic retreat, and the ruins of the thirteenth-century Augustinian abbey of Enlli or St Mary are still visible.[29] On the eastern coast of the isle of Anglesey (*Ynys Môn*), Penmon and Priestholm were associated with St Seiriol. Like Cuthbert he is said to have established the monastery at Penmon (the name meaning 'tip of Môn') but later also moved to a hermitage on a nearby island: Priestholm, otherwise known as Puffin Island or, in Welsh, *Ynys Lannog* or *Ynys Seiriol*, where he was buried. A monastery was also established here in the twelfth century. This, again like the Farne Islands, is noted for its seabird colonies of cormorants, guillemots, razorbills, kittiwakes, and puffins as well as its seal colony. Another island, also off the coast of Anglesey but on the west, became known as Holy Island or *Caergybi*, associated with St Cybi, a Cornish saint, who is said to have regularly met up with Seiriol at Clorach.[30] Another early saint, St Dwynwen, the daughter of King Brychan, is said to have lived a hermit's life (after an unhappy relationship) on a narrow strip of land in south-west Anglesey which becomes an island at high tide known as Llanddwyn Island. St Tudwal's (East), *Ynys Tudwal Fac*, is another tiny island off the Lleyn peninsula associated with St Tudwal, a Breton monk. The island is believed to be the original hermitage of St Tudwal and the remains of a priory, referred to in the 1291 tax rolls, can be found on its eastern side. Off the coast of south Wales, there were eremitic monasteries at Caldey, *Ynys Bŷr*, off Dyfed, at Barry Island — actually a peninsula but known as *Ynys y Barri* in Welsh — off Glamorgan, and at both Steep Holm, 'steep island', and Flat Holm, 'fleet island', in the Bristol Channel, the former associated in legend with St Gildas in the sixth century and the site of a twelfth-century priory, and the latter reputedly used by the disciples of St Cadoc as a retreat, also in the sixth century.[31]

Several early monastic sites, a number of early ones founded by Celtic monks, have been suggested in Cornwall. The twelfth-century Lammana Priory, on Looe Island, apparently stood above earlier Romano-British chapels built of wood but was traditionally thought to date from the sixth century. A later monastery was founded upon St Michael's Mount near Penzance in the eighth–eleventh century. There may have been an early monastery, too, on the island of Tintagel, where recent excavations have revealed what is likely to have been a palace site in the post-Roman/early medieval period when the site was an important trading centre.

There are fewer small islands off south-eastern and eastern England, but Hermitage Rock off Jersey attracted a devout hermit named St Helier who

---

29 RCHM, Wales and Monmouthshire, *An Inventory of the Ancient Monuments in Caernarvonshire*, II, 17.
30 Henken, *Traditions of the Welsh Saints*, p. 235.
31 According to Watts, *CDEPN*, p. 232, this was named after Viking fleets who used the island as a base, its original name (*æt*) *Bradan Relice* AD 918.

is said in an eleventh-century *Life* to have been martyred by pirates in 555, shielding the villagers of Jersey from a Viking raid.[32] In Kent there were early Saxon minsters on the isles of Sheppey, or 'sheep island', and Thanet — genuine islands in the Anglo-Saxon period — the former founded on the northern shore of the island *c.* 670 by Seaxburh, widow of King Eorcenberht of Kent. There were several on Thanet founded in the late seventh and eighth centuries, but, like so many other monastic sites in eastern England, they were probably destroyed by successive Danish attacks. Other communities were established on coastal headlands that were likewise surrounded on three sides by water and relatively remote. In what had been originally South Saxon territory, the precise location of Selsey minster, founded *c.* 710–20 on the end of the Manhood peninsula and Selsey Bill in West Sussex, is not known but it may have been at Church Norton on the shore of Pagham Harbour or on a site now submerged by the sea. Hoo monastery, to the north-east of Rochester in Kent, was said to have been founded *c.* 686–87 on an island by St Werburgh — it lies on the peninsula jutting out into the sea to the north of the Medway.

Fens and Marshlands

Clearly, islands and isolated headlands could not be the only places chosen for the establishment of early monasteries and were more suited to those seeking an eremitic existence. Equally attractive were the fenland and marshland areas of England. While it is not possible to note all monastic foundations here, many other early monasteries and hermitages were established in such watery locations, many noted by Bede in his *HE*, completed by 731. Ely was founded in *c.* 672/73 by Æthelthryth (St Etheldreda, daughter of the East Anglian King Anna), and its site resembled an island as it was surrounded by marshes or water: 'regio Elge undique est aquis ac paludibus circumdata, neque lapides maiores habet' (for the district of Ely is surrounded on all sides by waters and marshes and has no large stones).[33] Bede also notes how the monastery at Melrose, founded by St Aidan of Lindisfarne shortly before his death in 651, was also described as being almost encircled by water: 'quod Tuidi fluminis circumflexu maxima ex parte clauditur' (which is almost encircled by a bend in the river Tweed).[34] These marshlands and fenlands not only represented a world set apart but provided resources such as fish, eels, and waterfowl for the monks with the help of constructed weirs and fish-traps.

The fenland marshes continued to attract hermits and monastic communities, usually situated upon islands of slightly higher land amidst the marshes such as Ely and Crowland. The first community at Thorney, 'thorn-tree island', a

---

32 Mayhew-Smith, *Britain's Holiest Places*, pp. 28–29.
33 Bede, *HE*, IV.19, pp. 394–95. Ely, according to Bede, means 'eel district' but is further discussed by Watts, *CDEPN*, p. 215.
34 Bede, *HE*, V. 12, pp. 488–89.

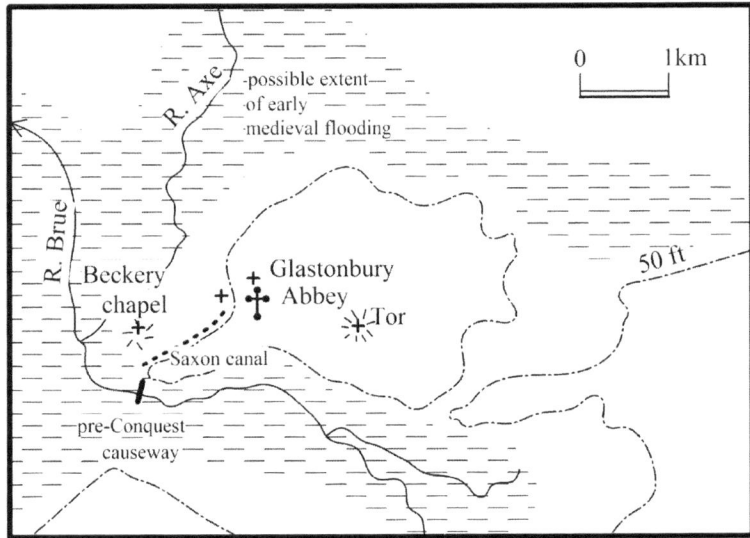

Figure 1.3. Glastonbury monastic centre (also showing churches and areas liable to flood). The dotted line following the foot of the upland to the west of the tor represents the dug canal which has now produced evidence of an Anglo-Saxon date. Map by author (based upon Blair, 'Transport and Canal Building on the Upper Thames', p. 267, fig. 61).

few miles to the south-east of Crowland, was established as a hermitage in the seventh century, as its original name *Ancarig igland* 'hermit-island' indicates,[35] only to be destroyed by Viking raids in the ninth century and re-established as a Benedictine house in the 970s. It is associated with three local martyrs, St Tancred, St Torthred, and St Tova. Further south, Chatteris was the site of a hermitage associated with another local saint, St Huna. Ramsey Abbey also stood upon an island in the fens and was founded by St Oswald of Worcester in 969. The cathedral of Peterborough was first built in 655 at a place then known as *Medeshamstede* and was situated on higher ground.

Recent archaeological discoveries at Glastonbury have cast a new light upon early monasticism (Figure 1.3). With enormous areas of wetland in what is now Somerset but was in Anglo-Saxon times part of the kingdom of Wessex, a number of religious centres were surrounded by marshes. Glastonbury Abbey lay within the marshy lowlands of the Somerset Levels below Glastonbury Tor (the fictional 'Isle of Avalon') close to the River Brue; its Welsh name, not certainly predating the late twelfth century, was *Inswytrin* (c. 601), sometimes interpreted as 'glassy island' due to the shine of its waters.

---

35 Reaney, *The Place-Names of Cambridgeshire and the Isle of Ely*, p. 280.

Avalon, if authentic, is a Brittonic name containing *aβall* 'apple' (singular *aβallen*) while Glastonbury is partly Brittonic, probably *glastan*, from British **glasto-tanna*, becoming confused 'through folk-etymology and obscured by mistranslation'.[36] Coates notes (personal communication) that **glasto-* 'is a colour term in the region blue/ green/ grey', while *glastan* is a Cornish derivative (plural form). Padel and Coates have further suggested the etymon of the folk-name *Glæstingas* from an uncertain British form, that is, the ancestor of Cornish *glastan* 'fruiting oaks'.[37] Padel has also considered the evergreen 'holm-oak', but this is a tree of Mediterranean regions and is not normally regarded as native to Britain. Watts also mentions 'woad' from the Gaulish *glastum* (one of the meanings of Latin *vitrum*), but this is a very much later recorded word, at least in Welsh. He concludes that the name probably means 'the fortified place of the Glæstingas', with interpretations as 'glass' a folk-etymological translation.

A cemetery of at least fifty individuals at Glastonbury — all but three adult males, with a nuclear grave which was perhaps that of a hermit[38] — was probably that of the monks of Beckery Chapel. This was built on rising ground within a great bend of the River Brue, at twelve metres (39 ft) only just above floodwaters, to the south-west of the present Glastonbury town centre. In May 2016 seven burials, following further excavations, were subject to carbon dating and found to date to the fifth or sixth century AD, making this, which Rahtz had considered to be a daughter monastery of Glastonbury Abbey,[39] the earliest archaeological evidence for monasticism in Britain, effectively a large hermitage with simple wattle-and-daub buildings. Not only that, but of the site, Richard Brunning, the site director, commented: 'It's on a small island just off Glastonbury so it's surrounded by wetlands and cut off from normal life'.[40] Such an early use of the site could be perhaps why Glastonbury with its legendary 'Isle of Avalon' was to play such a large part in Arthurian legend. The early use of the site even predates Iona Abbey in Scotland which was founded in the late sixth century (above), with Glastonbury Abbey itself founded in the seventh/eighth century. Legends link the chapel with an Irish connection, in particular St Bridget who, according to tradition, visited Beckery in 488.[41] Moreover, Beckery is now recognized as a wholly Goidelic name, *Becc-Ériu* 'little Ireland', although this occurs in a forged tenth-century charter surviving in an eleventh-/twelfth-century document.[42] Monastic

---

36 Watts, *CDEPN*, pp. 251–52; Rahtz, *English Heritage Book of Glastonbury*, p. 33; Coates, 'Gazeteer', pp. 330, 333.
37 Padel, *Cornish Place-Name Elements*, p. 104.
38 Rahtz, *English Heritage Book of Glastonbury*, p. 122.
39 Rahtz, *English Heritage Book of Glastonbury*, pp. 11, 118–22.
40 BBC News, 5 December 2016, and *The Times*, 6 December 2016.
41 Rahtz, *English Heritage Book of Glastonbury*, p. 119.
42 Coates, 'Liscard and Irish Names in Northern Wirral', p. 262, also Coates, 'Gazeteer', p. 334; Sawyer, *Anglo-Saxon Charters*, S 783.

Figure 1.4. Glastonbury Tor rising above the surrounding lowlands. Photo by author.

use of the site may have ended in the late ninth century when Somerset was attacked by the Vikings, but a new chapel and monastic complex were built in the late Saxon period.

On the tor itself, a post-Roman settlement has been archaeologically identified which led onto a series of Anglo-Saxon churches by the seventh century and a possible monastic community and a hermitage retreat from the abbey which continued in use into the post-Conquest period when the church of St Michael was erected, the tower of which survives today (Figure 1.4). This site was always subsidiary to Beckery or Glastonbury Abbey.[43] Stephen Rippon notes how John of Glastonbury's mid-fourteenth-century *Cronica sive Antiquites Glastoniensis Ecclesie* continues to refer to the symbolic role of marshlands and wetland environments as specially suitable for monastic retreats:

> of all the places he [Patrick] might have chosen, he settled upon Glastonbury as the spot most apt for his triumphs over the devil, where he might be able to earn most fully the joys of a heavenly reward; for the place was then suited to heavenly vigils because of its remoteness from mankind, being almost inaccessible because of the marshes.[44]

---

43 Rahtz and Watts, *Glastonbury*, pp. 67–84.
44 Rippon, 'Marshlands and Other Wetlands', p. 93. Irish pilgrims, whose books had been apparently studied by Dunstan, the tenth-century abbot of Glastonbury, claimed that St Patrick came to Glastonbury after the conversion of Ireland in the fifth century where he encountered holy men living as anchorites (perhaps, according to legend, on the tor), who

Muchelney, 'at the great island', was situated further south on an island of dry ground in the marshes between the Rivers Isle and Yeo, its abbey founded in the seventh or eighth century. A later foundation, Athelney Abbey, founded by King Alfred *c.* 888 at the place at which he had sought refuge from the Danes, its name meaning 'the island of the princes', was again situated on a small island in swampland just north of the River Tone in the parish now known as East Lyng, a place 'surrounded on all sides by very great swampy and impassable marshes', surely a reference to the ancient habit of founding minsters on islands.[45]

In eastern England, St Guthlac sought seclusion at Crowland in Lincolnshire, an isolated location in the heart of the fens, as previously noted above. A minster had been founded in his memory in the early eighth century by Æthelbald, king of Mercia, who had found refuge there with Guthlac when fleeing from his cousin King Ceolred of Mercia.[46] This was, however, another community to be entirely destroyed by the Danes in 866. It was refounded in the reign of King Edred as a monastery of the Benedictine Order about twenty years later but was again destroyed by fire in 1091 and had to be rebuilt yet again by Abbot Joffrid. As if this was not enough to try the faith of the monks, the greater part of the abbey and church was once more burnt down in 1170. Notwithstanding, it was to be rebuilt yet again under Abbot Edward, after which it was to enjoy almost unbroken prosperity down to the time of the Dissolution.

This was but one of the early minsters to suffer from Viking attacks. The first wave of raids began in the late eighth century, such as that upon Lindisfarne in 793, described thus by the Northumbrian scholar Alcuin of York: 'never before has such terror been seen in Britain as we have suffered from heathen people'.[47] The monastery at Iona was attacked in 795 and was burnt in AD 802, with sixty-eight monks being killed in another raid in AD 806. This was followed by further raids on England in the ninth century which came not from Norway — the Norwegians having moved on around Scotland to Ireland — but mostly from Danes from around the entrance to the Baltic, first attacking the Isle of Sheppey, in Kent, in 835 but then moving northwards.[48] Many other monasteries in Scotland and northern England simply disappear

---

elected him their superior. A cult of St Patrick flourished at Glastonbury in the tenth century. This was reiterated in forged medieval documents, and the monks claimed that Patrick had been buried here. There is little historical evidence to substantiate this. A lost cross marking the so-called 'grave of Arthur' is also likely to have been a medieval forgery, presumably to bring in increased revenue from pilgrims (or because the monks genuinely believed in the legend): Rahtz and Watts, *Glastonbury*, pp. 58–59, 62–63, and 73.

45 Havinden, *The Somerset Landscape*, p. 93, citing Asser's *Life of Alfred* in *English Historical Documents, c. 500–1042*, ed. by Whitelock, p. 273.
46 Mayhew-Smith, *Britain's Holiest Places*, p. 282.
47 Cavill, *Vikings, Fear and Faith in Anglo-Saxon England*, pp. 6–7.
48 Stenton, *Anglo-Saxon England*, pp. 239–43.

from the record. Raiding by Danish armies was clearly disastrous for many early minster foundations, especially in eastern England, with many destroyed in the late eighth-/ninth-century raids, and although most were later to be refounded, some were never replaced and others suffered yet again a further wave of raids in the tenth century.

Riverine Locations

Apart from marshlands, many river flood plains were often undrained and liable to flooding in the early medieval period. Nevertheless, such 'watery' locations may have had more to offer in a spiritual sense. John Blair has drawn attention to the siting of some early Christian minsters in Oxfordshire being established in places almost encircled by water; 'enabling the minsters to be *in* the world but not quite *of* it'.[49] Along the Upper Thames and its tributaries were a number of Anglo-Saxon minsters, all located upon the riverside gravel terraces close to crossing points of the river: St Frideswide's minster at Oxford was allegedly an early foundation which stood on the site of the present cathedral beside the many-channelled River Thames including the Trill Mill Stream which may have been recut in the eighth century for either defence, mill power, or navigation; the tenth-century minster of Bampton stood on a gravel promontory in a bend of the Shill Brook, a stream feeding into the Thames, not far from the possible pagan religious site which may have given rise to the place-name; and Eynsham minster, perhaps founded by the ninth century, stood beside another Thames tributary, the Chil Brook. Another early foundation was that of Abingdon, founded in the late seventh century, most likely on the site of St Helen's Church at the confluence of the Thames and the Ock.[50] In present-day Surrey, Chertsey Abbey, 'Cerot's island', founded in 666, lay close to the Bourne, a tributary of the Thames, while other Thames-side minsters included Battersea, 'Beaduric's island', and Bermondsey, 'Beornmund's island' — the latter again a reference to 'dry ground in marshland beside the Thames before the river was embanked'.[51] Canterbury in Kent, beside the River Stour, had been a Christian centre since before the arrival of St Augustine in 597, but interestingly, the Roman place-name *Durovernum Cantiacorum, Duro Averno Cantiacorum*, is derived from British *douro + *uerno*- 'alder fort, walled town by the alder swamp' (of the *Cantwara*).[52] Within the Hwiccan kingdom — later a part of Mercia — the majority of early minsters lay beside rivers: at Worcester, the site of the seventh-century diocesan see, with a monastery and at least five Anglo-Saxon

---

49 Blair, *The Church in Anglo-Saxon Society*, p. 194, citing Gittos, 'Sacred Space in Anglo-Saxon England', pp. 39–42.
50 Blair, *Anglo-Saxon Oxfordshire*, pp. 61–66.
51 Watts, *CDEPN*, p. 52.
52 Watts, *CDEPN*, p. 114; the British *uerno- becoming Welsh *gwern* 'alder'.

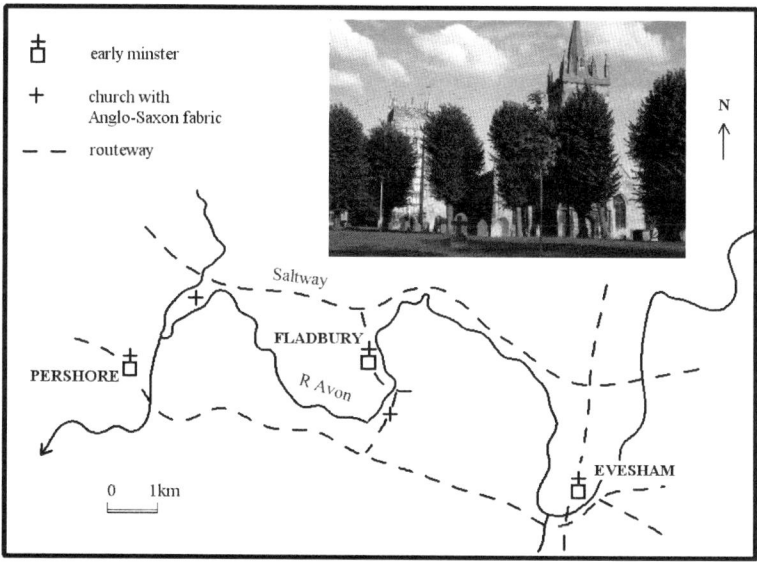

Figure 1.5. River Avon minsters in Worcestershire with an inset photograph of Evesham Abbey today. Map and photo by author.

churches; Kempsey; Twyning, 'between' rivers (see below); Tewkesbury c. 715; Deerhurst, late seventh-century; Gloucester Cathedral Abbey, before 679; and Gloucester St Oswald's Priory (660), together with Berkeley (before 807) beside the Severn, Kidderminster (c. 735) beside the Stour, and Pershore (c. 689), Fladbury (691–93), Evesham (c. 701), and Stratford (693–717) beside the Avon (Figure 1.5).

Beside lesser rivers lay Wootton Wawen (723–37) and probably Tredington (Warwickshire), Tetbury (before 680), Cheltenham (eighth-century), Daylesford (?718 × 727), Winchcombe (787), Cirencester (before 839), and Withington (?674–704) (Gloucestershire). The Roman site at Bath, beside the Bristol Avon, was to acquire a Christian minster probably by 675, founded by Osric, king of the Hwicce, but was to be taken into Wessex in 781.[53] Kempsey minster (Worcestershire) had probably been founded before 820, and its name, 'Cemmi's island', refers to the slightly raised patch of dry land upon which the present church stands, still an island in periods of flood (Figure 1.6). The minster at Evesham (Worcestershire), probably founded c. 703,[54] was set within a great bend of the River Avon. An early name for this site, *Cronuchomme*, refers to 'land hemmed in by water, land in a river bend' and, in this case, one frequented by cranes. Also within that kingdom, Twyning (Gloucestershire)

---

53 Sawyer, *Anglo-Saxon Charters*, S 1257.
54 Cox, *The Church and Vale of Evesham*, pp. 8–10.

Figure 1.6. Kempsey church beside the River Severn in Worcestershire. An early minster site on raised ground in the river meadows. Photo by author.

was the site of an early minster recorded in 814 but perhaps founded before c. 770 set 'between rivers' (*Tweoneaum*, i.e. *betwēonan* + **ēum*), here the Severn and its tributary the Avon.

Further north, in the kingdoms of Northumbria and Lindsey, several early minsters were established at similar 'island' sites, many first recorded by Bede c. 731: in Lincolnshire, Bardney, where a minster was established before 697 to the east of the River Witham, is 'Bearda's island', and Partney is 'Peata's island'; Lastingham in North Yorkshire, by Bede named as *Læstingae*, was named after a folk called the **Læstingas* with *ēg*; an early name for Hexham (founded 674) was *Inhagustaldaesae, Inhegustaldesiae*, perhaps **Hehstaldes ēg*, 'Hagustald's island';[55] and the monastery at Hartlepool was said by Bede to lie at *Heruteu* 'the island of the hart', a rocky peninsula connected to the mainland only by a narrow neck of blown sand, with the harbour to the south of this.[56]

Another reason for riverine siting has been suggested by Stocker and Everson. They note that places in Lincolnshire where votive offerings in rivers had been particularly abundant, especially along the River Witham and its tributaries, might have influenced the siting of some early churches and monasteries such as Kirkstead and Bardney. The monasteries there often assumed the responsibility for maintaining the earlier causeways that had

---

55 Watts, *CDEPN*, p. 301.
56 Bede, *HE*, III. 24, pp. 292–93.

provided access to pools along the course of the river where votive offerings had been made since prehistoric times, signalling the triumph of Christianity.[57]

Holy Water

With water essential for liturgical purposes, springs also seem to have continued to have a symbolic significance in Christian times.[58] Heathen shrines might be sanctified by dedication to a Christian saint, and water was an essential part of Christian baptism — water was seen increasingly as a means of purification and blessing.[59] Wells minster in Somerset, built on the site of a Roman mausoleum, became an ecclesiastical centre when it became the centre of a new diocese in AD 909, and the freshwater springs that gave rise to the place-name still flow to the south-east of the cathedral.[60] The cathedral at Winchester, built over a Roman mausoleum, was also built to incorporate ancient wells, one of the earliest in the Old Minster possibly dating from the seventh century.[61] Hagiographical literature is full of references to springs, especially those rising at the spot of a saint's martyrdom, as in the case of St Kenelm where a spring rose beneath the church constructed on the site at Romsley near Clent in Worcestershire.[62]

'Holy wells' also occur occasionally in Anglo-Saxon charter bounds. The south-eastern part of the boundary of Withington in Gloucestershire takes in a small area beyond the River Coln, recording here a *halgan wyllan* (Figure 1.7a–b). This lay not far from a Roman villa site in the adjacent parish of Chedworth, and Roman finds have also been found in the vicinity of the spring.[63] The villa site contained temples dedicated to various Roman gods but also a spring-fed pool dedicated to the water-nymphs (a nymphaeum) which supplied the villa with water. A Christian chi-rho symbol had been scratched on one of the rim-stones, probably in the fourth century AD. This testifies to the Christianization of pagan sites in the complex and perhaps even turning it into a Christian baptistery.[64] This may indicate the reuse of a pagan shrine for the 'new' religion, offering a similar scenario for the nearby 'holy well'.[65]

---

57 Stocker and Everson, 'The Straight and Narrow Way', pp. 277–79.
58 Hooke, 'Rivers, Wells and Springs in Anglo-Saxon England'.
59 Gittos, *Liturgy, Architecture, and Sacred Places*, p. 202; See Carolyn Twomey, 'Rivers and Rituals: Baptism in the Early English Landscape', in this volume.
60 Rodwell, 'From Mausoleum to Minster'.
61 Kjølbye-Biddle, 'Anglo-Saxon Baptisteries of the 7th and 8th Centuries'.
62 *Vita et miracula Sancti Kenelmi*, in *Three Eleventh-Century Anglo-Latin Saints' Lives*, ed. by Love, pp. 49–90.
63 Sawyer, *Anglo-Saxon Charters*, S 1556; RCHM, England, *Ancient and Historical Monuments in the County of Gloucester*, p. 25.
64 Wilson, *A Guide to the Roman Remains in Britain*, p. 197.
65 For other sites recorded in pre-Conquest charters, see Hooke, 'Rivers, Wells and Springs in Anglo-Saxon England'.

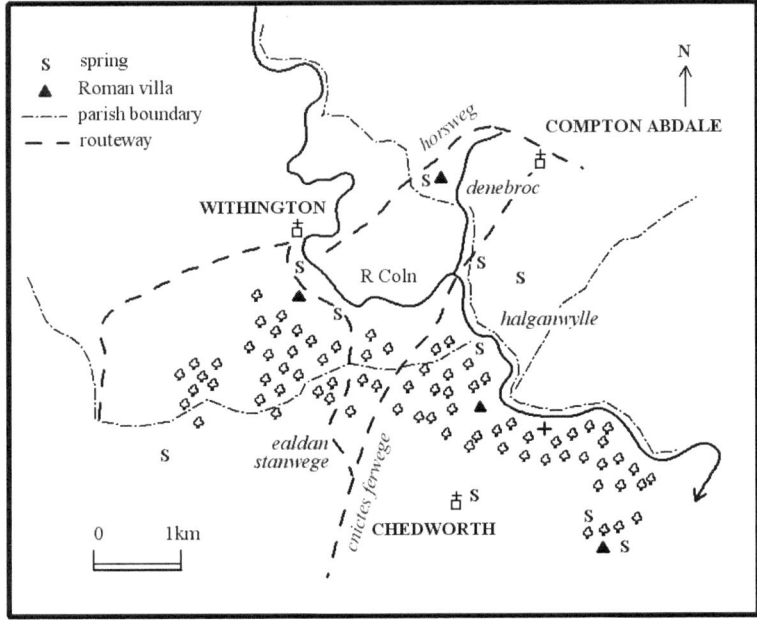

Figure 1.7a. Withington holy well in a boundary clause (S 1556) contained in an eleventh-century manuscript. Map by author.

Figure 1.7b. Withington holy well today. Photo by author.

Hundreds of churches have springs close by, but it is rarely possible to know if the spring had any earlier religious significance.

Commerce and Markets

Rivers also have offered a means of transport for the building materials required, and those in coastal locations were particularly well situated for communication and commerce. However, sites beside rivers were also to become commercially useful as so many were at the crossing places of the rivers, giving such a location facilities to attain 'central-place' status.

While the siting of early minsters may have had something of a symbolic quality, there is no doubt that riverine and coastal sites were thus enormously beneficial as the established Church began to prosper in the world of commerce and trade. Many early monastic communities began as hermitages, but others were established by kingly and elite families, as well as bishops, not only in order to spread the Christian faith but also as a means of asserting lordship over sizeable areas and strengthening their power base.[66] Indeed, some early minsters were subsumed into more powerful ones, such as Stratford in Warwickshire taken over by the church of Worcester, and conflict might continue over the ownership of estates even until the time of Domesday Book and beyond, such as over the ownership of Bengeworth and Hampton in Worcestershire which was contested between the church of Worcester and the abbey of Evesham.[67]

Many rivers were navigable by small craft, and the sea was a major highway for commerce. Canals were even being constructed by the eleventh century to help ships to navigate past difficult sections of rivers, as on the River Itchen where a new cutting was made to allow boats to pass as far as Bishopstoke, thereby moving goods closer to Winchester.[68] At Abingdon Abbey (Figure 1.8), which had been refounded after its virtual destruction by the Danes in the ninth century, the abbey's chronicles record how the citizens of Oxford petitioned for the course of the Thames to be diverted near Thrupp (where the river was particularly shallow), through the church's meadow, in order to ease access by rowers; this was on the understanding that a hundred herrings should be paid from each of their ships to the monks' cellarer.[69] Thus the river was canalized to facilitate downriver traffic from Oxford between 1052 and 1066, possibly by enlarging the channel of the Swift rivulet to the south of Andersey Island.[70]

---

66 Sargent, 'Lichfield and the Lands of St Chad', pp. 291–93.
67 *VCH Worcestershire*, pp. 397–99, 404–05.
68 Blair, 'Transport and Canal Building on the Upper Thames', pp. 258–59; Gardiner, 'Inland Waterways and Coastal Transport'.
69 Blair, 'Transport and Canal Building on the Upper Thames', p. 258, citing *Chronicon Monasterii de Abingdon*, ed. by Stevenson, I, 480–81.
70 Blair, 'Transport and Canal Building on the Upper Thames', p. 268.

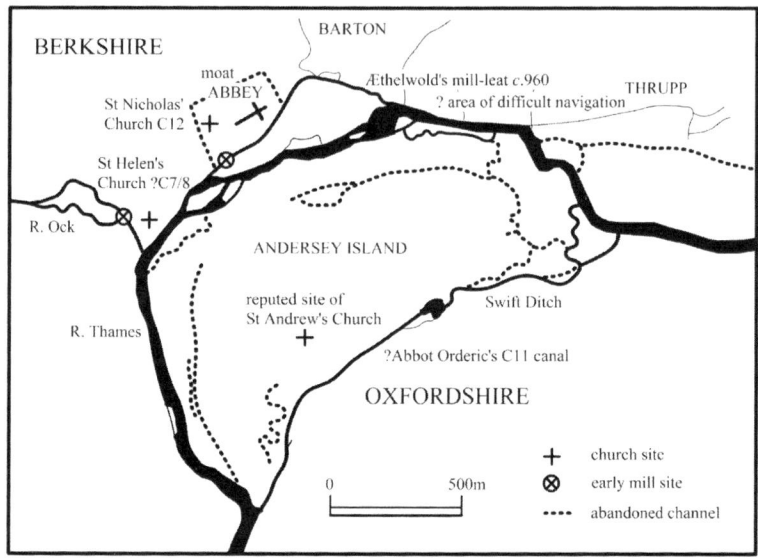

Figure 1.8. The site of Abingdon Abbey, its early water channels and canal. Map after Bond, 'The Reconstruction of the Medieval Landscape' and Blair, 'Transport and Canal Building on the Upper Thames'.

Several of the nearby streams also powered the abbey's mills, two recorded on the manor of Barton in Domesday Book by 1086.[71] The river meadows played an important role in the abbey's life. At that time Abingdon was on the eastern boundary of Berkshire while the adjacent Andersey Island lay in Oxfordshire in the parish of Culham. The chronicle notes how ownership of the meadow beside the flooded Thames was in dispute between the monks of Abingdon and the inhabitants of Oxfordshire in the mid-ninth century, perhaps due to the changing course of the river over time. This was settled by taking a sheaf of corn and a lighted taper, which, placed upon a round shield, was launched into the brook surrounding the meadow, its course drifting in the river deciding the issue in favour of the monks.[72] At Glastonbury, an early canal running north-eastwards from the Brue valley along the foot of the higher land of Wearyall Hill towards the abbey may have been constructed in the pre-Conquest period to transport building material during the abbey's rebuild in the tenth century during the abbacy of Dunstan,[73] when the abbey church was enlarged and the cloisters added (see Figure 1.3).

---

71 *Domesday Book*, vol. v: *Berkshire*, ed. by Morgan, 7.6.
72 *Chronicon Monasterii de Abingdon*, ed. by Stevenson, I, 88–90. Gelling, *The Place-Names of Berkshire*, p. 454.
73 Hollinrake and Hollinrake, 'Glastonbury's Anglo-Saxon Canal', pp. 235–38.

Clearly rights to transport along rivers and loading places were of considerable commercial value. With the expansion of Mercia in the eighth century, Milred, bishop of Worcester, was granted the toll of two ships at London.[74] Other similar grants included tolls for ships landing at certain ports in Kent were to Deneheah, abbot of Reculver, in AD 747 (at Fordwich) and to Sigeburga, abbess, and her *familia* at St Peter's Minster, Thanet, *c.* AD 761 (at Fordwich and Sarre).[75] Somewhat later, in 898/99, the bishop of Worcester and the archbishop of Canterbury were both granted the right to moor ships along the width of their properties at *Æðeredes hyd*, the later Queenhithe, on the Thames — the earliest evidence, if dated correctly, for commercial activity in the newly refounded city of London.[76]

Markets were being established by monasteries and abbeys, often at the gates of their precincts or outside the church, and 'attendance at religious ceremonies might have provided a convenient opportunity to meet and exchange goods within the territory of the church'.[77] At first, such gatherings may have been relatively informal, especially as the growing communities needed to purchase increasing amounts of food at a time when their estates were also increasing in extent. In Mercia, Æthelred, ealdorman, granted the church of St Peter at Worcester in 884 × 901 'half of all the rights which belong to their lordship, whether in the market or in the street […] excepting the wagon-shilling and the load-penny at Droitwich go to the king as they have always done'.[78] Evesham Abbey, founded in the early eighth century, contributed towards the growth of the town, and its market grew up directly outside the gates; this is firmly documented in 1055 when King Edward is said to have granted the town the status of a 'borough', thus granting it the legal right to profit by regulating the conduct of the market, especially by taking payments from those wishing to transact business within it, with produce of goods and livestock at first probably drawn from the abbey's estates.[79] These are two of the earliest documented markets in England.

Conclusions

Although it is not always possible to confirm the date for the earliest foundations of many Anglo-Saxon minsters, or to disentangle fact from legend in corroborating their association with particular Christian saints, their location

---

74 Sawyer, *Anglo-Saxon Charters*, S 98; Hooke, 'Water in the Landscape', p. 37.
75 Sawyer, *Anglo-Saxon Charters*, S 1612, S 29.
76 Sawyer, *Anglo-Saxon Charters*, S 1628.
77 Vince, 'The Growth of Market Centres and Towns in the Area of the Mercian Hegemony', p. 187.
78 Sawyer, *Anglo-Saxon Charters*, S 223.
79 Cox, *The Church and Vale of Evesham*, p. 74, citing *Chronicon Abbatiae de Evesham*, ed. by MacRay, m. 75.

is usually known. It is not perhaps surprising that hermits chose to establish themselves in remote places such as islands in the sea or in wide areas of fenland and marsh, but many of these eventually became monastic communities in their own right. Even then, other early minsters often seem to have chosen a site which was virtually surrounded by water, even within braided rivers or marshy floodplains. It is remarkable how many early minsters were located at places with Old English *ēg* / *īeg* place-names, signifying places 'surrounded by water',[80] and no other place-name term matches this in frequency. It appears that such sites provided some degree of separation from the everyday secular world. However, in time these very locations provided excellent sites for trade and commerce, and the monasteries were seldom slow to make the most of this. Indeed, many later monasteries were deliberately sited by powerful kingly and elite families or by bishops not only to spread Christianity and parochial care but to establish their lordship over sizeable areas and thus to extend their power base. Inextricably connected to the rivers, springs, and wetlands around them, the Anglo-Saxon Church thrived and profited spiritually and financially from their watery landscapes.

Works Cited

Primary Sources

The Anglo-Saxon Chronicle, ed. and trans. by Michael Swanton (London: Dent, 1996)
Anglo-Saxon Prose, ed. and trans. by Michael Swanton (London: Dent, 1975)
Bede, *Bede's Ecclesiastical History of the English People*, ed. and trans. by Bertram Colgrave and R. A. B. Mynors (Oxford: Clarendon Press, 1969)
―――, *Bede's Ecclesiastical History of the English People*, trans. by L. Sherley-Price, rev. by R. E. Latham (London: Penguin Books, 1990)
―――, *Vita S. Cuthberti prosaica*, in *Two Lives of Saint Cuthbert: A Life by an Anonymous Monk of Lindisfarne and Bede's Prose Life*, ed. and trans. by Bertram Colgrave (Cambridge: Cambridge University Press, 1940), pp. 141–307
Chronicon Abbatiae de Evesham, ed. by William Dunn MacRay, Rolls Series, 29 (London: Longman, Green, Longman, Roberts and Green, 1863; repr. Cambridge: Cambridge University Press, 2012)
Chronicon Monasterii de Abingdon, ed. by Joseph Stevenson, Rolls Series, 2 vols (Cambridge: Cambridge University Press, 2012)
Domesday Book, vol. v: *Berkshire*, ed. by Philip Morgan (Chichester: Phillimore, 1979)
English Historical Documents, c. 500–1042, ed. by Dorothy Whitelock (London: Eyre and Spottiswoode, 1955)

---

80 Hadcock and Knowles, *Medieval Religious Houses*; study by D. Hooke in progress.

The Exeter Book, ed. by George Philip Krapp and Elliott Van Kirk Dobbie, Anglo-Saxon Poetic Records, 3 (New York: Columbia University Press, 1936)

Three Eleventh-Century Anglo-Latin Saints' Lives, ed. by Rosalind Love (Oxford: Clarendon Press, 1996)

Vita S. Cuthberti auctore anonymo, in *Two Lives of Saint Cuthbert: A Life by an Anonymous Monk of Lindisfarne and Bede's Prose Life*, ed. and trans. by Bertram Colgrave (Cambridge: Cambridge University Press, 1940), pp. 59–139

Secondary Works

Blair, John, *Anglo-Saxon Oxfordshire* (Stroud: Alan Sutton, 1994)

——, *The Church in Anglo-Saxon Society* (Oxford: Oxford University Press, 2005)

——, 'Transport and Canal Building on the Upper Thames', in *Waterways and Canal-Building in Medieval England*, ed. by John Blair (Oxford: Oxford University Press, 2007), pp. 254–86

Bond, James, 'The Reconstruction of the Medieval Landscape: The Estates of Abingdon Abbey', *Landscape History*, 1 (1979), 59–75

Bonner, Gerald, 'St Cuthbert at Chester-le-Street', in *St Cuthbert, his Cult and his Community to AD 1200*, ed. by Gerald Bonner, David Rollason, and Clare Stancliffe (Woodbridge: Boydell, 1989), pp. 397–411

Brown, Michelle, *Pagans and Priests: The Coming of Christianity to Britain and Ireland* (Oxford: Lion Hudson, 2006)

Cavill, Paul, *Vikings, Fear and Faith in Anglo-Saxon England* (London: HarperCollins, 2001)

Coates, Richard, 'Gazeteer', in *Celtic Voices, English Places*, ed. by Richard Coates and Andrew Breeze with David Horovitz (Stamford: Shaun Tyas, 2000), pp. 263–345

——, 'Liscard and Irish Names in Northern Wirral', in *Celtic Voices, English Places*, ed. by Richard Coates and Andrew Breeze with David Horovitz (Stamford: Shaun Tyas, 2000), pp. 260–62

——, 'Un-English Reflections in Lindisfarne', in *Celtic Voices, English Places*, ed. by Richard Coates and Andrew Breeze with David Horovitz (Stamford: Shaun Tyas, 2000), pp. 241–59

Cox, David, *The Church and Vale of Evesham, 700–1215: Lordship, Landscape and Prayer* (Woodbridge: Boydell, 2015)

Gardiner, Mark, 'Inland Waterways and Coastal Transport: Landing Places, Canals and Bridges', in *Water and the Environment in the Anglo-Saxon World*, ed. by Maren Clegg Hyer and Della Hooke, Exeter Studies in Medieval Europe (Liverpool: Liverpool University Press, 2017), pp. 152–66

Gelling, Margaret, *The Place-Names of Berkshire*, Part II: *The Hundreds of Kintbury, Lambourn, Shrivenham, Ganfield, Ock, Hormer, Wantage, Compton, Moreton*, English Place-Name Society, 50 (Cambridge: English Place-Name Society, 1974)

Gittos, Helen, *Liturgy, Architecture, and Sacred Places in Anglo-Saxon England* (Oxford: Oxford University Press, 2013)

———, 'Sacred Space in Anglo-Saxon England: Liturgy, Architecture and Place' (unpublished doctoral thesis, University of Oxford, 2002)

Hadcock, David, and Richard Neville Knowles, *Medieval Religious Houses: England and Wales* (London: Longman, Green, 1953)

Havinden, Michael, *The Somerset Landscape* (London: Hodder & Stoughton, 1981)

Henken, Elissa R., *Traditions of the Welsh Saints* (Cambridge: D. S. Brewer, 1987)

Hollinrake, Charles, and Nancy Hollinrake, 'Glastonbury's Anglo-Saxon Canal', in *Waterways and Canal-Building in Medieval England*, ed. by John Blair (Oxford: Oxford University Press, 2007), pp. 235–43

Hooke, Della, 'Rivers, Wells and Springs in Anglo-Saxon England: Water in Sacred and Mystical Contexts', in *Water and the Environment in the Anglo-Saxon World*, ed. by Maren Clegg Hyer and Della Hooke, Exeter Studies in Medieval Europe (Liverpool: Liverpool University Press, 2017), pp. 107–35

———, 'Water in the Landscape: Charters, Laws and Place-Names', in *Water and the Environment in the Anglo-Saxon World*, ed. by Maren Clegg Hyer and Della Hooke, Exeter Studies in Medieval Europe (Liverpool: Liverpool University Press, 2017), pp. 33–67

Hunt, John, *Warriors, Warlords and Saints* (Alcester: West Midlands History, 2016)

Kjølbye-Biddle, Birthe, 'Anglo-Saxon Baptisteries of the 7th and 8th Centuries: Winchester and Repton', in *Acta XIII Congressus Internationalis Archaeologiae Christianae*, Studi di antichità cristiana, 54 (Vatican City: Pontificio Istituto di Archeologia Cristiana, 1998), pp. 757–78

Learey, Jim, and Matthew Symonds, 'The Many Faces of Silbury Hill', *Current Archaeology*, 293 (2014), 12–18

Mayhew-Smith, Nick, *Britain's Holiest Places: The All-New Guide to 500 Sacred Sites* (Bristol: Lifestyle Press, 2011)

Neville, Jennifer, *Representations of the Natural World in Old English Poetry* (Cambridge: Cambridge University Press, 1999)

Padel, Oliver, *Cornish Place-Name Elements*, English Place-Name Society, 56/57 (Cambridge: Cambridge University Press, 1985)

Pickles, Thomas, 'Anglo-Saxon Monasteries as Sacred Places: Topography, Exegesis and Vocation', in *Sacred Text, Sacred Space: Architectural, Spiritual and Literary Convergences in England and Wales*, ed. by Joseph Sterrett and Peter Thomas (Leiden: Brill, 2011), pp. 35–56

Rahtz, Philip, *English Heritage Book of Glastonbury* (London: Batsford/ English Heritage, 1993)

Rahtz, Philip, and Lorna Watts, *Glastonbury: Myth and Archaeology* (Stroud: Tempus, 2003)

RCHM [Royal Commission on Historical Monuments], England, *Ancient and Historical Monuments in the County of Gloucester*, vol. I: *Iron Age and Roman-British Monuments in the Gloucestershire Cotswolds* (London: Her Majesty's Stationery Office, 1976)

RCHM [Royal Commission on Historical Monuments], Wales and Monmouthshire, *An Inventory of the Ancient Monuments in Caernarvonshire*, vol. II: *Central* (London: Her Majesty's Stationery Office, 1960)

——— , *An Inventory of the Ancient Monuments in Caernarvonshire*, vol. III: *West* (London: Her Majesty's Stationery Office, 1964)

Reaney, P. H., *The Place-Names of Cambridgeshire and the Isle of Ely*, English Place-Name Society, 19 (Cambridge: Cambridge University Press, 1943)

Rippon, Stephen, 'Marshlands and Other Wetlands', in *Water and the Environment in the Anglo-Saxon World*, ed. by Maren Clegg Hyer and Della Hooke, Exeter Studies in Medieval Europe (Liverpool: Liverpool University Press, 2017), pp. 89–106

Rodwell, Warwick, 'From Mausoleum to Minster: The Early Development of Wells Cathedral', in *The Early Church in Western Britain and Ireland: Studies Presented to C. A. Ralegh Radford*, ed. by Susan Pearce, BAR British Series, 102 (Oxford: BAR, 1982), pp. 49–59

Sargent, Andrew, 'Lichfield and the Lands of St Chad' (unpublished doctoral thesis, University of Keele, 2012)

Sawyer, Peter H., *Anglo-Saxon Charters: An Annotated List and Bibliography*, Royal Historical Society Guides and Handbooks, 8 (London: Royal Historical Society, 1968)

Society of Antiquaries of London, 'The Hut that St Columba Built', *Salon*, 390 (July 1997), <http://www.sal.org.uk>

Stenton, F. M., *Anglo-Saxon England*, 3rd edn (Oxford: Clarendon Press, 1971)

Stocker, David, and Paul Everson, 'The Straight and Narrow Way: Fenland Causeways and the Conversion of the Landscape in the Witham Valley, Lincolnshire', in *The Cross Goes North: Processes of Conversion in Northern Europe, AD 300–1300*, ed. by Martin Carver (Woodbridge: Boydell & Brewer, 2003), pp. 271–88

VCH Worcestershire: *The Victoria History of the Counties of England, Worcester*, vol. II, ed. by J. W. Willis-Bund and William Page (London: Constable, 1906)

Vince, Alan, 'The Growth of Market Centres and Towns in the Area of the Mercian Hegemony', in *Mercia: An Anglo-Saxon Kingdom in Europe*, ed. by Michelle P. Brown and Carol A. Farr (London: Leicester University Press, 2001), pp. 183–93

Watson, William J., *The History of the Celtic Place-Names of Scotland*, repr. with an introd. by Simon Taylor (Edinburgh: Birlinn, 2004)

Watts, *CDEPN*: Watts, Victor, *The Cambridge Dictionary of English Place-Names* (Cambridge: Cambridge University Press, 2004)

Wilson, Roger G., *A Guide to the Roman Remains in Britain* (London: Constable, 2002)

CAROLYN TWOMEY

Rivers and Rituals: Baptism in the Early English Landscape

The Baptism of Christ in the Jordan River from the late tenth-century Benedictional of Æthelwold is a popular and paradigmatic image of baptism from early England (Figure 2.1). The waters of the river swell around the immersed figure of Christ and surge by the ankles of John the Baptist, with the dove of the Holy Spirit descending from above holding cruets of holy oil in its beak. The source of the flow is the jar of the horned river god Jordan while angels attend above and to the sides of the scene with new garments and crowns for the Son of God. All baptisms imitate this original act of theophany and reify the baptism of Christ in the Jordan River within the reborn bodies of new Christians. The physical practice of baptism in early England also imitated this famous scene in a more literal sense. In this chapter, I explore the evidence for and experience of river baptisms during the early Christian missions to England between the late sixth and ninth centuries.

Rivers were important features of the early English landscape for defining local communities as sites of assemblies, burials, and ritual depositions as well as powerful focal points for Christian cultural and spiritual metaphors. In choosing to baptize in rivers, early missionaries incorporated the localized meanings of such watery places into the moment of Christian initiation in order to achieve the gradual and successful conversion of the English. While river baptisms were an expedient way to baptize large numbers of new adult converts, as our earliest sources for the conversion period attest, this was not just a practical but also a strategic decision. The early mission to the English adapted a multifaceted ritual landscape for baptism that reused ancestral places important to community identity and emphasized continuity with pre-Christian ritual geography. The benedictional's Jordan River — with its new and old gods — depicted one specific sacred place in Christian history while it recalled the many multivalent watery landscapes of early England. After exploring how early medieval authors such as the Venerable Bede

Carolyn Twomey • (ctwomey@stlawu.edu) is a Visiting Assistant Professor of European History at St Lawrence University. She researches and teaches the history of medieval religion and the material world.

Meanings of Water in Early Medieval England, ed. by Carolyn Twomey and Daniel Anlezark, Studies in the Early Middle Ages, 47 (Turnhout: Brepols, 2021), pp. 59–84

Figure 2.1. 'Baptism of Christ', Benedictional of St Æthelwold, London, British Library, Add. MS 49598, fol. 25ʳ. AD 963–84. Image © The British Library Board.

(*c.* 673–735) explained contemporary river baptisms, I will show how the waters of local rivers defined early English communities and were selected for the sacrament of initiation by the missionary Church.

From the days of the early Church, it is evident that baptizers used natural settings for the rites of initiation. The earliest Christian liturgical instructions for baptism describe a preference for water outdoors, inherited from a Jewish tradition of ritual ablutions and in imitation of the biblical baptisms of Christ (Matt. 3. 13–17; Mark 1. 9–11; Luke 3. 21–22) and the eunuch (Acts 8. 36–39).[1] The *Didache* and *Apostolic Tradition* require actively flowing, clean water.[2] This was both a late antique public health concern to avoid stagnant water as well as exegetical references to the 'living water' of Christ (John 4. 10–14, 7. 38, 19. 34), the Fountain of Life and the four rivers of Paradise (Gen. 2. 10; Rev. 7. 17, 22. 1), the streams after which the hart pants (Psalm 42), and the miraculous water from the rock in the wilderness (Num. 20. 11; Exod. 17. 6).[3] Brief references to river baptisms in Bede's *Historia Ecclesiastica* and saints' lives from the seventh and eighth centuries suggest that it was common for missionaries to baptize early medieval converts in the open air. Bede provides scholars with the most often cited examples of river baptisms in the early Middle Ages; however, these episodes ought to be viewed critically, with an eye towards Bede's overall agenda.

Most references to baptism in Bede's *HE* are too short to provide any details of the specific setting, although there are some notable scenes involving rivers. Kings like Æthelberht of Kent, Sigeberht and Swithhelm of the East Saxons, Peada of the Middle Angles, and others are simply 'baptizatus est' (baptized), the quick word sometimes the only descriptor of what was a complex process of education, ritual, and politics.[4] The martyrdom of the Roman saint Alban is the first outdoor baptismal event mentioned early on in the *HE*, although Alban's baptism was in blood as a martyr and not in water.[5] Nevertheless, the saint works watery miracles that prefigure his baptism — causing a clear path to form across a running river and summoning a related spring to the top of a paradisal hill covered in flowers — acts which testified to the power of God to command nature and to the holiness of the professed Christian himself. The behaviour of the miraculous water in this passage takes on sacramental language, as water had the power to perform the 'ministerium' (duty), 'deuotio' (pious service), and 'officium' (ministry) asked of it, actions suggestive of baptism which Alban himself thought 'oportunum esse' (fitting) for the occasion of his death and rebirth.[6]

---

1 Davies, *The Architectural Setting of Baptism*, p. 2.
2 *Didache*, 7. 1, and *Apostolic Tradition*, 21. 2, in *Documents of the Baptismal Liturgy*, ed. and trans. by Whitaker, pp. 2 and 7; Jensen, 'Archaeology of Christian Initiation', pp. 255–56; Ferguson, *Baptism in the Early Church*, pp. 201–06 and 341; and Dix, *The Treatise of the Apostolic Tradition*, p. 33 and n. 2. For other discussions of fresh running water at baptism, see Crerar, 'Contextualising Romano-British Lead Tanks', p. 143.
3 Drewer, 'Fisherman and Fish Pond', pp. 542–43.
4 For more on this process, see Twomey, 'Kings as Catechumens'.
5 'de quo nimirum constat quia, etsi fonte baptismatis non est ablutus, sui tamen est sanguinis lauacro mundatus ac regni caelestis dingus factus ingressu' (In his case it is clear that though he was not washed in the waters of baptism, yet he was cleansed by the washing of his own blood and made worthy to enter the kingdom of heaven). Bede, *HE*, I. 7.
6 Bede, *HE*, I. 7.

Mass river baptisms follow royal baptisms in Bede's *HE*. After a comparatively lengthy description of the conversion of Edwin of Northumbria to Christianity in Book II, Bede describes in quick terms the baptisms of the king, his family, and his counsellors at Easter 626 in the church at York. But Bede then elaborates on the setting of the baptisms of the common people to follow: Bishop Paulinus baptized 'confluentem eo de cunctis uiculis ac locis plebem' (the crowds who flocked to him from every village and district) in the River Glen nearby the royal Northumbrian site of Yeavering.[7] A personal account of river baptism and a detailed description of Paulinus's physical appearance then follows from one of Bede's veritable witnesses. Deda, a monk at Partney Abbey, retells the personal story of an old man who described being baptized in the River Trent by the 'uenerabilis simul terribilis aspectu' (venerable and awe-inspiring) Paulinus at noon and in the presence of King Edwin and 'multam populi turbam' (a great crowd of people) near the city of Tiowulfingacæstir (Littleborough).[8] River settings for baptism were seemingly appropriate for the initiations of common people, if not for the individual conversion stories of Bede's early English kings.

In his travels throughout Northumbria accompanying the itinerant royal court, Paulinus also baptized many in the River Swale near Catterick; however, a clarifying statement to justify the natural setting quickly follows Bede's account of baptisms in the river.[9] Paulinus baptized the crowds in the river because 'nondum enim oratoria uel baptisteria in ipso exordio nascentis ibi ecclesiae poterant aedificari' (they were not yet able to build chapels or baptisteries there in the earliest days of the church).[10] Bede then continues to say that Paulinus 'attamen' (nevertheless) built a wooden chapel at the royal residence at *Campodonum*, presumably for baptismal purposes, though it later burnt down.[11] Importantly in that second clause, Bede makes it clear that such built structures were preferable but not required for the baptisms of Paulinus. Bede appears to be uncomfortable with the idea of river baptisms in this qualifying statement, but he was perhaps reluctant to omit these baptismal settings entirely due to the first-hand experience of others, the details of which were provided by his trustworthy informant Deda.

A baptistery of wood or stone may have been Bede's ideal baptismal setting, but little evidence exists for the widespread use of the baptistery model in England. Such practice did not suit either the itinerant nature of royal households or the particular ecclesiastical landscape of England. John Blair has shown how the English system of rural pastoral organization and monastic leadership did not favour the establishment of baptisteries, which

---

7   Bede, *HE*, II. 14.
8   Bede, *HE*, II. 16.
9   Bede, *HE*, II. 14.
10 Bede, *HE*, II. 14; Blair, *The Church in Anglo-Saxon Society*, p. 70.
11 *Campodonum* was a Roman site near Dewsbury in Yorkshire. Bede, *HE*, II. 14 and n. 3.

were more commonly associated with the bishop and his episcopal city. Despite an established late antique tradition of baptisteries as dedicated architectural spaces — a tradition for which we have evidence in Roman Britain as well as across the Mediterranean world — the baptistery was not adopted wholesale in the early Middle Ages.[12] The limited extant evidence for baptisteries at elite sites like Canterbury, Winchester, and Repton tells us more about the material deployment of royal patronage and the pastoral role of the saints than definitive baptismal practice. It is also likely that Bede's reference to a baptistery was a deliberate Romanization of Christianity's establishment in England, one that recalled structures such as the Lateran baptistery and suited Bede's focus on the role of Rome in the conversion of the English as a whole. Though the potential baptismal use of church *porticus* (small side rooms), towers, and west ends of church naves has been suggested alongside other functions and liturgical uses, the flexibility and diversity of these places ought to challenge any ideas of a uniform model of baptismal architecture during the evangelization of England.

Bede had another opportunity to reframe river baptism in his prose and poetic versions of an anonymous earlier eighth-century life of St Cuthbert. In a scene from the *Anonymous Life* (c. 699–705), Cuthbert sets out from his monastery at Melrose to baptize rural people along the River Teviot: 'Alia die proficiscebat iuxta fluuium Tesgeta tendens in meridiem inter montana docens rusticanos et baptizabat eos' (On a certain day, he was going along the river Teviot and making his way southward, teaching the country people among the mountains and baptizing them).[13] Throughout this life, Cuthbert teaches, baptizes, confirms, and performs many healing miracles involving water and chrism, later also preaching along the River Tweed of which the Teviot is a tributary.[14] Bede's versions of the life of Cuthbert recount many of the same events in greater detail; however, they do not record any river baptisms. His verse life (c. 705) introduces one of the saint's miracles by stating briefly how Cuthbert 'tendebat populos vitae renovare fluentis' (set out to renew the populace with the rivers of life) after his ordination to the priesthood.[15] Bede's prose version (721) omits even these metaphorical references to baptism in its accounts of Cuthbert's mission, focusing instead on general descriptions of pastoral preaching, correcting error, and confirming the already baptized.[16] These narrative decisions told the story of an existing population of baptized Christians and exaggerated

---

12 Blair, *The Church in Anglo-Saxon Society*, pp. 201–02; Lynch, *Christianizing Kinship*, pp. 48–55.
13 VCA, II. 5.
14 VCA, III. 5–6 and IV. 3–7.
15 Bede, *Vita metrica S. Cudbercti*, ed. by Lapidge, ch. 10. I am grateful to Matthew Delvaux for this reference.
16 Bede, VCP, chs 9, 12–14, 29, 30, and 32–34. For more on Bede's use of the anonymous *Life*, see Lutterkort, 'Beda Hagiographicus', and Shockro, 'Reading Bede as Bede Would Read', p. 81.

the hostility of the landscape that had guided Cuthbert in his mission along the river in the *Anonymous Life*. Bede's desire for churches as the appropriate setting for the sacraments remains evident in the prose life's description of the necessity to erect temporary tents and wooden booths for Cuthbert's preaching and confirming when no church could be found in the mountains.[17] Like in his later descriptions of river baptisms in the *HE*, Bede projected a more complete picture of pastoral care into the past, one that can obscure our view of early baptismal practice.

Noteworthy examples of baptism in rivers and streams also survive from accounts of Irish missionaries in western Britain and Ireland. In one of the many miracles of Adomnán's *Vita S. Columbae* (*c.* 690), St Columba, the founder of the monastery at Iona, baptized a visiting pagan ruler in a 'fluvius' (stream) which the saint summoned with the end of his staff:

> Cum per aliquot dies in insula demoraretur Scia vir beatus, alicujus loci terrulam mari vicinam baculo percutiens, ad comites sic ait, 'Mirum dictu, O filioli! hodie in hac hujus loci terrula quidam gentilis senex, naturale per totam bonum custodiens vitam, et baptizabitur, et morietur, et sepelietur.' Et ecce, quasi post unius intervallum horae, navicula ad eundem supervenit portum; cujus in prora quidam advectus est decrepitus senex, Geonae primarius cohortis, quem bini juvenes, de navi sublevantes, ante beati conspectum viri deponunt. Qui statim, verbo Dei a Sancto per interpretem recepto, credens, ab eodem baptizatus est, et post expleta baptizationis ministeria, sicuti Sanctus prophetizavit, eodem in loco consequenter obiit, ibidemque socii, congesto lapidum acervo, sepeliunt. Qui hodieque in ora cernitur maritima; fluviusque ejusdem loci in quo idem baptisma acceperat, ex nomine ejus, Dobur Artbranani usque in hodiernum nominatus diem, ab accolis vocitatur.
>
> [When St Columba was staying for a few days on the island of Skye, he struck with his staff a patch of ground by the seashore in a particular place, and said to his companions: 'Strange to tell, my dear children, today, here in this place and on this patch of ground, an old man — a pagan but one who has spent his whole life in natural goodness — will receive baptism, and will die and be buried.' Only an hour later — look! — a little boat came in to land on the shore, bringing in its prow a man worn out with age. He was the chief commander of the warband in the region of Cé. Two young men carried him from the boat and set him down in front of the blessed man. As soon as he had received the word of God from St Columba, through an interpreter, he believed and was baptized by him. When the rite of baptism was finished, as the saint has predicted, the old man died on the same spot and they buried him there and raised a mound of stones over the place. It is

---

17 Bede, *VCP*, ch. 32.

still visible there today by the seashore. The stream in which he had received baptism is even today called by the local people 'the water of Artbranan'.][18]

This scene establishes the sanctity of St Columba, namely, his ability to command nature and summon fresh water from the earth as well as his prescient knowledge of the arrival of the local leader to be baptized. The use of the stream of Artbranan for the baptism is notable for the author only in its miraculous origin and subsequent elite memorialization by the stone burial mound; the setting of the stream itself is unremarkable. Elsewhere, Columba — like Cuthbert himself[19] — summoned springs from the earth through prayer in order to have water for baptism, as seen on another occasion of a young boy brought to the saint for baptism by his parents.[20] These acts testified to Columba's saintly status and his role as a new Moses, who sprung water from the rock, a biblical event that would have been implicitly understood by Adomnán's readers as a prefiguration of baptism.[21] Indeed, the frequency of wells and pools used for baptism in Muirchú and Tíerchán's seventh-century lives of St Patrick has led to the assumption that such outdoor baptisms were limited to the Irish or 'Celtic' Church alone.[22] Holy men such as Aidan and Cedd played important roles in evangelizing the English territories, and these missionaries along with their texts travelled throughout the island. However, it is important to emphasize that — along with many other similarities between so-called 'national' Churches — we cannot ascribe river baptism to the Irish Church alone.[23] And while Bede tells us that there were disagreements in the practice of baptism both between British Christians and the Roman mission of Augustine — and indeed between the Ionan and Roman reckonings of Easter, the traditional time for baptism — there is no indication that the use of natural features in the landscape for baptism was behind them.

River baptisms might not have taken place in the exact ways that Adomnán, Bede, or the anonymous author of Cuthbert's *vita* described, in either their own times or the historical times about which they wrote. Yet we do know that they occurred. These scattered references to river baptisms are well known in the scholarship, but remain unconnected to the many cultural meanings of

---

18 Adomnán, *Life of St Columba*, ed. by Reeves, I. 27; Adomnán, *Life of St Columba*, trans. by Sharpe, I. 33.
19 St Cuthbert summoned 'fontem aquae uiue' (living water) from the rock on Inner Farne. *VCA*, III. 3.
20 Adomnán, *Life of St Columba*, ed. by Reeves, II. 10.
21 Num. 20. 11; Exod. 17. 6. O'Reilly, 'Reading the Scriptures in the Life of Columba', pp. 87–88.
22 Ray, *The Origins of Ireland's Holy Wells*, pp. 26–32; Turner, *Making a Christian Landscape*, pp. 178–80.
23 In lieu of national models, Blair emphasizes the early English Church as a 'melting pot of complex cultural elements, and its distinctive forms of religious life were a product of many influences both native and imported'. Blair, 'Churches in the Early English Landscape', p. 6.

water in the landscape of early England.[24] By bringing together these textual accounts with archaeological and landscape studies of riverine assembly places, burials, and ritual depositions, we can see why missionaries to the early English chose rivers as baptismal places. Contrary to what Bede suggests, river baptisms were not simply accidents of expediency or convenience. Rather, rivers were highly symbolic choices.

Rivers were central to defining early medieval community. They were landscape features essential to the human and agricultural lifecycles and places with a long pre-Christian history of water veneration. Deliberately locating baptism within rivers was an intentional act of spatial reuse that incorporated the many meanings of the water into the ritual of baptism. Archaeologists and environmental historians have emphasized the importance of the landscape in the formation of community identity as the 'medium *for* and the outcome *of* action and previous histories of action'.[25] Indeed, such places can be read as texts layered with human agency:

> The landscape is conceptualized not merely as a by-product of the economic and social activities and processes that unfolded upon it, but also as a dense and complex system of signs and symbols that can be decoded and deciphered. It is widely compared with a parchment and palimpsest, a porous surface upon which each generation inscribes its own values and preoccupations without ever being able to erase entirely those of the preceding one. It is a surface onto which cultures project their deepest concerns and recurring obsessions, a medal struck in the image of their mental structures.[26]

In viewing the environment as a type of material culture that can shape human lives while also being shaped by humans, we can see how the layers of meaning placed upon certain natural places continue over time to act in constructing the social and conceptual worlds for those inhabiting them.[27] While water in general has often been shown by anthropologists to be a universal mediator of social interaction, it is the particular local watery landscapes of early medieval England that we must read to understand how rivers embodied and constructed community relationships in the past.[28] The coming of Christianity to England

---

24 Blair, *The Church in Anglo-Saxon Society*, pp. 226, 377–80, 463, 476–78; Foot, '"By Water in the Spirit"', p. 182; Morris, 'Baptismal Places', pp. 19–20; Whitfield, 'A Suggested Function for the Holy Well?'.
25 Tilley, *A Phenomenology of Landscape*, p. 23, his emphasis; Blair also notes how 'anthropological studies have shown that people can invest landmarks with qualities — divine, magical, demonic, or social — which, though widely recognized, are both individual and flexible'. Blair, *The Church in Anglo-Saxon Society*, p. 474.
26 Walsham, *The Reformation of the Landscape*, pp. 4–7, at p. 6.
27 'Places themselves may be said to acquire a history, sedimented layers of meaning by virtue of the actions and events that take place in them'. Tilley, *The Materiality of Stone*, p. 27.
28 Strang, 'Aqua Culture', p. 79; Strang, 'Common Senses'. See also Hastrup, 'Water and the Configuration of Social Worlds', pp. 64–65. For a similar discussion of the sea acting to form communal identity, see Klein, 'Navigating the Anglo-Saxon Seas', p. 8.

added more layers of significance onto existing pre-Christian landscapes as both ancestral and ritual places. Rather than abolish their use, the Church's long-standing tradition of cultural reuse saw missionaries and the early English Church adopt rivers as places for the Christian practice of baptism.

Scholars have noted how water sources and river patterns were 'social arteries' that determined early medieval settlement and provided necessary physical conduits for the lives and livelihoods of the population.[29] Everything from everyday drinking and washing needs, the transport of goods, procurement of food, and agricultural cultivation depended on one's proximity to and use of rivers, even supposedly marginal territories such as fenland.[30] It was an intimate relationship to the natural world in which 'people were the landscape, and landscape was people'.[31] Understanding this experience of water in the past is hampered not only by our modern perceptions of separation between nature and society, but also by the gradual effects of climate change and modern human interventions of dams, weirs, and irrigation systems in a historic landscape.[32] Between the seventh and tenth centuries, water was more present as a whole in the English environment with a warmer climate and greater rainfall leading to higher water levels, unregulated flooding, and broader zones of fenland.[33] Wryly capturing the practical realities of living in such a landscape, Richard Jones reminds us, 'Bede's feet were wet'.[34]

In this watery world, navigable rivers and seas were fast and convenient modes of transport. We can see the types of vessels used to traverse them in surviving archaeology from ship burials along coastal regions to the more everyday boats recovered from previously underwater areas.[35] Clinker-built vessels such as those excavated whole from the elite seventh-century grave

---

29 Williamson, *Sutton Hoo and its Landscape*, p. 96.
30 Pelteret, 'The Role of Rivers and Coastlines in Shaping Early English History', pp. 43–45. The fenland of St Guthlac's Lincolnshire retreat was, at times, an unnavigable bog, but frequently a profitable source of grasses, thatch, saltpans, and wildlife such as eels, fish, and birds. Darby, *The Medieval Fenland*, pp. 21–85; see also Oosthuizen, *The Anglo-Saxon Fenland*.
31 Jones and Semple, 'Making Sense of Place in Anglo-Saxon England', p. 14. For how different regional identities in England were 'based on this mutual interaction of land and human', see Wickham-Crowley, 'Living on the *Ecg*', p. 91.
32 Pelteret, 'The Role of Rivers and Coastlines in Shaping Early English History', pp. 24–27; Williamson and Jones also remind us to avoid the phenomenological trap of equating modern experience of the landscape with that of the past. Williamson, *Sutton Hoo and its Landscape*, pp. 25 and 65; Jones and others, 'Living with a Trespasser'.
33 Jones, 'Responding to Modern Flooding'; Wickham-Crowley, 'Living on the *Ecg*', pp. 86–89; Hill, *An Atlas of Anglo-Saxon England*, p. 11.
34 Jones, 'Anglo-Saxon Water Consciousness'. See also the two-year Leverhulme-funded project, 'Flood and Flow: Place-Names and the Changing Hydrology of River-Systems', <https://waternames.wordpress.com/> [accessed 1 February 2020], and Jones and others, 'Living with a Trespasser'.
35 Williamson, 'The Environmental Contexts of Anglo-Saxon Settlement', p. 152. Estimated travel by sea is considerably faster than over land in the early medieval period. Ferguson, 'Re-Evaluating Early Medieval Northumbrian Contacts', p. 295.

monuments at Sutton Hoo and Snape (Suffolk), or found in pieces among the more modest sixth- and seventh-century grave goods of Kentish men and women, may have taken many frequent journeys to trade up or down rivers and along the Channel and North Sea coasts by sail or oar.[36] Smaller vessels such as the prehistoric sewn-plank boats of North Ferriby and Kilnsea and log boats surviving from early medieval contexts indicate other potential classes of easy-to-build, sturdy boats that would have been essential to daily life and transport in coastal and riverine communities.[37] There were likely many more of these small wooden boats manoeuvring down English waterways laden with rowers and goods, daily criss-crossing the rivers, lakes, and coasts of early England, such as the rafts (*ratis*) transporting wood on the River Tyne that nearly swept some Tynemouth monks out to sea, were it not for the prayers of St Cuthbert.[38] These journeys could be perilous as well as profitable.

By shifting our typical focus from the positions of churches, towns, or other features in the landscape to the perspective of the river, we begin to see how water was a central and liminal place that enabled cultural and social cohesion. Like the Roman *limes*, rivers have often been interpreted as borders and boundaries that divided different political states and/or cultures. Bede used rivers to delineate the peripheries of English kingdoms throughout the *HE*.[39] On a smaller scale, boundary clauses described particular springs and streams in order to demarcate the extent of landownership. Part of the seventh-century Old English bounds of the estate at Egham (Surrey), for example, outlined the property:

> fram þere `hore´ aepeldure to þe kneppe bi þe quelmes · fram þe quelmes binuþe þere stonie helde and sua goinde adun bi tigel bedde burne adun upe þat eigt þe stant in þere temes aet Loddere lake.
>
>> [from the hoar apple-tree to the hillock by the springs from the springs beneath the stony slope and so going down by tile bed bourn (stream) down as far as the eyot that is situated in the Thames at beggars' watercourse.][40]

---

36 See section on coastal transport in Carver, 'Travels on the Sea and in the Mind', pp. 26–29; and also Fleming, 'Elites, Boats and Foreigners'; Mason, 'Listening to the Early Medieval Dead', p. 57; Ferguson, 'Re-Evaluating Early Medieval Northumbrian Contacts', pp. 285–86.

37 Mason, 'Listening to the Early Medieval Dead', p. 57; Van de Noort, *The Humber Wetlands*, pp. 81–87; Fleming, 'Elites, Boats and Foreigners', pp. 421–22; McGrail and Switsur, 'Medieval Logboats'.

38 Bede, *VCP*, ch. 3; Ferguson, 'Re-Evaluating Early Medieval Northumbrian Contacts', p. 289.

39 Rivers delineated the extent of the lands of the Southern Angles (Bede, *HE*, I. 25, II. 5), the East Saxons (II. 3), the Mercian borders by the Rivers Idle and Trent (II. 12; III. 24), and the jurisdiction of the bishop who ministered to those west of the River Severn (v. 23). Synods also occurred at fords of rivers, including the one that elected Cuthbert (IV. 28) and readmitted Wilfrid (v. 19) to the episcopate.

40 *LangScape*, ed. by Jenkyns, Stokes, and Nelson, L1165.3.000 (S 1165), *c*. 670s. For more on boundary clauses as a source of watery place-names, see the bounds of Barnhorn (Sussex) AD 772 (S 108) in Jones, 'Responding to Modern Flooding'; as well as Pelteret, 'The Role of

Different springs of water, streams, and larger rivers like the Thames described in this boundary clause would have been important shared markers in the landscape that denoted communal territory and rights of ownership. The routes of rivers continue to determine the jurisdiction of modern counties and the extent of archaeological excavation in modern England, a practice that conceals overarching political and cultural relationships in the past.[41] Rather than reconstructing hard lines through the landscape, we ought to conceive of territorial boundaries as wider 'liminal zones' of activity where water in particular served a variety of purposes beyond the simplistic formulation of boundaries.[42] What might appear on a modern map as distant edges were central to members of local settlements, in both physical and metaphorical ways.[43] When we look from the water itself, these seemingly peripheral spaces on the outskirts of kingdoms become central places in the landscape.

Existing sites of assembly near rivers provided practical opportunities for Christian baptizers to convert large numbers of people and obtain access to local leaders. Rivers near royal centres on the borderlands of kingdoms were chosen specifically for baptism as they were existing meeting places accessible to those travelling with the king for other business.[44] The itinerant nature of the Northumbrian court, for example, provided many occasions for mass baptisms in the Rivers Swale, Trent, and Glen as King Edwin and his thegns stopped at the vills of Catterick, Tiowulfingacæstir, and Yeavering. Richard Morris has suggested that borders along the edges of territories provided a suitable neutral setting, particularly in the highly political case of a king serving as godparent for the baptism of another leader, as in the examples of the Northumbrian king Oswald standing sponsor for Cynegisl of Wessex and the Mercian king Wulfhere for Æthelwealh of the South Saxons.[45] Traditional sites used as gathering places for such crowds of people were natural places for missionaries to target — up to ten thousand baptized in a single day according to one (likely exaggerating) letter of Gregory the Great — but they were not static locations, and each carried with it a long local history of activity.[46]

Royal perambulations such as those described by Bede circulating between vill sites were often associated with former Roman and prehistoric

---

Rivers and Coastlines in Shaping Early English History', pp. 42–43; and Turner, 'Boundaries and Religion', p 54.

41 Mason, 'Listening to the Early Medieval Dead', p. 57. For example, contrary to how it may appear on modern maps, the kingdom of Lindsey was an independent political entity. Symonds, 'Territories in Transition', p. 36.

42 Pantos, '"On the Edge of Things"', pp. 40–41; Hooke, *The Landscape of Anglo-Saxon England*, pp. 76–77.

43 Mason, 'Listening to the Early Medieval Dead', p. 69.

44 Morris, 'Baptismal Places', pp. 20–22; see also Morris, *Churches in the Landscape*, pp. 63–72.

45 Bede, *HE*, III. 7 and IV. 13. See Morris, 'Baptismal Places', pp. 20–21. For diplomatic sponsorship at baptism in general, see Lynch, *Christianizing Kinship*, ch. 12.

46 Markus, *Gregory the Great and his World*, p. 179, citing *Ep.* viii. 29 to Eulogius, patriarch of Alexandria. See also Morris, 'Baptismal Places', p. 24.

landscapes.⁴⁷ Howard Williams and Sarah Semple have argued that early burial places, such as the cremation cemetery at Loveden Hill (Lincolnshire) and elite sites at Sutton Hoo and Yeavering served as dramatic gathering places for a range of social activities associated with burial traditions and regional displays of power.⁴⁸ Focused around existing prehistoric structures of hill forts and barrow mounds, the cremation and/or internment of the dead would have occasioned large social events for the living, involving multiple days of funerary feasting and gift giving, and likely the administration of local justice and trade. Yeavering, in particular, with its amphitheatre-like structure and visible landscape of prehistoric monuments in earth and stone, has shown how royal residences could be long-standing places of assembly and community ritual with an intended audience.⁴⁹ It is in this funerary landscape that we find the locations of early assembly places and, I argue, baptismal places.

Later meeting places for judicial and administrative gatherings such as the hundred courts also noticeably occurred at the transport junctions of rivers and roadways.⁵⁰ These were frequently located along the border zones of more than one parish and/or hundred unit and were polyfocal sites that extended over a considerable distance in the landscape.⁵¹ Occurring at only certain times of the year, these meetings were places for the exchange of oaths, goods, and hostages, as well as resolving disputes, feasting, festivals, and games.⁵² These might have been temporary sites of occupation with tents for hundreds of people, such as the high-status, middle Saxon gathering site at Dorney (Buckinghamshire) in the Thames valley that is marked only with pits containing pottery, bone, and small finds of combs, knives, and loom weights.⁵³ In addition to the trees and stones seen in boundary clauses and place-names such as *Augustinæs Ác* (Augustine's Oak), bridges and fords were also popular features in the landscape for meeting places, including ecclesiastical synods.⁵⁴

In addition to the benefits of easy access along watercourses for transportation, the perceived neutrality of the watery sites may also have encouraged their use for assemblies and baptisms. River crossings and bridges were often

---

47 Morris, 'Baptismal Places', p. 20.
48 Williams, 'Assembling the Dead'; Semple, 'Locations of Assembly in Early Anglo-Saxon England'.
49 Williams, 'Assembling the Dead', p. 110. See also the manmade and natural features of Yeavering in Hope-Taylor, *Yeavering*.
50 Meaney, 'Hundred Meeting-Places in the Cambridge Region'; Pantos, 'The Location and Form of Anglo-Saxon Assembly-Places'.
51 Pantos, '"On the Edge of Things"', p. 41.
52 Symonds, 'Territories in Transition', pp. 28–29; Pantos, '"On the Edge of Things"', p. 47.
53 Foreman, Hiller, and Petts, *Gathering the People, Settling the Land*, pp. 69–72, for the finds, see pp. 35–55 and 65–69, discussion of site dating, see p. 60.
54 Bede, *HE*, II. 2; Pantos, '"On the Edge of Things"', pp. 43–47; Cubitt, *Anglo-Saxon Church Councils*, pp. 27–39; see also Morris, 'Baptismal Places', p. 23.

meeting places for negotiations between opposing groups in the late antique and early medieval West.[55] Julia Barrow has argued that this practice explained a 973 event at Chester when the English king Edgar joined eight kings of Wales, the Isle of Man, Scotland, and Strathclyde in a boat in the middle of the River Dee.[56] The watery setting was physically convenient as well as metaphorically appropriate for the coming together of local communities and their leaders at ritual events such as recognizing overlordship. Aliki Pantos suggests, following Barrow, that 'it may even be that water, particularly running water, was seen as having some specifically neutral or liminal quality since it was continually changing'.[57] Certainly, something about the watery locations was attractive to organizers, whether from the point of view of convenience, commodities, or ceremony.

While the first textual references to the monthly meetings of the hundred courts began in the tenth century,[58] scholars have argued that such sites were already in place in early centuries and repeatedly reused in the landscape due to the appeal of continuity around particular landscape features.[59] Such sites were located at important locations like prehistoric hill forts, stones, and barrows, places Semple has shown to have held much more than a practical role in assemblies. Aside from the advantages of elevated positions for seeing and being seen when addressing a crowd,[60] such prehistoric places were appreciated by the early English as liminal places between natural and supernatural worlds, and they desired to bury the dead and assemble there.[61] This practice continued uninterrupted by the arrival of Christianity until the movement for enclosed burial in the tenth century marginalized such sites, now considered the domain of demons and criminals.[62] Before this, however, these places helped to form shared social identities through a variety of judicial, funerary, and feasting activities, as well as through the reuse of liminal landscape features of the pre-Christian past.[63] Placing churches on top of barrows was one way to 'assimilate' these important places and diffuse their potential otherworldly danger.[64] Guthlac famously converted his watery barrow into a monastery, and the church at Goodmanham — supposedly the site of the pagan temple profaned by Coifi as Bede tells us — sits on top of

---

55 The Roman emperor Valens was rowed out to meet the Gothic king Athanaric in a neutral boat in the middle of the River Danube, a practice continued into the ninth century by Carolingian rulers to emphasize peaceful co-rulership. Barrow, 'Chester's Earliest Regatta?', p. 84.
56 Barrow, 'Chester's Earliest Regatta?', pp. 88–89; Thornton, 'Edgar and the Eight Kings'.
57 Pantos, '"On the Edge of Things"', p. 46.
58 Loyn, *The Governance of Anglo-Saxon England*, pp. 140–41; Stenton, *Anglo-Saxon England*, pp. 298–301; Turner, *Making a Christian Landscape*, pp. 107–09.
59 Reynolds, *Later Anglo-Saxon England*, pp. 75–76.
60 Pantos, 'The Location and Form of Anglo-Saxon Assembly-Places', p. 171.
61 Semple, 'Locations of Assembly in Early Anglo-Saxon England', p. 139.
62 Semple, 'A Fear of the Past', p. 121.
63 Williams, 'Placing the Dead', p. 80; Pantos, '"On the Edge of Things"', p. 47.
64 Semple, 'A Fear of the Past', p. 123.

a similar, though likely artificial, mound.[65] The use of these places over time across the timeline of the Christianization of England indicates the importance of place and the reuse of specific landscapes for important community events.

Buried ancestors and their memory are an integral part of understanding the meaning of rivers in the early English landscape. The distribution of burial traditions across England and their prominent locations, particularly barrow and cemetery burials, indicate how watersheds and water views defined early communities. Rivers were central to the conceptual world of the early English through the many symbolic layers of meaning placed on rivers seen through the treatment of the dead.[66] Evidence from cemeteries corresponds to Tom Williamson's 'river and wold' model, which remade the map of settlement zones in early England wherein the watersheds of major rivers formed distinct cultural communities.[67] Rather than transposing the later divisions of kingdoms back into the sixth and seventh centuries, this model focused on how arable land and accessible water sources in the environments of river valleys formed three provinces based around the drainage basins of the North Sea, the English Channel, and the Irish Sea, each separated from one another by high areas with poor water sources and little settlement.[68] The rivers at their centres would have provided commercial and cultural contacts between valley inhabitants who used the rivers as connective highways, and these 'drainage basins and river systems tended, over time, to approximate social territories'.[69] This perspective reinterprets the historical use of the Rivers Thames and Lea in the 886 division of England between the viking Guthrum and English king Alfred not as arbitrary boundaries, but rather distinct cultural regions. Guthrum's territory encompassed the extent of the North Sea Province watershed, which was an area that had already been in active cultural communication with the Danes connected by the sea.[70]

The burial traditions within these areas corresponded to the idea that rivers held a particular role in defining regional cultural expressions and identities in the fifth through seventh centuries. Trends in prone burial and

---

65 Bede, *HE*, II. 13; Blair, 'Churches in the Early English Landscape', p. 8. See also Della Hooke, 'The Sacred Nature of Rivers, Wells, Springs, and Other Wetlands in Anglo-Saxon England' and Helen Appleton, 'Water, Wisdom, and Worldliness in the Anglo-Saxon Prose Lives of Guthlac' in this volume.
66 Williamson, *Sutton Hoo and its Landscape*, p. 23. See also Mason, 'Listening to the Early Medieval Dead', ch. 2.
67 Williamson, *Sutton Hoo and its Landscape*, pp. 94–96 and 134–35; Williamson, 'The Environmental Contexts of Anglo-Saxon Settlement', p. 147; Everitt, 'River and Wold'.
68 Williamson, 'The Environmental Contexts of Anglo-Saxon Settlement', pp. 147–50.
69 Williamson, 'The Environmental Contexts of Anglo-Saxon Settlement', p. 147. See also Mason, 'Listening to the Early Medieval Dead', p. 69.
70 Williamson, 'The Environmental Contexts of Anglo-Saxon Settlement', p. 153. See also Klein, 'Navigating the Anglo-Saxon Seas', p. 9, and Carver, 'Travels on the Sea and in the Mind', for how an east-west division of Britain more suitably reflects the island's cultural links by sea.

cremation cemeteries coordinated to Williamson's watershed zones.[71] Prone burials — in which individuals were interred lying face down in the grave due to many possible reasons, including their status as social or judicial outsiders, fear of the dead, or penitential practices — were statistically few in number, but noticeably more frequent in the North Sea Province river valleys and coasts.[72] This practice spread along rivers such as the Tees, Wear, and Tyne and continued across the conversion period, notably found among the dead of Bede's double monastery at Wearmouth and Jarrow.[73] The practice of cremating horses with the dead and depositing certain types of grave goods in fifth- and sixth-century burials also followed this riverine cultural map of the North Sea.[74] Horse cremations stretched from Sancton (E. Yorks) to Spong Hill (Norfolk) and a majority of metal wrist clasps and brooches of cruciform and annular styles found in female graves from East Anglia, the East Midlands, and Yorkshire defined the burial preferences of people living in the riverine community of the North Sea zone.[75] Rivers were thus highways that transported not only goods but also cultural traditions across areas previously thought to be entirely separate polities.

The individual burials corroborate Williamson's watersheds and indicate the visual importance of the river in the cognitive landscape. Though the famous seventh-century Mound 1 ship burial at Sutton Hoo has been frequently compared to the prominent monument on the cliff edge erected over the grave of the epic hero Beowulf, none of the phases of burial at Sutton Hoo, including these late sixth- and seventh-century monumental barrows, would have been very visible from afar.[76] They did, however, provide an excellent amphitheatre-like view of the sprawling Deben River below.[77] Williamson argues that the view *from* the grave rather than the view *of* the grave indicated how 'the river would have been the centre of the imaginative and experienced world of those who dwelt beside it'.[78] Similarly, cemeteries at Spong Hill, Castle Acre and West Acre, Walsingham, Brettenham, and Pewter Hill all

---

71 Mason, 'Listening to the Early Medieval Dead', pp. 58–69, following Svanberg, *Decolonizing the Viking Age*.
72 Reynolds, *Anglo-Saxon Deviant Burial Customs*, pp. 68–76 and 183; Mason, 'Listening to the Early Medieval Dead', pp. 58–63.
73 Mason, 'Listening to the Early Medieval Dead', pp. 63–64; McNeil and Cramp, 'The Wearmouth Anglo-Saxon Cemetery', p. 82.
74 Mason, 'Listening to the Early Medieval Dead', pp. 64–69. For the significance of horse cremation and burial, see Fern, 'Horses in Mind'.
75 Mason dismisses Bede's claim (*HE*, I. 15) that this represents a uniformly 'Anglian' region derived from an ethnic 'Germanic' migration. Mason, 'Listening to the Early Medieval Dead', pp. 64 and 66–67; Williamson, 'The Environmental Contexts of Anglo-Saxon Settlement', pp. 147–51; Fern, 'Early Anglo-Saxon Horse Burial'.
76 Williamson, *Sutton Hoo and its Landscape*, pp. 104–05, citing *Beowulf* ll. 2802–07 and ll. 3156–60; see also Williams, 'Placing the Dead'.
77 Williamson, *Sutton Hoo and its Landscape*, p. 107.
78 Williamson, *Sutton Hoo and its Landscape*, p. 108.

overlook rivers.⁷⁹ This concern for the water also likely influenced the burial places of secular and ecclesiastical powers in seventh- to eighth- century coastal Northumbria. The Bowl Hole cemetery near Bamburgh Castle commands a view of the ocean and St Cuthbert's island retreat, and the burials at the female Cross Close cemetery at the monastery of Hartlepool were aligned north–south with the River Tees.⁸⁰ The physical and visual relationship to water was what mattered to the ancestors buried within the barrows and those orchestrating the burying of the dead both before and after the arrival of Christianity.

The care with which dead members of the community were interred in the riverine landscape, in places associated with views of water and determined by watersheds, suggests a supernatural concern for the afterlife associated with water. As both spiritual and physical highways, rivers forged connections between communities of the living and the otherworld.⁸¹ Water, like other landscape features such as barrows, groves, and trees, was an in-between place where gods and humans communicated; the earliest forms of pre-Christian temples in Britain were likely open-air enclosures.⁸² We can see this reflected in Christian tradition in the physical placement of Christ near the water's edge when speaking to the apostles in the *Heliand* (*c.* 830),⁸³ and the occurrence of miracles on the edges of watery places such as the miraculous dolphin flesh provided for St Cuthbert and his starving companions at the seashore on Epiphany (the feast day commemorating Christ's baptism).⁸⁴ Departed ancestors connected individuals not only to their living communities but also to the divine across the water. From the few textual references to pagan deities, we know that kings traced their lineages to members of a pantheon of gods like Woden, Tiw, and Thunor, whose names marked associated sacred places in the landscape.⁸⁵ And as we have seen in a telling example from the Continent, Radbod, the ruler of the Frisians, refused to be baptized once St Wulfram of Sens told him his ancestors were in hell; the king would not be separated from his family in the afterlife.⁸⁶

---

79 Williamson, *Sutton Hoo and its Landscape*, p. 107.
80 Mason, 'Listening to the Early Medieval Dead', pp. 85–104; see also Groves, 'The Bowl Hole Burial Ground' and Daniels, *Anglo-Saxon Hartlepool and the Foundations of English Christianity*, pp. 74–96, at 81–82.
81 Heide, 'Holy Islands and the Otherworld'.
82 Williamson, *Sutton Hoo and its Landscape*, p. 100. Blair, 'Churches in the Early English Landscape', p. 8–9.
83 Lund, 'At the Water's Edge', p. 57; see also Hayward, 'Contextualizing the Gospel among the Saxons'.
84 *VCA*, II. 4, and Bede, *VCP*, ch. 11.
85 Stenton, 'The Historical Bearing of Place-Name Studies'; Wilson, *Anglo-Saxon Paganism*, pp. 5–21.
86 *Vita Vulframni*, ed. by Levison, ch. 9; Lynch, *Christianizing Kinship*, p. 72; see also Wood, *The Merovingian Kingdoms*, pp. 318–19.

In a literal sense, the identity of people in the landscape was bound up with the river. The names of people living in riverine communities originated from the names of the rivers themselves: Blythingas, the people of the River Blyth, and the Beningas, the people of the River Beane.[87] Ultimately, Williamson argues, against Bede's etymology, that the Wuffingas of Sutton Hoo received their name from the River Deben, or 'the winding one'.[88] A similar origin has been argued for Deira from the River Derwent.[89] Other English names for watery places indicated a great knowledge of and appreciation for how water behaved in the environment and represented different relationships between communities and water in the cultural landscape.[90] For example, the use of the Old English term *wella* (for a spring, well, and/or stream) was 'associated with the property of communities and individuals' and only became commonly used after *c.* 730, when the eighth century saw a shift toward nucleated settlement.[91] Moreover, river names were more likely to survive over time compared to other types of landscape names.[92] The landscape project directed by Jones approaches naming conventions behind riverine landscapes as holding traditional ecological knowledge that reveals river behaviour necessary for local inhabitants to learn in order to manoeuvre the reaches of their river.[93] This approach challenges modern organizations to utilize Old English place-name evidence to best prepare today's communities for flooding in an age of global warming, such as the late 2015 York floods which restored the King's Fishpond to its medieval dimensions and inundated the city. From these watery perspectives, early English rivers worked to form cultural and social territories, in which rivers were not political borders between peoples, but definitions of community in themselves.[94]

The burial of not only people but also objects along rivers indicated the importance of water and the ritual reuse of local riverine places. Like the meanings attached to rivers and oceans seen in the spatial evidence of

---

87 Williamson, *Sutton Hoo and its Landscape*, p. 109.
88 Williamson, *Sutton Hoo and its Landscape*, pp. 109–10 and 116–18.
89 Yeates, 'Still Living with the Dobunni', p. 89.
90 Gardiner, 'Oral Tradition, Landscape and the Social Life of Place-Names', p. 21; see also, for the physical naming and oral bounding of territories preserved in Old English and Latin charters, Geary, 'Land, Language and Memory in Europe'.
91 Jacobsson, *Wells, Meres, and Pools*, p. 220. For the general discussion of *wella* and its use as a gloss for *fons, fluvius,* and *flumen*, see pp. 219–27. For the uses of *mere* for natural freshwater lakes, flooded river valley and marsh, and man-made ponds along roadways, see pp. 208–18. See also Cole, 'Distribution and Use of *Mere* as a Generic in Place-Names'.
92 Gelling, 'The Place-Names of the Isle of Man', pp. 143–44; Gelling and Cole, *The Landscape of Place-Names*, chs 1 and 2 for watery place-names.
93 Jones and others, 'Living with a Trespasser'. Gelling grouped place-names into three groups: topographic, habitative, and folk names in Gelling, *Signposts to the Past*; see Jones, 'Responding to Modern Flooding' for alternative categorizations of Old English place-names according to a Traditional Ecological Knowledge (TEK) framework.
94 Williamson, *Sutton Hoo and its Landscape*, p. 95.

assemblies and human burials in early medieval England, particular studies of the Rivers Witham and Severn have explored the long history of ritual depositions in their waters. Deposits of votive metalwork clustered near bridges and crossing points of the marshy valley of the River Witham in Lincolnshire are objects that date from the late Bronze Age to the later medieval period. This extensive continuous use was uninterrupted by Christianity; the only changes to the river after evangelization were that the buildings of churches now marked the natural crossing points where depositions were popular.[95] Similarly, the River Severn saw continuous use of its waters despite changing cultural and political categories from the prehistoric to the early medieval era.[96] Regardless of the precise function that these acts served, such repeated depositions in rivers from the prehistoric Iron Age to the medieval Christian period indicated an important cultural connection with water. River deposits continued to be 'the proper thing to do'.[97]

Similarly, we should not view the adaptation of ritually charged watery places for baptism as a stubborn holdover from pagan practices, which would eventually be rooted out in favour of 'proper' baptismal places in churches and baptisteries. The use of rivers continued unopposed in a new Christian landscape. Traditions of burial and votive deposits along rivers indicate that the early English landscape encountered by Christian missionaries was already marked with ritual watery foci. These central places for defining community and social identity had been reused repeatedly in many cases since the Iron Age. The status of rivers persisted not due to the stubbornness of 'paganism' after the conversion, but because of the roles of water perpetuated in the landscape as focal places for ritual.[98] The waters of the river provided an apt setting for further rituals of community identity and water, a landscape palimpsest overlaid with a new baptismal gloss.

Before and after the conversion, the early English maintained a complex conceptual relationship to rivers. Local communities encoded layers of use and meaning onto a landscape of ritual water marked by pre-Christian and Christian depositions, burials, and meeting places as well as everyday interactions with water as central economic and social places. The act of baptism emphasized long-term continuity with ancestral foci and ritual practices in the landscape in a new form of social and spiritual initiation.

---

95   Stocker and Everson, 'The Straight and Narrow Way'; Hooke, 'Rivers, Wells and Springs in Anglo-Saxon England'. For the broader context of watery deposits from pre-Roman and Roman watery places, see Merrifield, *The Archaeology of Ritual and Magic*, pp. 23–30 and 45–48. Similar to the Witham River, the River Thames functioned as a 'unifier' and highway through the use of ferry and crossing points aligned with roadways. Cohen, 'Boundaries and Settlement', p. 18.
96   Yeates, 'Still Living with the Dobunni'.
97   Merrifield, *The Archaeology of Ritual and Magic*, p. 115.
98   For this continuity as 'repeated renegotiation', see Semple, 'Sacred Spaces and Places', p. 757.

This was a message of inculturation captured in Pope Gregory the Great's seventh-century instructions to Bishop Mellitus to reuse pagan temples as Christian churches with a sprinkling of 'aqua benedicta' (holy water) to ensure the successful conversion of the English.[99] Our image of the Baptism of Christ in the river with which we began would have been a familiar scene of baptism purposefully repeated in the Swale, Thames, Trent, and Severn: men and women up to their waists or ankles in the flowing water, awed by the presence of both old and new gods, wearing new clothes and indelible crowns of heaven.

Though by the time the manuscript artist was layering the rising blue waters of the River Jordan into the parchment of the Benedictional of Æthelwold, river baptisms in general were disappearing. Unlike other evidence for the Christianization of the landscape, the practice of baptism in rivers is difficult to chart over time, though it likely lasted longer than Bede would have us think. In the same way that prehistoric barrows became associated with the evil otherworld, watery places over time were marginalized as the church building and its complex became increasingly the centre of the Christianized landscape. We can look to the formation of minor estates beginning in the tenth century and the growth of churches attached to them for competing places for baptism other than in watery places. The increasing desire to manage burial in the landscape within new churchyards is likely paralleled in this more ephemeral practice of baptism.[100] As the Church sought to centre all activity within the 'holy cities' of churches, rivers likely faded out of popular use.[101]

The early English had both local tradition and biblical authority to place baptism in the river. As landscape features dedicated to mediating social relationships and imbued with supernatural significance in many cultural traditions, rivers would have easily accommodated new meanings as the sites of the Christian ritual of initiation into the social and spiritual body of the Church and the promise of heaven. By choosing to adapt familiar ritual places into Christian use, missionaries hoped to accomplish a gradual conversion of the English embedded in the existing ritual and cognitive landscape.

---

99 Bede, *HE*, I. 30. Key contributors to the scholarly discussion of this famous letter include Meyvaert, 'Diversity within Unity'; Markus, 'Gregory the Great and a Papal Missionary Strategy'; Cubitt, 'Unity and Diversity in the Early Anglo-Saxon Liturgy'; Demacopoulos, 'Gregory the Great and the Pagan Shrines of Kent'; and Abrams, *Bede, Gregory, and Strategies of Conversion*, pp. 23–24.
100 Blair, *The Church in Anglo-Saxon Society*, pp. 463–71 and 472–75; Gittos, 'Is There Any Evidence for the Liturgy of Parish Churches in Late Anglo-Saxon England?'.
101 Gem, 'Church Buildings'.

Works Cited

Primary Sources

Adomnán of Iona, *Life of St Columba*, trans. by Richard Sharpe (London: Penguin, 1995)

——, *The Life of St Columba, Founder of Hy, Written by Adamnan, Ninth Abbot of That Monastery*, ed. by William Reeves (Edinburgh: Edmonston and Douglas, 1874)

Bede, *Bede's Ecclesiastical History of the English People*, ed. and trans. by Bertram Colgrave and R. A. B. Mynors (Oxford: Clarendon Press, 1969)

——, *Vita metrica S. Cudbercti*, in *Bede's Latin Poetry*, ed. by Michael Lapidge, Oxford Medieval Texts (Oxford: Oxford University Press, 2019), pp. 181–314

——, *Vita S. Cuthberti prosaica*, in *Two Lives of Saint Cuthbert: A Life by an Anonymous Monk of Lindisfarne and Bede's Prose Life*, ed. and trans. by Bertram Colgrave (Cambridge: Cambridge University Press, 1940), pp. 141–307

Documents of the Baptismal Liturgy, ed. and trans. by Edward Whitaker, rev. by Maxwell E. Johnson, 3rd edn, Alcuin Club Collections, 79 (London: Society for Promoting Christian Knowledge, 2003)

LangScape. The Language of Landscape: Reading the Anglo-Saxon Countryside, ed. by Joy Jenkyns, Peter Stokes, and Janet L. Nelson, version 0.9, <http://langscape.org.uk> [accessed 15 February 2020]

Vita S. Cuthberti auctore anonymo, in *Two Lives of Saint Cuthbert: A Life by an Anonymous Monk of Lindisfarne and Bede's Prose Life*, ed. and trans. by Bertram Colgrave (Cambridge: Cambridge University Press, 1940), pp. 59–139

Vita Vulframni episcopi Senonici, ed. by Wilhelm Levison, in *Passiones vitaeque sanctorum aevi Merovingici*, ed. by Bruno Krusch and Wilhelm Levison, vol. III, MGH Scriptores rerum Merovingicarum, 5 (Hannover: Hahnsche Buchhandlung, 1910), pp. 647–73

Secondary Works

Abrams, Lesley, *Bede, Gregory, and Strategies of Conversion in Anglo-Saxon England and the Spanish New World*, Jarrow Lecture, 2013 (Newcastle: Parish Church Council of St Paul's Church, Jarrow, 2015)

Barrow, Julia, 'Chester's Earliest Regatta? Edgar's Dee-Rowing Revisited', *Early Medieval Europe*, 10 (2001), 81–93

Blair, John, *The Church in Anglo-Saxon Society* (Oxford: Oxford University Press, 2005)

——, 'Churches in the Early English Landscape: Social and Economic Contexts', in *Church Archaeology: Research Directions for the Future*, ed. by John Blair and Carol Pyrah (Walmgate: Council for British Archaeology, 1996), pp. 6–19

Carver, Martin, 'Travels on the Sea and in the Mind', in *The Maritime World of the Anglo-Saxons*, ed. by Stacey S. Klein, William Schipper, and Shannon Lewis-Simpson, Essays in Anglo-Saxon Studies, 5 (Tempe: ACMRS, 2014), pp. 21–36

Cohen, Nathalie, 'Boundaries and Settlement: The Role of the Thames', in *Boundaries in Early Medieval Britain*, ed. by David Griffiths, Andrew Reynolds, and Sarah Semple, Anglo-Saxon Studies in Archaeology and History, 12 (Oxford: Oxbow Books, 2003), pp. 9–21

Cole, Ann, 'Distribution and Use of *Mere* as a Generic in Place-Names', *Journal of the English Place-Name Society*, 26 (1993), 38–50

Crerar, Belinda, 'Contextualising Romano-British Lead Tanks: A Study in Design, Destruction and Deposition', *Britannia*, 43 (2012), 135–66

Cubitt, Catherine, *Anglo-Saxon Church Councils, c. 650–c. 850* (London: Leicester University Press, 1995)

——, 'Unity and Diversity in the Early Anglo-Saxon Liturgy', *Studies in Church History*, 32 (1996), 49–50

Daniels, Robin, *Anglo-Saxon Hartlepool and the Foundations of English Christianity: An Archaeology of the Anglo-Saxon Monastery*, Monograph Series, 3 (Hartlepool: Tees Archaeology, 2007)

Darby, H. C., *The Medieval Fenland* (Newton Abbot: David & Charles, 1974)

Davies, J. G., *The Architectural Setting of Baptism* (London: Barrie & Rockliff, 1962)

Demacopoulos, George, 'Gregory the Great and the Pagan Shrines of Kent', *Journal of Late Antiquity*, 1.2 (2008), 353–69

Dix, Gregory, *The Treatise of the Apostolic Tradition of St Hippolytus of Rome*, rev. and ed. by Henry Chadwick, 2nd edn (London: Society for Promoting Christian Knowledge, 1968)

Drewer, Lois, 'Fisherman and Fish Pond: From the Sea of Sin to the Living Waters', *Art Bulletin*, 63 (1981), 533–47

Everitt, Alan, 'River and Wold: Reflections on the Historical Origins of Regions and Pays', *Journal of Historical Geography*, 3 (1977), 1–19

Ferguson, Christopher, 'Re-Evaluating Early Medieval Northumbrian Contacts and the "Coastal Highway"', in *Early Medieval Northumbria: Kingdoms and Communities, AD 450–1100*, ed. by David Petts and Sam Turner, Studies in the Early Middle Ages, 24 (Turnhout: Brepols, 2011), pp. 283–302

Ferguson, Everett, *Baptism in the Early Church: History, Theology, and Liturgy in the First Five Centuries* (Grand Rapids, MI: William B. Eerdmans, 2009)

Fern, Christopher, 'Early Anglo-Saxon Horse Burial of the Fifth to Seventh Centuries AD', in *Early Medieval Mortuary Practices*, ed. by Sarah Semple and Howard Williams, Anglo-Saxon Studies in Archaeology and History, 14 (Oxford: University of Oxford, School of Archaeology, 2007), pp. 92–109

——, 'Horses in Mind', in *Signals of Belief in Early England: Anglo-Saxon Paganism Revisited*, ed. by Martin Carver, Alexandra Sanmark, and Sarah Semple (Oxford: Oxbow Books, 2010), pp. 128–57

Fleming, Robin, 'Elites, Boats and Foreigners: Rethinking the Rebirth of English Towns', *Città e campagna nei secoli altomedievali: Spoleto, 27 marzo – 1 aprile 2008*, Settimane di studio del Centro italiano di studi sull'alto medioevo, 56 (Spoleto: Presso la sede della Fondazione, 2009), pp. 393–425

Foot, Sarah, '"By Water in the Spirit": The Administration of Baptism in Early Anglo-Saxon England', in *Pastoral Care Before the Parish*, ed. by John Blair and Richard Sharpe (Leicester: Leicester University Press, 1992), pp. 171–92

Foreman, Stuart, Jonathan Hiller, and David Petts, *Gathering the People, Settling the Land: The Archaeology of a Middle Thames Landscape, Anglo-Saxon to Post-Medieval*, Thames Valley Landscapes Monograph, 14 (Oxford: Oxford Archaeology, 2002)

Gardiner, Mark, 'Oral Tradition, Landscape and the Social Life of Place-Names', in *Sense of Place in Anglo-Saxon England*, ed. by Richard Jones and Sarah Semple (Donington: Shaun Tyas, 2012), pp. 16–30

Geary, Patrick J., 'Land, Language and Memory in Europe, 700–1100', *Transactions of the Royal Historical Society*, 6th series, 9 (1999), 169–94

Gelling, Margaret, 'The Place-Names of the Isle of Man', in *Language Contact in the British Isles: Proceedings of the Eighth International Symposium on Language Contact in Europe, Douglas, Isle of Man, 1988*, ed. by P. Sture Ureland and George Broderick (Tübingen: Max Niemeyer Verlag, 1991), pp. 141–55

———, *Signposts to the Past: Places, Names and the History of England* (Chichester: Phillimore, 1978)

Gelling, Margaret, and Ann Cole, *The Landscape of Place-Names* (Stamford: Shaun Tyas, 2000)

Gem, Richard, 'Church Buildings: Cultural Location and Meaning', in *Church Archaeology: Research Directions for the Future*, ed. by John Blair and Carol Pyrah (Walmgate: Council for British Archaeology, 1996), pp. 1–6

Gittos, Helen, 'Is There Any Evidence for the Liturgy of Parish Churches in Late Anglo-Saxon England? The Red Book of Darley and the Status of Old English', in *Pastoral Care in Late Anglo-Saxon England*, ed. by Francesca Tinti (Woodbridge: Boydell & Brewer, 2005), pp. 63–82

Groves, Sarah E., 'The Bowl Hole Burial Ground: A Late Anglian Cemetery in Northumberland', in *Burial in Later Anglo-Saxon England, c. 650 to 1100 AD*, ed. by Jo Buckberry and Annia Cherryson (Oxford: Oxbow Books, 2010), pp. 114–25

Hastrup, Kirsten, 'Water and the Configuration of Social Worlds: An Anthropological Perspective', *Journal of Water Resource and Protection*, 5.4A (2013), 59–66

Hayward, Douglas, 'Contextualizing the Gospel among the Saxons: An Example from the Ninth Century of the Cultural Adaptation of the Gospel as Found in *The Heliand*', *Missiology: An International Review*, 22 (1994), 439–53

Heide, Eldar, 'Holy Islands and the Otherworld: Places beyond Water', in *Isolated Islands in Medieval Nature, Culture and Mind*, ed. by Gerhard Jaritz and Torstein Jørgensen, Central European University Medievalia Series, 14 (Budapest: Central European University Press, 2011), pp. 57–80

Hill, David, *An Atlas of Anglo-Saxon England* (Oxford: Basil Blackwell, 1981)

Hooke, Della, *The Landscape of Anglo-Saxon England* (London: Leicester University Press, 1998)

———, 'Rivers, Wells and Springs in Anglo-Saxon England: Water in Sacred and Mystical Contexts', in *Water and the Environment in the Anglo-Saxon World*, ed. by Maren Clegg Hyer and Della Hooke, Exeter Studies in Medieval Europe (Liverpool: Liverpool University Press, 2017), pp. 107–35

Hope-Taylor, Brian, *Yeavering: An Anglo-British Centre of Early Northumbria* (Swindon: English Heritage, 1977)

Jacobsson, Mattias, *Wells, Meres, and Pools: Hydronymic Terms in the Anglo-Saxon Landscape*, Acta Universitatis Upsaliensis, Studia Anglistica Upsaliensia, 98 (Uppsala: Reklam & Katalogtryck AB, 1997)

Jensen, Robin M., 'Archaeology of Christian Initiation', in *A Companion to the Archaeology of Religion in the Ancient World*, ed. by Rubina Raja and Jörg Rüpke (Chichester: Wiley-Blackwell, 2015), pp. 253–67

Jones, Richard, 'Anglo-Saxon Water Consciousness' (unpublished conference paper, Water in Anglo-Saxon England, Institute of Historical Research Colloquium, London, 2015)

———, 'Responding to Modern Flooding: Old English Place-Names as a Repository of Traditional Ecological Knowledge', *Journal of Ecological Anthropology*, 18.1 (2016), 25 pp.

Jones, Richard, Rebecca Gregory, Susan Kilby and Ben Pears, 'Living with a Trespasser: Riparian Names and Medieval Settlement on the River Trent Floodplain', *European Journal of Post-Classical Archaeologies*, 7 (2017), 33–64

Jones, Richard, and Sarah Semple, 'Making Sense of Place in Anglo-Saxon England', in *Sense of Place in Anglo-Saxon England*, ed. by Richard Jones and Sarah Semple (Donington: Shaun Tyas, 2012), pp. 1–15

Klein, Stacey S., 'Navigating the Anglo-Saxon Seas', in *The Maritime World of the Anglo-Saxons*, ed. by Stacey S. Klein, William Schipper, and Shannon Lewis-Simpson, Essays in Anglo-Saxon Studies, 5 (Tempe: ACMRS, 2014), pp. 1–20

Loyn, H. R., *The Governance of Anglo-Saxon England, 500–1087* (London: Hodder Arnold, 1984)

Lund, Julie, 'At the Water's Edge', in *Signals of Belief in Early England: Anglo-Saxon Paganism Revisited*, ed. by Martin Carver, Alexandra Sanmark, and Sarah Semple (Oxford: Oxbow Books, 2010), pp. 49–66

Lutterkort, Karl, 'Beda Hagiographicus: Meaning and Function of Miracle Stories in the *Vita Cuthberti* and the *Historia Ecclesiastica*', in *Beda Venerabilis: Historian, Monk and Northumbrian*, ed. by Luuk A. J. R. Houwen and A. A. MacDonald (Groningen: E. Forsten, 1996), pp. 81–106

Lynch, Joseph H., *Christianizing Kinship: Ritual Sponsorship in Anglo-Saxon England* (Ithaca, NY: Cornell University Press, 1998)

Markus, R. A., *Gregory the Great and his World* (Cambridge: Cambridge University Press, 1997)

———, 'Gregory the Great and a Papal Missionary Strategy', *Studies in Church History*, 6 (1970), 29–38

Mason, Austin, 'Listening to the Early Medieval Dead: Religious Practices in Eastern Britain, 400–900 CE' (unpublished doctoral thesis, Boston College, 2012)

McGrail, Sean, and Roy Switsur, 'Medieval Logboats', *Medieval Archaeology*, 23 (1979), 229–31

McNeil, Susan, and Rosemary Cramp, 'The Wearmouth Anglo-Saxon Cemetery', in *Wearmouth and Jarrow Monastic Sites*, vol. 1, ed. by Rosemary Cramp, English Heritage Archaeological Monographs (Swindon: English Heritage, 2005), pp. 77–90

Meaney, Audrey L., 'Hundred Meeting-Places in the Cambridge Region', in *Names, Places and People: An Onomastic Miscellany in Memory of John McNeal Dodgson*, ed. by A. R. Rumble and A. D. Mills (Stamford: Shaun Tyas, 1997), pp. 195–233

Merrifield, Ralph, *The Archaeology of Ritual and Magic* (New York: New Amsterdam, 1988)

Meyvaert, Paul, 'Diversity within Unity: A Gregorian Theme', *Heythrop Journal*, 4 (1963), 141–62

Morris, Richard, 'Baptismal Places 600–800', in *People and Places in Northern Europe, 500–1600*, ed. by Ian Wood and Niels Lund (Woodbridge: Boydell & Brewer, 1991), pp. 15–24

———, *Churches in the Landscape* (London: J. M. Dent & Sons, 1989)

Oosthuizen, Susan, *The Anglo-Saxon Fenland* (Oxford: Windgather Press, 2017)

O'Reilly, Jennifer, 'Reading the Scriptures in the Life of Columba', in *Studies in the Cult of Saint Columba*, ed. by Cormac Bourke (Dublin: Four Courts Press, 1997), pp. 80–116

Pantos, Aliki, 'The Location and Form of Anglo-Saxon Assembly-Places: Some "Moot Points"', in *Assembly Places and Practices in Medieval Europe*, ed. by Sarah Semple and Aliki Pantos (Dublin: Four Courts, 2004), pp. 155–80

———, '"On the Edge of Things": The Boundary Location of Anglo-Saxon Assembly Sites', in *Boundaries in Early Medieval Britain*, ed. by David Griffiths, Andrew Reynolds, and Sarah Semple, Anglo-Saxon Studies in Archaeology and History, 12 (Oxford: Oxbow Books, 2003), pp. 38–49

Pelteret, David A. E., 'The Role of Rivers and Coastlines in Shaping Early English History', *Haskins Society Journal*, 21 (2009), 21–46

Ray, Celeste, *The Origins of Ireland's Holy Wells* (Oxford: Archaeopress Archaeology, 2014)

Reynolds, Andrew, *Anglo-Saxon Deviant Burial Customs* (Oxford: Oxford University Press, 2009)

———, *Later Anglo-Saxon England: Life & Landscape* (Stroud: Tempus, 2002)

Semple, Sarah, 'A Fear of the Past: The Place of the Prehistoric Burial Mound in the Ideology of Middle and Later Anglo-Saxon England', *World Archaeology*, 30 (1998), 109–26

———, 'Locations of Assembly in Early Anglo-Saxon England', in *Assembly Places and Practices in Medieval Europe*, ed. by Sarah Semple and Aliki Pantos (Dublin: Four Courts, 2004), pp. 135–54

———, 'Sacred Spaces and Places in Pre-Christian and Conversion Period Anglo-Saxon England', in *The Oxford Handbook of Anglo-Saxon Archaeology*, ed. by Helena Hamerow, David A. Hinton, and Sally Crawford (Oxford: Oxford University Press, 2011), pp. 742–63

Shockro, Sally, 'Reading Bede as Bede Would Read' (unpublished doctoral thesis, Boston College, 2008)

Stenton, F. M., *Anglo-Saxon England*, 3rd edn (Oxford: Oxford University Press, 1971)

——, 'The Historical Bearing of Place-Name Studies: Anglo-Saxon Heathenism', *Transactions of the Royal Historical Society*, 4th series, 23 (1941), 1–24

Stocker, David, and Paul Everson, 'The Straight and Narrow Way: Fenland Causeways and the Conversion of the Landscape in the Witham Valley, Lincolnshire', in *The Cross Goes North: Processes of Conversion in Northern Europe, AD 300–1300*, ed. by Martin Carver (Woodbridge: Boydell & Brewer, 2003), pp. 271–88

Strang, Veronica, 'Aqua Culture: The Flow of Cultural Meanings in Water', in *Water: Histories, Cultures, Ecologies*, ed. by M. Laybourne and A. Gaynor (Crawley: University of Western Australia Press, 2006), pp. 68–80

——, 'Common Senses: Water, Sensory Experience and the Generation of Meaning', *Journal of Material Culture*, 10 (2005), 92–120

Svanberg, Fredrik, *Decolonizing the Viking Age* (Stockholm: Almqvist & Wiksell, 2003)

Symonds, Leigh A., 'Territories in Transition: The Construction of Boundaries in Anglo-Scandinavian Lincolnshire', in *Boundaries in Early Medieval Britain*, ed. by David Griffiths, Andrew Reynolds, and Sarah Semple, Anglo-Saxon Studies in Archaeology and History, 12 (Oxford: Oxbow Books, 2003), pp. 28–37

Thornton, David E., 'Edgar and the Eight Kings, AD 973: *Textus et Dramatis Personae*', *Early Medieval Europe*, 10 (2001), 49–79

Tilley, Christopher, *The Materiality of Stone* (Oxford: Berg, 2004)

——, *A Phenomenology of Landscape: Places, Paths and Monuments*, Explorations in Anthropology (Oxford: Bloomsbury Academic, 1994)

Turner, Sam, 'Boundaries and Religion: The Demarcation of Early Christian Settlements in Britain', in *Boundaries in Early Medieval Britain*, ed. by David Griffiths, Andrew Reynolds, and Sarah Semple, Anglo-Saxon Studies in Archaeology and History, 12 (Oxford: Oxbow Books, 2003), pp. 50–57

——, *Making a Christian Landscape: The Countryside in Early Medieval Cornwall, Devon and Wessex* (Exeter: University of Exeter Press, 2006)

Twomey, Carolyn, 'Kings as Catechumens: Royal Conversion Narratives and Easter in Bede's *Historia Ecclesiastica*', *Haskins Society Journal*, 25 (2014), 1–18

Van de Noort, Robert, *The Humber Wetlands: The Archaeology of a Dynamic Landscape* (Sheffield: Windgather, 2004)

Walsham, Alexandra, *The Reformation of the Landscape: Religion, Identity, and Memory in Early Modern Britain and Ireland* (Oxford: Oxford University Press, 2011)

Whitfield, Niamh, 'A Suggested Function for the Holy Well?', in *Text, Image, Interpretation: Studies in Anglo-Saxon Literature and its Insular Context in Honour of Éamonn Ó Carragáin*, ed. by Alastair Minnis and Jane Roberts (Turnhout: Brepols, 2007), pp. 495–563

Wickham-Crowley, Kelley M., 'Living on the *Ecg*: The Mutable Boundaries of Land and Water in Anglo-Saxon Contexts', in *A Place to Believe In: Locating Medieval Landscapes*, ed. by Clare A. Lees and Gillian R. Overing (University Park: Pennsylvania State University Press, 2006), pp. 85–110

Williams, Howard, 'Assembling the Dead', in *Assembly Places and Practices in Medieval Europe*, ed. by Sarah Semple and Aliki Pantos (Dublin: Four Courts, 2004), pp. 109–34

——, 'Placing the Dead: Investigating the Location of Wealthy Barrow Burials in Seventh-Century England', in *Grave Matters: Eight Studies of First Millennium AD Burials in Crimea, England and Southern Scandinavia*, ed. by Martin Rundkvist, BAR International Series, 781 (Oxford: Archaeopress, 1999), pp. 57–86

Williamson, Tom, 'The Environmental Contexts of Anglo-Saxon Settlement', in *The Landscape Archaeology of Anglo-Saxon England*, ed. by N. J. Higham and Martin J. Ryan (Woodbridge: Boydell & Brewer, 2010), pp. 133–56

——, *Sutton Hoo and its Landscape: The Context of Monuments* (Oxford: Windgather Press, 2008)

Wilson, David, *Anglo-Saxon Paganism* (London: Routledge, 1992)

Wood, Ian N., *The Merovingian Kingdoms, 450–751* (London: Longman, 1994)

Yeates, Stephen James, 'Still Living with the Dobunni', in *Perspectives in Landscape Archaeology*, ed. by Sarah Semple and Helen Lewis, BAR International Series, 2103 (Oxford: Archaeopress, 2010), pp. 78–93

SIMON TRAFFORD

Swimming in Anglo-Saxon England

Swimming comes very close to being a human cultural universal; although by no means everyone can swim, the habit is encountered in every continent, evidenced in every age, and unites cultures that in other ways are wildly diverse. In the last century, the ubiquity of swimming even encouraged the formation of the 'aquatic ape hypothesis', which postulated a stage in the process of human evolution in which our ancestors adopted a predominantly aquatic existence. According to the theory, this period found its legacy in a number of adaptations to life in water — such as hairlessness or the ability to hold our breath — that are characteristic of humans but absent in other primates.[1]

Despite considerable popular interest, the aquatic ape hypothesis has failed to win any discernible academic acceptance; nevertheless, it usefully focused attention on the status of swimming as a behaviour to which humans are to some degree physiologically adapted but which also has an indispensable culturally acquired component. The majority of mammals swim extremely well even on the first occasion that they are immersed in water. Many apes, on the other hand — and uninstructed humans — are instinctively afraid in deeper water, tending to panic and drown extremely easily.[2] However, large brains and a highly developed capacity to learn allow humans to swim extremely well — once they have been taught or otherwise acquired the knack — and,

---

1 The idea was put forward (separately and in ignorance of one another) by Westenhöfer, *Der Eigenweg des Menschen* and Hardy, 'Was Man More Aquatic in the Past?', but attracted relatively little attention until it was popularized by Morgan in *The Descent of Woman* and *The Aquatic Ape*. Heavily criticized in some quarters as pseudoscience, the hypothesis has never been taken seriously by academic palaeo-anthropologists, although various evolutionary adaptations that suggest a waterside existence for early humans continue to be the subject of research. On the thesis and the debate surrounding it, the papers gathered in Roede and others, *The Aquatic Ape* are a useful introduction, although they do not cover more recent developments. For more up-to-date sceptical approaches to the hypothesis, see, for instance, Gee, *The Accidental Species*.
2 Wind, 'The Non-Aquatic Ape', pp. 273–77.

Simon Trafford • (simon.trafford@sas.ac.uk) is Lecturer in Medieval History at the Institute of Historical Research, University of London.

Meanings of Water in Early Medieval England, ed. by Carolyn Twomey and Daniel Anlezark, Studies in the Early Middle Ages, 47 (Turnhout: Brepols, 2021), pp. 85–107

indeed, to cover distances that would confound most other terrestrial mammals. Furthermore, two characteristics that, though not unique to humans, are well developed in us are the ability to hold breath voluntarily and the 'diving reflex' (which diverts bodily supply of oxygen to core functions and lowers the heart rate in response to immersion in cold water) which permits diving for periods of a minute or, with training, considerably longer. This in turn allows the gathering of food and other resources from relatively deep water in the sea, lakes, and other bodies of water, a behaviour which is unusual in terrestrial mammals.[3]

Justification for the expenditure of time and effort in learning to swim is not hard to find. Aquatic environments offer enormous opportunities and advantages: most obviously as an essential source of hydration and year-round food but also as a means of sanitation, transport, power, and defence, alongside many other uses. The ability to swim is best seen, therefore, first and foremost, as a cultural adaptation to avoid death by drowning, a risk that must be tolerated and managed in pursuit of the immense benefits bestowed by the proximity of water. In this respect, swimming sits alongside a number of other adaptive strategies — most obviously the building of boats and bridges — that allow humans to negotiate and exploit aquatic and semi-aquatic environments despite their innate physical incompetence and vulnerability in water. Despite this crucial role in self-preservation, it should immediately be said that this has never meant that all members of society do in fact acquire the skill; even some of those for whom it would seem most essential never learn. One thinks here of the oft-repeated though not necessarily well-evidenced assertion that sailors were rarely able to swim before the twentieth century.

Besides avoiding drowning, swimming offers a multitude of other opportunities for those who acquire the skill: although it is too slow and effortful to be viable as a means of *transport* over any notable distance, it can be vital as a means of *traverse* — across rivers, in order to reach moored boats, and in a range of other contexts. It also has clear military application, especially given the frequent use of seas, rivers, and ditches/moats as a means of defence. It is, moreover, an excellent form of exercise, an opportunity for competitive sport, and a locus for displays of prowess. By no means least, swimming is widely and fondly regarded as a relaxing and highly pleasurable activity in its own right, and one that can be enjoyed both by the solitary participant and also in groups; it offers rich opportunities for social interaction in an element that encourages not just the casting-off of clothing but also of inhibitions and normal social conventions.

Nor does this exhaust the functional and symbolic richness of swimming, for it exists beside, and sometimes converges with, a variety of other immersions, each with their own meanings and associations but all united by the commonality of experience of humans finding themselves in an unfamiliar

---

3   Roede, 'Aquatic Man', p. 315.

element. Washing — a human activity with a range of utilitarian and social functions and a history of its own — can obviously shade into swimming; indeed, in common usage the English word 'bathing' can describe both concepts (bathing in the Roman manner, of course, blurs the distinction even further). Still more significant are various types of ritual immersion: in our period, Christian baptism and penitential immersions are fundamental and have attracted considerable study, but we should also note what seems to have been an important Irish monastic tradition of prayer whilst immersed in cold water, a tradition which was transmitted to Anglo-Saxon England, Northumbria in particular. The enchanting story recounted by Bede of St Cuthbert having his feet dried by two otters after such an immersion is by far the most famous, but other examples are abundant, including SS Patrick, Ciarán of Saighir, Comgall of Bangor, and Kevin of Glendalough in Ireland, and SS Wilfrid, Dryhthelm of Melrose, Aldhelm of Malmesbury, and Aelred of Rievaulx in England, as well as a number of Welsh and Breton saints.[4] Lastly in considering ritual immersion it is important to recognize the presence in this category of numerous examples of exposure of the human body to water for judicial and penal purposes.[5]

Considerable attention has been devoted in recent years to the many and varied types of human interaction with water in early medieval England, and to analysing the very substantial place that seas, lakes, and rivers seem to have had in the imaginary of the pre-Conquest English; the present volume is the most recent manifestation of this trend.[6] With the exception of one significant debate (of which more below), there has been relatively little consideration specifically of swimming, although it is arguably the closest and most intimate interaction with water that is possible. The prevailing silence about Anglo-Saxon swimming is largely maintained by that literature — itself by no means substantial — which takes the history of swimming through the ages as its principal subject; although there are a number of good popular works devoted to its history and appeal, the only full scholarly history of swimming in Britain remains that published over thirty years ago by the Renaissance specialist Nicholas Orme.[7] The first chapter of his book explores the period up to 1066, with the overwhelming majority of the space devoted to Rome.

---

[4] Bede, *VCP*, ch. 10, pp. 188–91. On Irish ascetic immersion, see Gougaud, *Devotional and Ascetic Practices in the Middle Ages*, pp. 159–63; Bonser, 'Praying in Water'; Reynolds, '*Virgines subintroductae* in Celtic Christianity'; Cooper, 'New Light on Aelred's Immersion'; Herity, 'Early Irish Hermitages in the Light of the *Lives* of Cuthbert', pp. 51–53; Ireland, 'Penance and Prayer in Water'.

[5] See, amongst others, Hill, 'The Weight of Love and the Anglo-Saxon Cold Water Ordeals' and Reynolds, *Anglo-Saxon Deviant Burial Customs*, pp. 21–24, 171, 210, 214, 225.

[6] See, for instance, the papers gathered in Klein, Schipper, and Lewis-Simpson, *The Maritime World of the Anglo-Saxons*, Blair, *Waterways and Canal-Building in Medieval England*, Sobecki, *The Sea and Englishness in the Middle Ages*, or Hyer and Hooke, *Water and the Environment*.

[7] Orme, *Early British Swimming*.

The Anglo-Saxons receive a magnificent four short pages, of which three are devoted to *Beowulf*, with all of the rest of the six Anglo-Saxon centuries covered in rather less than one page. This extreme brevity should not, though, be taken as any reproach to Orme's generosity in dealing with a period outside his own, but rather as indicative of the extreme poverty of our sources; as Orme says, mentions of swimming in English pre-Conquest materials are 'scanty' and 'few'. Perhaps wisely, he restricts his discussion to a simple affirmation that swimming was known and respected among the early medieval English, adding that the bare bones provided by the sources might be fleshed out a little with reference to the much richer Scandinavian record presented in the sagas.[8]

The intention of this article is to reopen discussion of swimming in Anglo-Saxon England, locating it alongside an array of other immersions and human interactions with water. It will consist of an initial attempt to bring together as much as possible of the admittedly sparse textual evidence for swimming, including a number of sources not considered by Orme, and to reconsider interpretation of the best-known of alleged swimming episodes — those found in *Beowulf* — in the light of decades of protracted debate. The not-inconsequential possibilities of non-textual sources of data will also be considered, and comparisons made with Scandinavian and Irish examples, both of which seem able to present a much richer picture of swimming.

Beowulf

The only section of the evidence for swimming in Anglo-Saxon England that has benefited from any sort of concerted scholarly discussion is that which is found in *Beowulf*, doubtless as a result of the poem's status as the most prominent work of Old English literature.[9] This is, therefore, the starting point for this enquiry, as the debate has introduced some important arguments and parallels which demand consideration. It should be noted, however, that the primary focus of these discussions has been on swimming as a literary theme or discursive device rather than as a lived practice; this is, of course, entirely reasonable in analysis of a work of literature, but it is a difference of emphasis from the present enquiry.

Traditionally there have been said to be three swimming episodes in the poem:[10]

1) The famous competition in Beowulf's youth with his friend Breca, lasting five days in the open ocean and incorporating a lengthy battle with sea monsters (ll. 506–81).

---

8 Orme, *Early British Swimming*, pp. 9–18.
9 *Klaeber's Beowulf*, ed. by Fulk, Bjork, and Niles.
10 From the earliest editions of the poem at the start of the nineteenth century through to the 1960s and later, all these episodes were uniformly taken as examples of swimming, and it remains a common translation today. See Fulk, 'Afloat in Semantic Space', pp. 458–59.

2) Beowulf's plunge into the mere, involving almost a day's worth of breath-holding, preliminary to the fight with Grendel's mother (ll. 1492–1512).
3) The retreat from Hygelac's disastrous raid on Frisia, after which Beowulf swam, on his own, hundreds of miles back to the land of the Geats, clutching, allegedly, thirty men's mail-coats on his arm (ll. 2359–62).

All three present difficulties of interpretation, and a substantial literature has grown up which has focused attention on a number of features.[11] Above all, the exaggeratedly superhuman character of these feats is much remarked upon and is the cause of some disquiet; to some commentators it strikes a fantastical and hyperbolic note that contrasts with the otherwise prevailing mood of understatement.[12] Profound doubt has been raised over whether the escapade with Breca was actually a swimming competition at all, but rather something that involved boats; after all, Beowulf himself says of the incident 'wit on sund reon' (we rowed on the sea) (l. 539b).[13] Opinion has divided over the extent to which this and other allusions to rowing are to be taken literally or as metaphorical expressions for swimming. Linked to this has been extensive discussion of the word *sund*, which, it is normally suggested, means either 'the sea' in poetic contexts or 'swimming' in prose; the *Beowulf* poet though at least sometimes seems to use it to describe swimming. More recently, it has been suggested that the word is deployed precisely because of the ambiguity it introduces, and that it may have been, in the words of R. D. Fulk, 'semantically indeterminate, floating among the senses "sea, swimming, boating, floating" and so forth'.[14]

A similar re-evaluation of the 'retreat from Frisia' episode has also taken place: again, the idea that the poet portrays Beowulf as literally swimming hundreds of miles home to the land of the Geats has been questioned, with debate this time centring around how to reconstruct a damaged section of text and, in particular, how to interpret the words *sundnytte* and *ofer-swam*, both of which, it has been suggested, may indicate human movement through or over the water without necessarily making clear if this was by boat or swimming.[15]

There remain stalwart defenders of both the 'rowing' and 'swimming' standpoints with regard to the Breca and the Frisian episodes (the plunge

---

11 Considerable attention has been devoted to the competing claims of 'Germanic' and 'Celtic' analogues to the aquatic episodes: Earl, 'Beowulf's Rowing Match'; Puhvel, 'The Swimming Prowess of Beowulf'; Puhvel, 'Beowulf and Celtic Under-Water Adventure'; Puhvel, 'The Aquatic Contest in *Hálfdanar saga Brönufóstra* and Beowulf's Adventure with Breca'; Jorgensen, 'Beowulf's Swimming Contest with Breca'; Russom, 'A Germanic Concept of Nobility in *The Gifts of Men* and *Beowulf*'.
12 Puhvel, 'The Swimming Prowess of Beowulf'; Orchard, *A Critical Companion to 'Beowulf'*, p. 125.
13 Wentersdorf, 'Beowulf's Adventure with Breca'; Earl, 'Beowulf's Rowing Match'.
14 Fulk, 'Afloat in Semantic Space', p. 472; cf. Frank, '"Mere" and "sund"'; Anlezark, 'All at Sea', pp. 227–32.
15 Wentersdorf, 'Beowulf's Withdrawal from Frisia'; see also Anlezark, 'All at Sea', pp. 233–35.

into the mere is unambiguously and incontrovertibly a bodily immersion — although a feat of breath-holding rather than of swimming as such — but textual ambiguities ensure that even in this case there is debate over its precise nature);[16] it is perhaps safest simply to acknowledge the vagueness and ambiguity of the poem in this respect. With the uncertainty that surrounds them, the aquatic episodes in *Beowulf* are limited in their usefulness for our present purposes. Even if we were to accept all of the swimming narratives, these anecdotes are, above all, examples of the superhuman prowess of a hero. They are introduced for specific purposes within the narrative, with each story related to a monster fight, and — as Daniel Anlezark has argued — make the point that 'Beowulf's strength and his ability to traverse the sea, and immerse himself in it, are inseparable aspects of his character'.[17] They may, perhaps, indicate that skill in swimming might be esteemed in a warrior (if that had ever been in doubt), but to extrapolate from that to the role swimming might play in everyday life is a long step.

The Literary Evidence

Turning away from the famous, but problematic, *Beowulf* episodes, it is soon evident that references to swimming (or at least to *humans* swimming: there are plenty of fish and other beasts) are actually few and only very occasionally encountered: they are chance occurrences, not elaborated upon or emphasized, and rarely with any particular significance invested in them.[18] Given that they are so few, it is useful to consider them individually, as their context may have some significance.

Perhaps most closely related to *Beowulf*, not least in that it appears in a poetic context, is a line (58) in *The Gifts of Men*, a poem preserved in the tenth-century Exeter Book. Listing a very large number of desirable attributes of all sorts, including keeping beer, piloting a ship, being quick with dice and being handsome, the poem also mentions those who are 'syndig' (skilled in swimming).[19] Geoffrey Russom has urged that these be seen alongside similar lists of *íþróttir* or high-status masculine accomplishments in a variety of Scandinavian poems; swimming, in particular, he identifies as 'an especially distinguished sport', described in the Eddic poem *Rígsþula* as characteristic of earls rather than freemen or thralls.[20]

Just as fleeting as the reference to swimming in *The Gifts of Men* is a mention in the *Prognostics* (London, British Library, MS Cotton Tiberius A.iii, of the

---

16 Anlezark, 'All at Sea', p. 233.
17 Anlezark, 'All at Sea', p. 239.
18 For an example of these swimming creatures, see Elizabeth A. Alexander, 'The Sailors, the Sea Monster, and the Saviour: Depicting Jonah and the *Ketos* in Anglo-Saxon' in this volume.
19 *The Exeter Anthology of Old English Poetry*, ed. by Muir.
20 Russom, 'A Germanic Concept of Nobility in *The Gifts of Men* and *Beowulf*', p. 7.

eleventh century) that to dream of swimming signifies anxiety or harm: 'In flumen n(a)tare. anxietatem. significat. | on flod swymman anxsumnesse <getacnað>' (l. 135); 'Natare se uidere dampnum. significat. | swimman hine geseon hearm <getacnað>' (l. 263).[21] Beyond the fairly redundant point that swimming was clearly sufficiently prominent in the minds of the pre-Conquest English that it might invade their subconscious, it is hard to know what can be made of this. In *Solomon and Saturn*, however, there is a more substantial allusion to swimming, appearing in a context which, with its mention of a great water-traveller named Wulf, Anlezark has linked with *Beowulf*. Solomon notes:

> Dol bið se gæð   on deop wæter,
> se ðe sund nafað,   ne gesegled scip,
> ne fugles flyht,   ne he mid fotum ne mæg
> grund geræcan;   huru se Godes cunnað
> full dyslica,   Dryhtnes meahta.
>
>> [Foolish is he who goes into deep water, he who can't swim, nor has a sail-rigged ship, nor the flight of a bird, who cannot reach the bed with his feet. Indeed, he very foolishly tests God, the Lord's might.][22]

Alongside these are a number of mentions of swimming that appear in translations or versions of stories from classical or biblical sources. One, notably, occurs in the letter of Alexander to Aristotle bound together with the manuscript of *Beowulf* in London, British Library, MS Cotton Vitellius A.xv and in the same hand as one of the scribes of *Beowulf*. It describes the fate of a section of Alexander's army that was attacked by a herd of hippopotamus:

> Þa het ic .CC. minra þegna of greca herige leohtum wæpnum hie gegyrwan, 7 hie on sunde to Þære byrig foron 7 swumman ofer æfter þære ea to þæm eglande. Þa hie ða hæfdon feorðan dæl þære ea geswummen, ða becwom sum ongrislic wise on hie.
>
>> [Then I ordered two hundred of my thegns from the Greek army to arm themselves with light weapons and go over to the village by swimming, and they swam over across the river to that island. And when they had swum about a quarter of the river, something terrible happened to them.][23]

Dragged to a watery death by monsters, it seems clear that there is a deliberate linkage of this text and *Beowulf* with its themes of aquatic monstrousness.[24]

---

21 *Anglo-Saxon Prognostics*, ed. and trans. by Liuzza, pp. 98–99 and 116–17.
22 *The Old English Dialogues of Solomon and Saturn*, ed. and trans. by Anlezark, ll. 48–51, pp. 80–81. See Anlezark, 'All at Sea', pp. 236–39.
23 *The Old English Letter of Alexander to Aristotle*, ed. by Orchard, pp. 234–35.
24 Orchard, *Pride and Prodigies*, pp. 25–27.

More substantially, two other translations into Old English both see situations in which noblemen encounter bodies of water which they need to cross.[25] Thus in the Old English translation of Orosius's *History*, one of Cyrus the Great's warriors attempts to cross a wide river: 'þa gebeotode an his ðegna þæt he mid sunde þa ea oferfaran wolde, mid twam tyncenum, ac hiene se stream fordraf' (then one of his thanes vowed that he would cross the river by floating (or swimming) with two small casks, but the current swept him away).[26] And in the eleventh-century Old English prose translation of *Apollonius of Tyre*, the hero is cast up by a shipwreck: 'Þæt scip eal tobærst. On ðissere egeslican reownesse Apollonius geferan ealle forwurdon to deaðe, and Apollonius ana becom mid sunde to Pentapolim þam Ciriniscan lande, and þar up eode on ðam strande' (the ship broke up completely. All of Apollonius' companions perished in this terrible tempest, and Apollonius alone, by swimming, came to Pentapolis in the Cyrenaican land and went ashore there).[27] In the original Latin, it is clear that Apollonius floats ashore with the aid of a plank from the boat; Nicholas Orme attempts to bolster his case for widespread swimming amongst the pre-Conquest English by taking the omission of the buoyancy aid as indicative of an expectation on the part of the English translator that the ability to swim was a standard attribute of noble males, but this seems a rather laboured argument, especially given that *mid sunde* need not mean unassisted swimming, but can refer to any type of watery traverse.[28] In sum, although the general presence of swimming in the mental furniture of the early medieval English is very clear from this handful of allusions in literary sources, it is very hard to craft it into anything more substantial.

Army (and Other) Drownings

By contrast, chronicle and narrative sources seem to provide some rather more promising material. In particular, it is worthwhile to note a number of occasions on which armies, or parts of armies, are drowned when attempting to cross water, normally when in retreat.[29] In the *Historia Ecclesiastica* I. 20,

---

25 See Wentersdorf, 'Beowulf's Adventure with Breca', pp. 156–57; Orme, *Early British Swimming*, pp. 12–13.
26 *The Old English Orosius*, ed. by Bately, p. 43; trans. by Wentersdorf, 'Beowulf's Adventure with Breca', p. 157.
27 *The Old English Apollonius of Tyre*, ed. by Goolden, pp. 16–17; trans. by Wentersdorf, 'Beowulf's Adventure with Breca', p. 157.
28 Orme, *Early British Swimming*, p. 13.
29 Whether warriors specifically trained in swimming has occasioned somewhat inconclusive comment: Weinhold, *Altnordisches Leben* believes it was commonplace in Scandinavia, but Wentersdorf, 'Beowulf's Adventure with Breca', pp. 147–48, points out the extreme poverty of the evidence for this. The principal late Roman military manual, Vegetius's *Epitoma Rei Militaris* (of the late fourth century) emphasizes the importance of swimming to armies, advice that was redoubled in Carolingian Francia by Hrabanus Maurus's commentary on the

Bede recounts the calamitous fate of the Saxon and Pictish army that attempted to flee after being terrified by St Germanus at the 'Alleluia!' battle: 'Passim fugiunt, arma proiciunt, gaudentes uel nuda corpora eripuisse discrimini; plures etiam timore praecipites flumen quod transierant deuorauit' (They fled hither and thither, casting away their weapons and glad even to escape naked from the danger. Many of them rushed headlong back in panic and were drowned in the river which they had just crossed).[30] The gruesome death of the pagans is of special significance, for earlier in the same chapter we see Germanus administering Lenten baptism to a large portion of the army of the Britons; the very different consequences of bodily exposure to water for the members of the respective hosts is very pointed. A comparable pairing of pagan drowning and Christian baptism is evident in *HE* IV. 13: when St Wilfrid enters the (pagan) kingdom of the South Saxons, it has been suffering from three years of drought, causing the starving population to abandon hope: 'xl simul aut l homines inedia macerati procederent ad praecipitium aliquod siue ripam maris, et iunctis misere manibus pariter omnes aut ruina perituri aut fluctibus absorbendi deciderent' (forty or fifty men, wasted with hunger, would go together to some precipice or to the sea shore where in their misery they would join hands and leap into the sea, perishing wretchedly either by the fall or by drowning).[31] But as soon as the people received baptism from Wilfrid a gentle rain fell, the land was replenished, and the newly Christian populace was saved.

In common with patristic writers, in Bede death by drowning is almost without fail something that happens only to pagans.[32] Indeed, the battle that marked the effective end of paganism as any sort of threat in England, the defeat of Penda and his allies at the Battle of the *Winwæd* in 655, is appropriately attended by another mass drowning of those put to flight:

> Et quia prope fluuium Uinued pugnatum est, qui tunc prae inundantia pluuiarum late alueum suum, immo onmes ripas suas transierat, contigit ut multo plures aqua fugientes quam bellantes perderet ensis.
>
> [The battle was fought near the river *Winwæd* which, owing to heavy rains, had overflowed its channels and its banks to such an extent

---

Epitoma, De Procinctu romanae Miliciae, ed. by Dümmler, which insisted that it was essential for armies of his own time to swim well (see Bachrach, *Early Carolingian Warfare*, p. 257). The extent to which Vegetius circulated in England before the Conquest is unclear: Bede quotes the *Epitoma* but without mentioning the title or author of the work (Jones, 'Bede and Vegetius', p. 248), but neither of the two surviving early copies now in England can conclusively be shown to have been there before the middle of the eleventh century (Gneuss and Lapidge, *Anglo-Saxon Manuscripts*). On Vegetius and swimming, see also Orme, *Early British Swimming*, p. 7.

30 Bede, *HE*, I. 20, pp. 63–65.
31 Bede, *HE*, IV. 13, pp. 373–75.
32 Daniell, *Death and Burial in Medieval England*, pp. 65–68.

that many more were drowned in flight than were destroyed by the sword in battle.]³³

Moving later in the period, the *Anglo-Saxon Chronicle* recounts an attack in 853 on the Viking base at Thanet in which the English were initially successful but later fled with significant loss on both sides through battle and drowning: '7 þy ilcan geare Alhhere mid Cantwarum 7 Huda mid Suþrigum gefuhton on Tænet wið hæþenne here, 7 þær wurdon feola ofslægene 7 adruncene on agðre hand 7 þa ealdormen begen ofsleagon' (And the same year Ealhhere with the Kentishmen and Huda with the men of Surrey fought in Thanet against a heathen host; and there were many slain and drowned on either side, and the ealdormen both slain).³⁴

A rather more successful English defence is recorded in 915:

> Þ(a) bestælon hi þeah nihtes up æt sumum twam cyrrum, æt oþrum cyrre beeastan Weced, <æt> oþrum cyrre æt Porlocan, þa sloh hi mon æt ægþerum cyrre þæt hyra feawa onweg comon, buton þa ane <þe> þær ut oðswymman mihton to þam scipum.
>
> [However, they landed secretly by night on two separate occasions: once east of Watchet and again at Porlock; and on each occasion the English struck them so that only those few escaped who were able to swim out to the ships.]³⁵

Exactly what lies behind this entry is hard to know: Were the escapees those with the strength to manage a swim in armour? To what extent, indeed, was swimming whilst encumbered with weapons and armour a possibility? It is probably safest to say that it depends upon a number of factors: the type of armour worn, the strength and preparedness of the swimmer, and the distance attempted. Videos of experiments by plucky re-enactors available online suggest that swimming in mail is possible but hard, although these rarely seem to have accounted for the substantial additional weight of a weapon.³⁶ This should be set against such evidence as the story recounted by Gregory of Tours of an unnamed friend of Count Guntram Boso who, whilst wading across the River Rhône wearing a heavy mail shirt, stepped into a pit in the riverbed dug as a defensive measure and was never seen again.³⁷ Those wearing lighter or non-metallic armour would, presumably, have had much less trouble in swimming.

It is equally interesting to conjecture about the background to the *Chronicle* entry for 1013, referring to Sweyn's army: 'Wendon þa þanon eastweard to

---

33 Bede, *HE*, III. 24, pp. 290–91.
34 *The Anglo-Saxon Chronicle* [D], ed. by Cubbin, *s.a.* 853; trans. by Garmonsway, pp. 65, 67.
35 *The Anglo-Saxon Chronicle* [D], ed. by Cubbin, *s.a.* 915; trans. by Garmonsway, p. 99.
36 For instance, Kaptorga – Visual History, 'Can You Swim in Chainmail?'. See also note 29, above.
37 Gregory of Tours, *Decem Libri Historiarum*, ed. by Krusch and Levison, VI. 26, pp. 265–66.

Lundene, 7 mycel his folces adrangc on Temese, for þam þe hi nanre brice ne cæpton' (Thence they went east to London, and a great part of his host was drowned in the Thames, because they did not bother to look for any bridge).[38] This would appear to be the chronicler's sardonic allusion to some sort of attempt to bypass or attack the heavily defended London Bridge.[39] The circumstances are completely unknown, and it is hard to know what light, if any, the incident sheds upon swimming; it is probably best interpreted alongside the *Chronicle* entry for 1016, which famously records Cnut's excavation of a trench or canal for his boats in order to circumvent the bridge and avoid the fate which had befallen his father's troops, or even alongside later Scandinavian traditions of an attack on the bridge in 1014.[40]

Moving towards the end of the period, in 1066 we are twice told of portions of armies drowning when in retreat from a defeat. Firstly the English at Fulford:

> Þa, ær þam þe se cynning Harold þyder cuman mihte, þa gegaderode Eadwine eorl 7 Morkere eorll of heora eorldome swa mycel werod swa hi begitan mihton 7 wið þone here gefuhton 7 mycel wæl geslogon, 7 þær wæs þæs engliscan folces mycel ofslagen 7 adrent 7 on felam bedrifen, 7 Normen ahton wælstowe gewald.

> [But before King Harold could arrive, Earl Edwin and Earl Morcar had gathered as great a force as they could from their earldom, and fought that host, and made great slaughter of them; but a great number of the English were either slain or drowned or driven in flight, and the Norwegians had possession of the place of slaughter.][41]

The description finds its mirror-image at Stamford Bridge a few days later, with the defeat and dispersal of Harald Hardrada's army: '7 þa engliscan hi hindan hetelice slogon, oð þæt hig sume to scype coman, sume adruncen, 7 sume eac forbærnde, 7 swa mislice forfarene þæt þær wæs lyt to lafe, 7 Engle ahton wælstowe geweald' (the English fiercely assailed their rear until some of them reached their ships; some were drowned, others burnt to death, and thus perished in various ways so that there were few survivors, and the English had possession of the place of slaughter).[42]

The descriptions of the two northern, riverine, battles of 1066 are paired, and the same formula of slaughter, death by water and complete disaster, befalls the respective losing side of each encounter. We are entitled to wonder if there is a hint of a rhetorical flourish in these descriptions that takes advantage of

---

38   *The Anglo-Saxon Chronicle* [D], ed. by Cubbin, *s.a.* 1013; trans. by Garmonsway, p. 143.
39   Cf. Pelteret, 'The Role of Rivers and Coastlines in Shaping Early English History', p. 38.
40   Hagland, 'Saxo-Norman London Bridge and Southwark', pp. 232–33; Hagland and Watson, 'Fact or Folklore'.
41   *The Anglo-Saxon Chronicle* [C], ed. by O'Brien O'Keeffe, *s.a.* 1066; trans. by Garmonsway, p. 196.
42   *The Anglo-Saxon Chronicle* [D], ed. by Cubbin, *s.a.* 1066; trans. by Garmonsway, p. 199.

the similar topography of the two battlefields to play up a detail that fitted with the mood of national disaster that characterizes the annal for the year.

As with our other types of evidence, the several stories of mass military drownings in the *Chronicle* only provide hints and glimpses of what must surely have been regular interactions of warriors with water. The ability to swim to and from moored boats is an essential prerequisite of marine hit-and-run tactics and seems to have been standard and unremarkable practice, although it would certainly have been a feat which would become far harder when defeated, pressed by pursuers, and wearing heavy armour. To be sure, the occasions on which we are told of drownings are those when things have gone drastically wrong for one side or the other. The special ghastliness of an ignoble death by water heightens the calamity of defeat or sharpens the victory as the invader is repulsed, and surely must also have echoes of the fate of Pharaoh's army in the Red Sea in the Book of Exodus. We might compare a Viking defeat in 891 at the River Dijle near Leuven in Belgium, described with no small relish by the Annals of Fulda:

> Nordmanni fuge praesidium querentes, flumen, quod antea eis a tergo pro muro habebatur, pro morte occurrebat. Nam instantibus ex altera parte cede christianis coacti sunt in flumen pracipitari. Coacervatim se per manus et colla cruribusque complectentes in profundum per centena vel milia numero mergebantur, ita ut cadaveribus interceptum alveum amnis siccum appareret. In eo proelio cesi sunt duo reges eorum, Sigifridus scilicet et Gotafridus; regia signa XVI ablata et in Baioaria in testimonium transmissa sunt.
>
> [The Northmen sought safety in flight and found that the river, which before they had thought of as a wall to their rear, was now their death. For with the Christians bringing death from the other side they were forced to throw themselves into the river, and, grasping at each other in heaps by hand, neck and limbs, they sank in hundreds and thousands, so that their corpses blocked the river bed and it seemed to run dry. In that battle two of their kings were killed, that is Sigifrid and Godafrid, and sixteen royal standards were carried off and were sent to Bavaria as a witness.][43]

The final example to bring forward here, though also a narrative source, is of a rather different type. This is to be found in the account of an Englishman named Saewulf of a pilgrimage to Jerusalem in 1103 (and thus admittedly only tenuously included in this survey). His ship and many others were at anchorage off Jaffa when they were caught in a terrible storm:

> Quidam stupore consumpti ibidem dimersi sunt; quidam a lignis proprie nauis, quod incredibile multis uidetur, adherentes, me uidente,

---

43 *Annales fuldenses*, ed. by Kurze, s.a. 891, pp. 120–21. The translation is from *The Annals of Fulda*, ed. and trans. by Reuter, pp. 122–23. On Viking drownings, see Clarke, 'The Vikings', p. 45.

ibidem sunt obtruncati: quidam autem a tabulis nauis euulsis iterum in profundum deportabantur. Quidam autem natare scientes sponte se fluctibus commiserunt, et ita quamplures perierunt. Perpauci quippe, propria uirtute confidentes, ad litus illesi peruenerunt.

> [Some, stupefied with terror, were drowned; some, as they were clinging, were decapitated by the timbers of their own ship. This may seem incredible to many, yet I saw it. Some, washed off from the decks of their ships, were carried out again to the deep. Some, who knew how to swim, voluntarily committed themselves to the waves, and thus many of them perished. Very few, who had confidence in their own strength, arrived safe on shore.][44]

Amidst the remarkable turn of events Saewulf describes, one thing that is almost lost is, by contrast, the complete ordinariness of swimming: an English commentator with no particular axe to grind accepts as normal that the ability to swim is widespread, but by no means ubiquitous among his peers. It is matter-of-fact and commonplace; indeed the lack of any particular interest in swimming on those rare occasions that it does appear tempts us to indulge the argument that it is precisely because the ability to swim was so widespread, quotidian, and unremarkable that it is hardly ever mentioned.

Beyond the Textual

Having noted the bare scraps that the documentary record provides, it is important to turn to other types of evidence. Although a number of pieces of surviving Anglo-Saxon artwork depict people in water, in practically all cases they are unambiguously receiving baptism rather than swimming as such. However accompanying Psalm 66 in the Bury Psalter, and immediately below the line 'transivimus per ignem et aquam' (we went through fire and water), there are depicted a number of human figures: from the left, the first is emerging from flames, and the next — clearly immersed in water — does indeed seem to be doing something most easily interpreted as swimming; indeed, the position of the legs looks very like an attempt to represent the frog kick of breast stroke.[45] Next to this figure another, on its back, is also apparently in water but reaches towards an apparently toppling tower marked 'opera iustitie' (works of justice), while at the far right, a final figure emerges

---

44 Saewulf, *Relatio de peregratione ad Hierosolymam*, trans. by Brownlow.
45 Vatican City, Biblioteca Apostolica Vaticana, MS Reg. lat. 12. fol. 71ᵛ. See Ohlgren, *Anglo-Saxon Textual Illustration*. On strokes, whilst it is quite impossible to know what approaches to propulsion were adopted by medieval swimmers, it should be noted that the first swimming manual to be published in England, Sir Everard Digby's *De Arte Natandi* of 1587 gives pride of place to breast stroke. Interestingly, despite describing over forty strokes, techniques, and manoeuvres, Digby does not ever mention anything that resembles front

Figure 3.1. Detail of a swimming figure from Psalm 66 in the Bury Psalter. Vatican City, Biblioteca Apostolica Vaticana, MS Reg. lat. 12. fol. 71ᵛ. Second quarter of the eleventh century. Reproduced with permission of the Biblioteca Apostolica.

onto dry land and is clasped in the arms of God (Figure 3.1). So far as I am aware this is the only depiction of human swimming from Anglo-Saxon England: although other psalters illustrate this psalm, they plump for wading figures rather than swimmers.

Moving to onomastic sources, it is probably over-optimistic to hope that place-name evidence might be able to add much to the picture. All that is needed to be able to swim is about a metre's depth of water in river, pond, lake, or sea, so potential locations for recreational swimming are spectacularly abundant throughout the British Isles. It seems inherently unlikely that any particular spot would have acquired a name indicating its especial suitability for swimming, although that is a possibility for a small scatter of places whose names incorporate the element *bæð*, such as Bathley in Nottinghamshire.[46] Another possibility is that some of those places that include the element *gelad*, 'a passage over water', such as Lechlade in Gloucestershire or Cricklade, Wiltshire, may indicate spots suitable for transit by swimming as opposed to a ford, although in most cases the shallowness of the river involved would seem to make that unlikely.[47]

Study of swimming through the archaeological record raises very similar problems to those that impoverish the toponymic: swimming can take place anywhere, requires no particular material components, and leaves no direct trace. There is, however, one significant avenue for archaeological research

---

crawl; Europeans only adopted it when confronted with the 'hand-over-hand' technique of indigenous Americans and Australians in the mid-nineteenth century and later. See Orme, *Early British Swimming*, pp. 85–88, and Sher, *Swim*, pp. 66–68.

46 Parsons and Styles, *The Vocabulary of English Place-Names (Á – Box)*, p. 62.
47 Cronan, 'Old English *gelad*'.

on swimming which, so far as I am aware, is yet to be exploited for medieval populations in Britain. This makes use of the phenomenon of external auricular exostosis, a growth of bone around the ear-canal which is stimulated by repeated exposure to cold water and is particularly prevalent amongst water-sports enthusiasts (hence its popular name, 'surfer's ear'). It is possible to recognize the condition in well-preserved skeletal material and thus to single out individuals who regularly immersed themselves. This technique has already been used to answer questions about marine resource exploitation in a variety of other contexts, and it seems at least possible that similar studies could be undertaken on medieval English material.[48] While by no means looking to these techniques as some sort of panacea for the gaping holes in our knowledge and understanding of immersions in Anglo-Saxon England, that there is potential for them to open entirely new lines of enquiry — for instance, into the gendering of swimming — is undeniable.

Comparanda: The Scandinavian and Irish Examples

In the absence — for the time being — of any expansion of the evidence for swimming in pre-Conquest England, what may be most instructive is to contextualize what little we do have by looking at comparable examples from elsewhere, where evidence of different kinds from that available for England at least seems to give a rather richer picture of swimming in an early medieval society. This is by no means a new approach; as we have seen, much discussion of the aquatic episodes in *Beowulf* has revolved around establishing parallels for them in so-called 'Germanic' or 'Celtic' literature, suggesting that they represent reflexes of an archetype of the aquatically skilled hero within the respective traditions of heroic literature. Some have even gone so far as to suggest direct influence of the *Beowulf* stories on these foreign materials, or vice versa. What is intended in the very limited space available here is rather more limited in ambition: simply to note the relative preponderance of references to swimming in neighbouring and very similar cultures to that of Anglo-Saxon England, and to conjecture as to whether it is reasonable to take the more informative and detailed picture that they provide as also representing the situation in England. Are we able safely to suppose that what was the case in Scandinavia or Ireland was also true of England, and that the relative paucity of references to swimming from the latter is merely an artefact of the quality or nature of the surviving source materials?

The Scandinavian material deserves first place in this discussion, for saga references to swimming and diving are famously abundant, and the

---

48 Villotte and Knüsel, 'External Auditory Exostoses and Prehistoric Aquatic Resources Procurement'. See, for example, Crowe and others, 'Water-Related Occupations and Diet in Two Roman Coastal Communities'.

impression conveyed is that it was ubiquitous and commonplace, an activity engaged in and, indeed, enjoyed by all sorts of characters in the stories. Of course, as we all know, considerable caution always needs to be exercised in applying the picture presented in these generally much later Icelandic tales to the first-millennium Continental and Insular societies that they purport to describe, and this may be even more the case with swimming. Then as now, Iceland had a peculiarly strong and highly developed culture of outdoor swimming in its proliferation of geothermally heated pools. Many saga writers and readers were used to spending an appreciable portion of their time in comfortably warm water — Snorri Sturluson's pool is a present-day tourist attraction in Reykholt — but it is impossible to know if and to what extent that may have encouraged unrealistic portrayals of protracted swims and socializing at length in the considerably less congenial rivers, lakes, and fjords of continental Scandinavia or of those Northern Atlantic islands not fortunate enough to be volcanic.[49]

That said, the cumulative weight of saga characters entering the water is very substantial and not easily set aside. It is neither possible nor desirable to list every aquatic episode here, but it is useful at least to convey an impression of their frequency and significance in the world view of the sagas. Very often swimming prowess is used — as perhaps in *Beowulf* — to demonstrate the heroic or sometimes superhuman qualities of the saga protagonists. Thus, Skallagrim and Egil in *Egil's Saga*, Kjartan and Bolli in *Laxdæla Saga*, Gunnar in *Njal's Saga*, and Grettir and Bjorn in *Grettir's Saga* are all portrayed as prodigious swimmers.[50] Nevertheless, swimming is by no means the preserve of the saga-heroes but can be found at all levels of society: *Orkneyinga Saga* sees farmhands displaying their ability, and in *Egil's Saga* we find Irish slaves escaping their servitude by swimming to an island off the Icelandic coast.[51] We even see, again in *Egil's Saga*, a woman swimming: Thorgerd Brak, Skallagrim's slave woman, attempts to evade him when he is pursuing her by jumping into the sea and swimming.[52]

The sagas also show us something of the social role of swimming, as a pleasurable activity to be enjoyed with friends or as an opportunity for competition, perhaps the most obvious example being a well-known episode

---

49 The point that saga accounts of lengthy immersions in frigid northern waters need not always be taken absolutely literally might equally be raised with regard to Thorkel Farserk's supposed two-mile swim off the coast of Greenland shortly before the year 1000 that is described in Landnámabók, but, as Holsinger, 'Thorkel Farserk Goes for a Swim' has explored, this has not prevented the episode from being cited incautiously in contemporary debates on anthropogenic climate change.
50 *Grettis saga Ásmundarsonar*, ed. by Guðni Jónsson; Puhvel, 'The Aquatic Contest in *Hálfdanar saga Brönufóstra* and Beowulf's Adventure with Breca', p. 135; Wentersdorf, 'Beowulf's Adventure with Breca', pp. 146–47; Orme, *Early British Swimming*, p. 15.
51 *Orkneyinga Saga*, trans. by Pálsson and Edwards, p. 110; *Egil's Saga*, trans. by Scudder, p. 149.
52 *Egil's Saga*, trans. by Scudder, p. 63.

in *Laxdæla Saga* in which the hero, Kjartan, accompanies a group of men from Nidaros in Norway to swim in the nearby River Nid. They engage in a ducking and breath-holding game — *kefja* — with Kjartan emerging triumphant over all apart from a mysterious stranger who turns out to be King Olaf Tryggvason.[53] The overall picture of the Scandinavian evidence is vivid and emphatic and portrays swimming not just as a widely available and respected skill, but as an enjoyable pastime in its own right, and a focus for social interaction, performance, and display. Is it reasonable to believe that a similar picture might prevail were sources of a similar type available for Anglo-Saxon England? That would be to venture into realms of speculation for which we are unprepared, but there is nothing inherently unlikely about it. For an additional comparison which may enhance our picture, however, we may perhaps turn to a number of references to swimming in Irish sources which, though by no means as plentiful or as descriptive as those in the sagas, have the distinct advantage of much closer geographical and temporal proximity to Anglo-Saxon material.

Swimming makes its appearance in Irish records largely as a consequence of a potent and frequently encountered hagiographical trope that sees those on or near the water being menaced by aquatic monsters.[54] Easily the most famous of these is Adomnán's story of St Columba's brush with the Loch Ness Monster (more strictly the River Ness Monster, for it is there, rather than in the loch, that he encounters it).[55] Crossing the river, Columba encounters the funeral of a Pict who, whilst swimming, had been savaged by the monster. Hearing this, Columba orders one of his companions to jump in and bring him back a boat from the other side of the river. No doubt regretting the day he ever met the saint, his monk Lugne takes off his clothes apart from his tunic and dives in, immediately attracting the attentions of the monster, which swims up from the bottom where it has been resting and rushes at him. Columba, of course, commands the beast to stop, and it swims off in fear. Far less well known, but just as germane for our purposes, is the story of the sixth-century evangelist of Scotland, St Moluag (or St Lugid), who, when only a boy, swims across a pond with two companions, where he encounters a water-monster the size of a ship. One of Lugid's friends looks at the creature and dies, but Lugid is able to resurrect him and, naturally, defeats the creature by blessing it; thereafter the water is no longer a danger.[56] It is, though, not just in hagiography that the status of swimming in Irish society is evident: to take just one example, the *Cáin Íarraith*, the eighth-century law of fosterage fee, states that fostered boys of noble family should be taught swimming (along with riding, marksmanship,

---

53   *Laxdæla Saga*, trans. by Kunz, pp. 347–48.
54   Borsje, 'The Movement of Waters as Symbolised by Monsters in Early Irish Texts'.
55   Adomnán, *Life of St Columba*, trans. by Sharpe, II. 27, pp. 175–76. Thomas, 'The "Monster" Episode in Adomnan's *Life* of St Columba'; Borsje, 'The Monster in the River Ness in *Vita Sancti Columbae*'.
56   *Vita St Lugidus*, ed. by Heist, p. 136.

and boardgames) by their foster fathers.[57] There is no space here to explore this subject in depth, but the impression that these Irish examples provide is of a world in which swimming is abundant, quotidian, and unremarkable.

To conclude, although this exploration has brought forward some new material, the written evidence on which to base our understanding of swimming in early medieval England is meagre indeed. For all the debate they have occasioned, *Beowulf*'s 'swimming' episodes tell us remarkably little about how swimming was practised or regarded; nor do any of the other literary sources add much. Narrative histories and chronicles fill out the picture, but only to a limited extent. It can be said with reasonable confidence that many early medieval military operations — and most notably Viking marine assaults — were predicated upon the ability to swim; the skill certainly also came in handy when circumstances necessitated the beating of a hasty retreat. At the same time, death by drowning is clearly a literary topos in these texts, conveying utter defeat and calamity, and it is above all in that sense that they must be read.

All-in-all, it is not a huge tally, and it compares particularly unfavourably with the relative wealth and details of accounts from neighbouring early medieval societies, particularly Scandinavia. A number of factors seem to suggest that it may be reasonable to understand this discrepancy not wholly as indicative of a relative absence of swimming in England compared with neighbouring societies, but to some degree as a peculiarity of the respective sources available and their interests. The ubiquity and utility of swimming that was explored at the start of this chapter would certainly tend to favour the idea of a well-established culture surrounding recreational bodily immersions, particularly given what a profoundly damp place Britain is; there was certainly no lack of opportunity or bodies of water in which to practise the skill. Added to this, though, must be the profoundly ordinary and unremarkable quality that characterizes every single one of the allusions that we do have to swimming in Anglo-Saxon England. That people sometimes successfully negotiate water by swimming and sometimes do not is not remarkable but a simple fact of life that needs no further comment.

To be sure, some doubts do remain: the absence of named individuals in historical or hagiographical sources expressing enjoyment of swimming is notable and contrasts with examples in texts of comparable type elsewhere: there is, for instance, no English swimming saint like Lugid. Nor, to take a famous Frankish example, is there a swimming king of the sort that Einhard tells us Charlemagne was.[58] Yet all-in-all, with all due caution and with the caveats we have noted, it seems not unreasonable to wonder if our picture of early medieval English society should include a rich set of social, recreational, and cultural roles for swimming of the kind that is portrayed so vividly to have been the case in Ireland and Scandinavia.

---

57 *Càin Íarraith*, ed. by Binchy; *Càin Íarraith*, ed. and trans. by Hancock and others, pp. 155–59.
58 Einhard, *The Life of Charlemagne*, trans. by Turner, XXII, p. 51.

Works Cited

Manuscript

Vatican City, Biblioteca Apostolica Vaticana, MS Reg. lat. 12 (Bury Psalter)

Primary Sources

Adomnán of Iona, *Life of St Columba*, trans. by Richard Sharpe (London: Penguin, 1995)
The Anglo-Saxon Chronicle, ed. and trans. by George N. Garmonsway (London: J. M. Dent & Sons, 1972)
The Anglo-Saxon Chronicle: A Collaborative Edition, vol. v: *MS C*, ed. by Katherine O'Brien O'Keeffe (Cambridge: D. S. Brewer, 2001)
The Anglo-Saxon Chronicle: A Collaborative Edition, vol. vi: *MS D*, ed. by Geoffrey Peter Cubbin (Cambridge: D. S. Brewer, 1996)
Anglo-Saxon Prognostics: An Edition and Translation of Texts from London, British Library, MS Cotton Tiberius A.iii, ed. and trans. by Roy M. Liuzza (Cambridge: D. S. Brewer, 2011)
Annales fuldenses: sive, Annales regni Francorum orientalis, ed. by Friedrich Kurze (Hannover: Impensis bibliopolii Hahniani, 1891)
The Annals of Fulda, ed. and trans. by Timothy Reuter, vol. ii of *Ninth-Century Histories*, Manchester Medieval Texts Series (Manchester: Manchester University Press, 1992)
Bede, *Bede's Ecclesiastical History of the English People*, ed. and trans. by Bertram Colgrave and R. A. B. Mynors (Oxford: Clarendon Press, 1969)
——, *Vita S. Cuthberti prosaica*, in *Two Lives of Saint Cuthbert: A Life by an Anonymous Monk of Lindisfarne and Bede's Prose Life*, ed. and trans. by Bertram Colgrave (Cambridge: Cambridge University Press, 1940), pp. 141–307
Cáin Íarraith, in *Ancient Laws of Ireland*, vol. ii, ed. and trans. by W. Neilson Hancock, Thaddeus O'Mahony, Alexander George Richey, and Robert Atkinson (Dublin: Her Majesty's Stationery Office, 1869), pp. 147–93
Cáin Íarraith, in *Corpus iuris Hibernici: ad fidem codicum manuscriptorum*, vol. v, ed. by Daniel A. Binchy (Dublin: Dublin Institute for Advanced Studies, 1978), pp. 1759.6–1770.14
Egil's Saga, trans. by Bernard Scudder, in *The Sagas of Icelanders*, ed. by Örnólfur Thorsson (Harmondsworth: Penguin, 1997), pp. 3–184
Einhard, *The Life of Charlemagne*, trans. by Samuel E. Turner (Ann Arbor: University of Michigan Press, 1960)
The Exeter Anthology of Old English Poetry: An Edition of Exeter Dean and Chapter MS 3501, ed. by Bernard J. Muir, 2 vols (Exeter: University of Exeter Press, 1994)
Gregory of Tours, *Decem Libri Historiarum: Gregorii episcopi Turonensis. Libri Historiarum X*, ed. by Bruno Krusch and Wilhelm Levison, MGH Scriptores rerum Merovingicarum, 1.1 (Hannover: MGH, 1951)

Grettis saga Ásmundarsonar, ed. by Guðni Jónsson (Reykjavik: Hid Íslenzka fornritafelag, 1936)

Hrabanus Maurus, *De Procinctu romanae Miliciae*, ed. by Ernst Dümmler, *Zeitschrift für deutsches Alterthum*, 15 (1872), 443–51

Klaeber's Beowulf, ed. by R. D. Fulk, Robert E. Bjork, and John D. Niles, Toronto Old English Studies, 4th edn (Toronto: University of Toronto Press, 2008)

Laxdæla Saga: The Saga of the People of Laxardal, trans. by Keneva Kunz, in *The Sagas of Icelanders*, ed. by Örnólfur Thorsson (Harmondsworth: Penguin, 1997), pp. 270–421

The Old English Apollonius of Tyre, ed. by Peter Goolden (London: Oxford University Press, 1958)

The Old English Dialogues of Solomon and Saturn, ed. and trans. by Daniel Anlezark, Anglo-Saxon Texts, 7 (Cambridge: D. S. Brewer, 2009)

The Old English Letter of Alexander to Aristotle, ed. by Andy Orchard in *Pride and Prodigies: Studies in the Monsters of the Beowulf-Manuscript* (Toronto: University of Toronto Press, 2003), pp. 224–53

The Old English Orosius, ed. by Janet Bately, EETS, s.s. 6 (Oxford: Oxford University Press, 1980)

Orkneyinga Saga: The History of the Earls of Orkney, trans. by Hermann Pálsson and Paul Edwards (Harmondsworth: Penguin, 1981)

Saewulf, *Relatio de peregratione ad Hierosolymam*, trans. by William R. Brownlow, *Saewulf (1102, 1103 A.D.)* (London: Palestine Pilgrims' Text Society, 1892)

Vita St Lugidus, in *Vitae sanctorum Hiberniae: ex codice olim Salmanticensi, nunc Bruxellensi*, ed. by William Watts Heist, Subsidia Hagiographica, 28 (Brussels: Société des Bollandistes, 1965), pp. 131–45

Secondary Works

Anlezark, Daniel, 'All at Sea: Beowulf's Marvellous Swimming', in *Myths, Legends and Heroes: Essays on Old Norse and Old English Literature in Honour of John McKinnell*, ed. by Daniel Anlezark (Toronto: University of Toronto Press, 2011), pp. 225–41

Bachrach, Bernard S., *Early Carolingian Warfare: Prelude to Empire* (Philadelphia: University of Pennsylvania Press, 2011)

Blair, John, ed., *Waterways and Canal-Building in Medieval England* (Oxford: Oxford University Press, 2007)

Bonser, Wilfrid, 'Praying in Water', *Folklore*, 48.4 (1937), 385–88

Borsje, Jacqueline, 'The Monster in the River Ness in *Vita Sancti Columbae*: A Study of a Miracle', *Peritia*, 8 (1994), 27–34

——— , 'The Movement of Waters as Symbolised by Monsters in Early Irish Texts', *Peritia*, 11 (1997), 153–70

Clarke, H. B., 'The Vikings', in *Medieval Warfare: A History*, ed. by Maurice Keen (Oxford: Oxford University Press, 1999), pp. 36–58

Cooper, Roger C., 'New Light on Aelred's Immersion', *Harvard Theological Review*, 69.3/4 (1976), 416–19

Cronan, Dennis, 'Old English *gelad*, "A passage across water"', *Neophilologus*, 71 (1987), 316–19

Crowe, Fiona, Alessandra Sperduti, Tamsin C. O'Connell, Oliver E. Craig, Karola Kirsanow, Paola Germoni, Roberto Macchiarelli, Peter Garnsey, and Luca Bondioli, 'Water-Related Occupations and Diet in Two Roman Coastal Communities (Italy, First to Third Century AD): Correlation between Stable Carbon and Nitrogen Isotope Values and Auricular Exostosis Prevalence', *American Journal of Physical Anthropology*, 142.3 (2010), 355–66

Daniell, Christopher, *Death and Burial in Medieval England, 1066–1655* (London: Routledge, 2005)

Earl, James W., 'Beowulf's Rowing Match', *Neophilologus*, 63 (1979), 285–90

Frank, Roberta, '"Mere" and "sund": Two Sea-Changes in *Beowulf*', in *Modes of Interpretation in Old English Literature*, ed. by Phyllis Rugg Brown, Georgia Ronan Crampton, and Fred C. Robinson (Toronto: University of Toronto Press, 1986), pp. 153–72

Fulk, Robert D., 'Afloat in Semantic Space: Old English *sund* and the Nature of Beowulf's Exploit with Breca', *Journal of English and Germanic Philology*, 104 (2005), 456–72

Gee, Henry, *The Accidental Species: Misunderstandings of Human Evolution* (Chicago: University of Chicago Press, 2013)

Gneuss, Helmut, and Michael Lapidge, *Anglo-Saxon Manuscripts: A Bibliographical Handlist of Manuscripts and Manuscript Fragments Written or Owned in England up to 1100*, Toronto Anglo-Saxon Series, 15 (Toronto: University of Toronto Press, 2014)

Gougaud, Louis, *Devotional and Ascetic Practices in the Middle Ages*, trans. by Gerald C. Bateman (London: Burns, Oates and Washbourne, 1927)

Hagland, Jan Ragnar, 'Saxo-Norman London Bridge and Southward: The Saga Evidence Reconsidered', in *London Bridge: 2000 Years of a River Crossing*, ed. by Bruce Watson, Trevor Brigham, and Tony Dyson (London: Museum of London Archaeology Service, 2001), pp. 232–33

Hagland, Jan Ragnar, and Bruce Watson, 'Fact or Folklore: The Viking Attack on London Bridge', *London Archaeologist*, 10 (2012), 328–33

Hardy, Alister, 'Was Man More Aquatic in the Past?', *New Scientist*, 7 (1960), 642–45

Herity, Michael, 'Early Irish Hermitages in the Light of the *Lives* of Cuthbert', in *St Cuthbert, his Cult and his Community to AD 1200*, ed. by Gerald Bonner, David Rollason, and Clare Stancliffe (Woodbridge: Boydell, 1989), pp. 45–63

Hill, Thomas D., 'The Weight of Love and the Anglo-Saxon Cold Water Ordeals', *Reading Medieval Studies*, 40 (2014), 34–41

Holsinger, Bruce, 'Thorkel Farserk Goes for a Swim: Climate Change, the Medieval Optimum, and the Perils of Amateurism', in *The Middle Ages in the Modern World: Twenty-First Century Perspectives*, ed. by Bettina Bildhauer and Chris Jones (Oxford: Oxford University Press, 2017), pp. 27–44

Hyer, Maren Clegg, and Della Hooke, eds, *Water and the Environment in the Anglo-Saxon World*, Exeter Studies in Medieval Europe (Liverpool: Liverpool University Press, 2017)

Ireland, Colin, 'Penance and Prayer in Water: An Irish Practice in Northumbrian Hagiography', *Cambrian Medieval Celtic Studies*, 34 (1997), 51–66

Jones, Charles W., 'Bede and Vegetius', *Classical Review*, 46.6 (1932), 248–49

Jorgensen, Peter A., 'Beowulf's Swimming Contest with Breca: Old Norse Parallels', *Folklore*, 89.1 (1978), 52–59

Kaptorga – Visual History, 'Can You Swim in Chainmail?' <https://www.youtube.com/watch?v=bwd2ZEav2vE> [accessed 12 January 2020]

Klein, Stacey S., William Schipper, and Shannon Lewis-Simpson, eds, *The Maritime World of the Anglo-Saxons*, Essays in Anglo-Saxon Studies, 5 (Tempe: ACMRS, 2014)

Morgan, Elaine, *The Aquatic Ape* (London: Souvenir Press, 1982)

——, *The Descent of Woman* (London: Souvenir Press, 1972)

Ohlgren, Thomas H., ed., *Anglo-Saxon Textual Illustration: Photographs of Sixteen Manuscripts with Descriptions and Index*, vol. 1 (Kalamazoo: Medieval Institute Publications, 1992)

Orchard, Andy, *A Critical Companion to 'Beowulf'* (Cambridge: D. S. Brewer, 2003)

——, *Pride and Prodigies: Studies in the Monsters of the Beowulf-Manuscript* (Toronto: University of Toronto Press, 2003)

Orme, Nicholas, *Early British Swimming, 55 BC – AD 1719* (Exeter: University of Exeter, 1983)

Parsons, David, and Tania Styles with Carole Hough, eds, *The Vocabulary of English Place-Names (Á – Box)* (Nottingham: Centre for English Name Studies, 1997)

Pelteret, David A. E., 'The Role of Rivers and Coastlines in Shaping Early English History', *Haskins Society Journal*, 21 (2009), 21–46

Puhvel, Martin, 'The Aquatic Contest in *Hálfdanar saga Brönufóstra* and Beowulf's Adventure with Breca: Any Connections?', *Neuphilologische Mitteilungen*, 99 (1998), 131–38

——, 'Beowulf and Celtic Under-Water Adventure', *Folklore*, 76.4 (1965), 254–61

——, 'The Swimming Prowess of Beowulf', *Folklore*, 82.4 (1971), 276–80

Reynolds, Andrew, *Anglo-Saxon Deviant Burial Customs* (Oxford: Oxford University Press, 2009)

Reynolds, Roger E., '*Virgines subintroductae* in Celtic Christianity', *Harvard Theological Review*, 61.4 (1968), 547–66

Roede, Machteld, 'Aquatic Man', in *The Aquatic Ape: Fact or Fiction?*, ed. by Machteld Roede, Jan Wind, John Patrick, and Vernon Reynolds (London: Souvenir Press, 1991), pp. 306–28

Roede, Machteld, Jan Wind, John Patrick, and Vernon Reynolds, eds, *The Aquatic Ape: Fact or Fiction?* (London: Souvenir Press, 1991)

Russom, Geoffrey R., 'A Germanic Concept of Nobility in *The Gifts of Men* and *Beowulf*', *Speculum*, 53 (1978), 1–15

Sher, Lynn, *Swim: Why We Love the Water* (New York: Public Affairs, 2012)

Sobecki, Sebastian I., ed., *The Sea and Englishness in the Middle Ages: Maritime Narratives, Identity and Culture* (Cambridge: D. S. Brewer, 2011)

Thomas, Charles, 'The "Monster" Episode in Adomnan's *Life* of St Columba', *Cryptozoology*, 7 (1988), 38–45

Villotte, Sébastien, and Christopher J. Knüsel, 'External Auditory Exostoses and Prehistoric Aquatic Resources Procurement', *Journal of Archaeological Science: Reports*, 30 (2016), 633–36

Weinhold, Karl, *Altnordisches Leben* (Berlin: Weidmann, 1856)

Wentersdorf, Karl P., 'Beowulf's Adventure with Breca', *Studies in Philology*, 72 (1975), 140–66

——, 'Beowulf's Withdrawal from Frisia: A Reconsideration', *Studies in Philology*, 68 (1971), 395–415

Westenhöfer, Max, *Der Eigenweg des Menschen* (Berlin: W. Mannstaede, 1942)

Wind, Jan, 'The Non-Aquatic Ape: The Aquatic Ape Theory and the Evolution of Human Drowning and Swimming', in *The Aquatic Ape: Fact or Fiction?*, ed. by Machteld Roede, Jan Wind, John Patrick, and Vernon Reynolds (London: Souvenir Press, 1991), pp. 263–82

REBECCA SHORES

Sounds of Salvation: Nautical Noise in Old English and Anglo-Latin Literature

In her introduction to *The Maritime World of the Anglo-Saxons*, Stacy Klein writes that the sea could be 'used to symbolize a host of different phenomena, from earthly life to psychological turmoil', and was often 'described through figural language and imagery'.[1] This chapter will supplement the foundational work of Klein and her colleagues by focusing not on the sea, nor even on water itself, but on the act of sailing, rowing, or floating on and across the riverways and seascapes of early medieval England. When our literary subjects are amid these bodies of water, their nautical movements divulge emotional conceptualizations and experiences of piety that are, I argue here, especially sonic. By studying these nautical noises, I hope to pivot from what Martin Foys identifies as the 'heavily visualist framework of language through which we predictably study and reconstitute the medieval past' to articulate how nautical undertakings engage aural, embodied, and emotional interactions with the sacred.[2] Being receptive to distinctly medieval descriptions of 'physical sensuality' allows us to reconsider ways in which Anglo-Latin and Old English literature were shaped by 'different indices of language and physical phenomena than those that have dominated the modern world'.[3] At a time when the seas are rising and warming at irreversible rates, it may do us good to ponder not only what we see, but also what we hear, as we near the waves ourselves.

The present chapter studies sonic elements of human, animal, and instrumental noise aboard boats at sea. The first section investigates the confluence of song, memory, and seascape in the anonymous *Vita Ceolfridi*

---

1 Klein, 'Navigating the Anglo-Saxon Seas', p. 18.
2 Foys, 'A Sensual Philology for Anglo-Saxon England', p. 470.
3 Foys, 'A Sensual Philology for Anglo Saxon England', p. 470.

Rebecca Shores • (arpshores@gmail.com) earned her PhD from the University of North Carolina at Chapel Hill. Her dissertation, 'Nautical Narratives in Anglo-Latin Hagiographies, ca. 700–1100', was completed in 2017. Preliminary thoughts on the topic of this chapter were developed at the University of Denver's 'Seafaring' conference in 2016.

and the Old English poem *The Seafarer*. In the second section, nautical noise of the classical *celeuma* links Bishop Wilfrid's shipwreck on the shores of Sussex to Moses's exodus through the Red Sea. Finally, the third section considers the reappearance of Aldhelm's *proreta* in Byrhtferth of Ramsey's *Vita Oswaldi* and finds that the twice-borrowed classicism is yet another example of the early medieval association among sailors' sounds, surroundings, and salvation. In poetry and prose, in Anglo-Latin and Old English, a variety of works depicts aural elements of seafaring as both physically and spiritually powerful. Nautical noise, it seems, can signify salvation.

Solitary Soundscapes

A soundscape is an acoustic ecology. It refers to the human perception of how the noises of our natural environment (such as birdsong and ocean waves) interact with human noises — those produced by our bodies and those produced by our machines. In broad terms, this is not so different from medieval ideas about the space of sound. Medieval writers understood that even the most fleeting sounds could have physical, lasting consequences. In his exploration of 'how aspects of medieval language can play a central role in the relations between bodily action and communicational media', Foys provides an example of just how important the materiality of sound could be: 'Air was an environment of spiritual, moral and corporeal contention: the *gæst* of Old English and the *spiritus* of Latin connected in the airy breath of the soul'.[4] The embodied and material air was a place where the frenetic battles of the highest stakes were played out, and where the very order of the universe was hidden. Even silence, Valerie Allen reminds us, was thought to be occupied by the music of the spheres, ever-present but imperceptible to the mortal ear.[5]

Like built environments, sonic environments both reflect (or in this case, echo) and redirect the sensory experience of human emotions. Even now, our vocabulary for emotion retains the physical remnants from a more etymologically transparent past. We are still 'touched' or 'moved' by a show of 'tenderness'; we can be 'stopped cold' by fear. To authors like Ambrose and Augustine, sacred song could *literally* move people — in mind, body, and spirit. The psalms enriched and even engendered communal unity: Ambrose wrote, 'psalmus dissidentes copulat, discordes sociat, offensos reconciliat' (a psalm joins those with differences, unites those at odds and reconciles those who have been offended).[6] Augustine considers the means by which this is

---

4 Foys, 'A Sensual Philology for Anglo-Saxon England', pp. 459, 463.
5 Allen, 'Broken Air', p. 311.
6 Ambrose, *Exposito Psalmi CXVIII*, ed. by Petschenig and Zelzer, I. 9, p. 8; see McKinnon, *Music in Early Christian Literature*, no. 276, pp. 126–27.

achieved: the psalter's 'iste dulcis sonus, ista suauis melodia, tam in cantico quam in intellectu, etiam monasteria peperit' (sweet sound, this lovely melody […] — as much as in the song as in the understanding — has even begotten monasteries).[7] Melody and lyrics could move the body to acts of devotion and the soul towards spiritual understanding. Music mattered to salvation.

But the sonic and emotive elements of holy music were not without complications. Augustine was torn 'inter periculum uoluptatis et experimentum salubritatis magisque' (between the pleasure [of listening] and the value of the experience).[8] On the one hand, he believed

> dum ipsis sanctis dictis religiosius et ardentius sentio moueri animos nostros in flammam pietatis, cum ita cantantur, quam si non ita cantarentur, et omnes affectus spiritus nostri pro sui diuersitate habere proprios modos in uoce atque cantu, quorum nescio qua occulta familiaritate excitentur.[9]
>
>> [that our souls are more piously and earnestly moved to the ardor of devotion by these sacred words when they are thus sung than when they are not thus sung, and that all the affections of our soul, by their own diversity, have their proper measures […] in voice and song, which are stimulated by I know not what secret correspondence.][10]

The danger, he writes, is in enjoying the sensory experience of the music without the understanding of the words. Nevertheless, the transformative power of the sacred song tips the scales:

> cum reminiscor lacrimas meas, quas fudi ad cantus ecclesiae in primordiis recuperatate fidei meae, et nunc ipsum cum moueor non cantu, sed rebus quae cantantur […] magnam instituti huius utilitatem rursus agnosco.[11]
>
>> [when I recall the tears which I shed at the song of the Church in the first days of my recovered faith, and even now as I am moved not by the song but by the things which are sung, […] I acknowledge again the great benefit of this practice.][12]

In this context, the River Wear creates an emotionally sonic environment in the early eighth-century *Vita Ceolfridi* during the anonymous author's rendering of Abbot Ceolfrith's ritualized departure from his seaside monastery. In perhaps the earliest English example of a shipside liturgy, this polyphonic

---

7   Augustine, *Enarrationes in Psalmos*, ed. by Dekkers and Fraipont, CXXXII. 2, p. 1927; McKinnon, *Music in Early Christian Literature*, no. 365, pp. 159–60.
8   Augustine, *Confessiones*, ed. by Verheijen, X. 33, 50, p. 182; McKinnon, *Music in Early Christian Literature*, no. 352, p. 155.
9   Augustine, *Confessiones*, ed. by Verheijen, X. 33, 49, p. 181.
10 McKinnon, *Music in Early Christian Literature* , no. 352, p. 154.
11 Augustine, *Confessiones*, ed. by Verheijen, X. 33, 50, p. 182.
12 McKinnon, *Music in Early Christian Literature*, no. 352, p. 155.

scene captures the monastic community at Wearmouth in a uniquely musical stage of its history. Ceolfrith had accompanied Benedict Biscop on a trip to Rome, where they asked John the Archcantor to return with them to England. He obliged, and his arrival caused 'Wearmouth [to gain] prestige as a centre for liturgical music in the late 7th century', which contributed to 'the flurry of musical activity in Northumbria'.[13] During his tenure at Wearmouth-Jarrow, Ceolfrith would travel six times to Rome to enrich the community he had helped to establish. His devotion to the dual monastery and its ritual practices inform his departure from the worldly matters of his monastic life, and his departure from earthly life on this final journey to Rome by boat. His separation from his monastic family is signalled by the dissolution of his native, terrestrial soundscape across the River Wear.

His journey begins with the departure liturgy at the Church of St Peter (Wearmouth), located on the River Wear where the river empties into the North Sea. Having been granted permission to go to Rome for a seventh and final time, and to live out his life among the holy places near the tomb of St Peter, Ceolfrith leads his final Mass for the community and:

> rogat pro se orare, dicit et ipse orationem, accendit thymiama, habensque in manu turribulum consistit in gradibus ubi legere consuerat, dat osculum plurimis, nam ne omnibus posset luctu et suo et ipsorum praepeditur.
>
>> [asked them to pray for him, himself said a prayer, lit the incense, and holding the thurible in his hand he stood on the steps where he was accustomed to read out loud; he gave a kiss to many, for he could not do it for all, as both his and their grief hindered them.][14]

Ceolfrith's presence at the liminal threshold of the church triggers memories of a life long-committed to the people and place he now leaves behind. For him and his brethren, the sight, smell, and sound of the scene beget debilitating grief. The sonic experience of the final service is so emotionally evocative that it becomes paralysing, but these corporeal reactions shape a scene of collective ritual into one of individual affect.

Archaeologists have found that the Wear estuary was then wider than it is today, and the church itself was surrounded by water on three sides: bound by 'a low limestone cliff that faced directly out to sea' to the east, a gently sloping shore leading down to 'a rocky beach about 160 m from the church' to the south, and 'a small stream or inlet' feeding into the Wear for the same distance to the south-west.[15] These measurements suggest that the monastery grounds were not much narrower than the river crossing itself, which would have been between two hundred and three hundred metres across. Moreover, the church would have been the focal point of this 'low promontory' at the

---

13 Haines, 'A Musical Fragment from Anglo-Saxon England', p. 219.
14 *Vita Ceolfridi*, ed. and trans. by Grocock and Wood, pp. 104–05.
15 Turner, Semple, and Turner, *Wearmouth and Jarrow*, p. 115.

mouth of the Wear, and was therefore a site of interaction and intervisibility with the sea.[16] Since the monastery also owned the property on the opposite bank of the river, its inhabitants might well have imagined the Wear as part of their sacred and communal space. It is therefore the water itself, not simply the distance, that separates the sounds and spirits of those present.

The anonymous author pays careful attention to localization and vocalization as the procession moves through the monastic complex. In their earliest stages, the monastery grounds were built out from St Peter's Church — the northernmost building reaching just over twenty metres east to west (excluding the porch). A long, 'elaborately decorated' covered walkway emerged from the lateral midpoint of the church and stretched due south for at least thirty-five metres, bisecting burial grounds directly to its east and west until it met a building or wall parallel to the church. Rosemary Cramp's excavations did not reveal if the covered walk 'terminated in a building or passed through buildings to the riverbank', but the entire complex seems not to have exceeded three hundred square metres.[17] While the procession of the abbot and his priests might not have covered a great distance, each movement they make as a worshipping body is accompanied by a shared and unifying sound.

Ceolfrith and his brethren walked from St Peter's Church through a beautiful portico from which they could have seen cemeteries on either side, and within which their prayers would have reverberated. The procession enacted Ceolfrith's pilgrimage in unison with the antiphons and psalms, which 'could be interpreted either in terms of the pilgrimage he was going to make to Rome, or of the journey he was about to make to the other world'.[18] These psalms, hymns, and prayers articulate a soundscape that undulates across both monastery and memory. The monks traverse the lands of their dead in a corridor of stone and song; just as Ambrose had imagined, the notes and tunes of familiar verses created a sonic unity that bound the brethren to their abbot in space and time. But in boarding the boat across the river, Ceolfrith brings these bonds to a breaking point.

After Ceolfrith exhorted his colleagues to maintain their unity with Jarrow, they resumed the sung liturgy until they reached the river bank. Again, 'itidimque singulis osculum pacis dat, intercepto saepius cantu prae lacrimis' (he gave the kiss of peace to each of them, while the singing was often interrupted because of their tears).[19] The comforting rhythm of their remembered verses is jolted out of cadence by the sobs of the brethren Ceolfrith leaves behind. His departure is so devastating that it disrupts both the order of the liturgy and the very fabric of the air.

---

16 Turner, Semple, and Turner, *Wearmouth and Jarrow*, p. 115.
17 Cramp, *Wearmouth and Jarrow Monastic Sites*, p. 98.
18 Ó Carragáin has identified the first antiphon as one 'based on Isaiah 26:7' and the second antiphon inspired by Psalm 83 (84). 8. He writes that the monks would have sung Psalm 66 (67) during the procession; *The City of Rome and the World of Bede*, p. 13.
19 *Vita Ceolfridi*, ed. and trans. by Grocock and Wood, pp. 104–07.

As the drama of this scene reaches nearly histrionic heights, the abbot boards the ship on the waters of the Wear with candles and a cross, literalizing the metaphor of the ship of the Church.[20] Familiar sounds follow him close behind:

> Currente trans fluuium naui, aspectans contra merentes suo abscessu fratres, audiensque sonum sublimem mixti cum luctu carminis, nullatenus ualuit ipse a singultu et lacrimis temperare.
>
> > [As the ship sailed swiftly across the river, he looked across at the brothers mourning his departure, and heard the glorious sound of their song mingled with their grief, and he could not prevent himself from giving way to sobs and tears.][21]

Awash with overwhelming sorrow aboard the liminal and ecclesiastical space of the ship, he is brought to the other side of the river. Before he disembarks, he answers the fractured, intermingled 'mixti' noise of his brethren with a loud 'crebri', repeated prayer: 'Christe, misere illi coetui!' (Christ have mercy on this company!).[22] By transposing his own voice over their broken utterances, he reinscribes the echo that returns across the river. He hears nothing back.

If his departure had been sonic and communal, his arrival is silent and solitary. He disembarks, bows to the cross, and rides away on a horse, without a prayer of thanks at his safe arrival or even a whisper of his adoration at the cross. This quiet isolation is surely only a narrative hush; Ceolfrith would have been travelling with a retinue, and certainly would have heard, and even interacted with, his attending colleagues. What marks him as a pilgrim — as a stranger in a not-so-foreign land — is the breathtaking silence through which he 'festinans ab ipsa quoque cognata sibi Anglorum gente' (hasten[s] from the race of the Angles) for the sound of the heavenly spheres.[23]

Communal sound is just as resonant for the plaintive seaman of *The Seafarer*, whose bodily and emotional experience are also borne of a sonically inspired, nautical environment. Found in an extraordinary Old English miscellany called the Exeter Book, the poem is one of many in a compilation of poetry that includes ninety-five riddles, and is otherwise more or less interested in various components of the Christian life. Within this remarkable manuscript, *The Seafarer* is among a small group of poems set against a maritime backdrop and concerned with the sense of isolation engendered by it. As Daniel Anlezark points out, 'the speaker of *The Wanderer* has crossed waterways, and has spent time contemplating the "binding of the waves"; [...] the speaker of *The Husband's Message* has journeyed across the water, and has been separated from his wife; the female speaker of *Wulf and Eadwacer* cannot leave her

---

20 Grocock and Wood remind the reader that 'he is only crossing the river' in *Vita Ceolfridi*, ed. and trans. by Grocock and Wood, p. 106, n. 133.
21 *Vita Ceolfridi*, ed. and trans. by Grocock and Wood, pp. 104–05.
22 *Vita Ceolfridi*, ed. and trans. by Grocock and Wood, pp. 106–07.
23 *Vita Ceolfridi*, ed. and trans. by Grocock and Wood, pp. 106–07.

island, and longs for Wulf who is on another'.[24] These archipelagic poems are often classified as elegies because they evoke emotional responses to suffering and loss. And while the seascape of *The Seafarer* is frequently understood as a metaphor for ascetic isolation, the soundscape of the poem offers a more profound register of interpretation.[25]

For the inveterate sailor of the poem, the maritime soundscape must be spoken back into existence, since at his opening line, it resounds only in his memory. When he recounts 'þær ic ne gehyrde butan hlimman sæ, iscaldne wæg' (there I heard nothing but the roaring sea, the ice-cold waves, ll. 18–19), the seafarer links his natural soundscape to the human sense of touch. This pattern occurs again in embodied, affective birdsong:

Hwilum ylfete song
dyde ic me to gomene,   ganetes hleoþor
ond huipan sweg   fore hleahtor wera,
mæw singende   for medodrince. (19b–22)[26]

> [Sometimes I took the song of the swan for my entertainment, the cry of the gannet and the call of the curlew instead of the laughter of men, the gull singing instead of the mead-gulp.][27]

Aural attention helps the modern reader perceive the physicality of this guttural scene. The ambiguity of 'gomene' can be slightly clarified by the suitability of its partner, 'song'. From this it may be inferred that the entertainment is musical, or even vocal, but not necessarily verbal. The near-pun of 'hleoþor' and 'hleahtor' confirms the match between gannet and curlew, and laughing men; it is a subtle touch that the human laughter is mirrored by two (perhaps conversational) birds. The assumption that each avian utterance is meant to juxtapose or replace a related, human noise suggests a more sensual and embodied 'mead-gulp', or perhaps just 'gulp', than the traditional and literal translations allow.

Further support for this sensorial approach to the text comes from the seafarer's manipulation of voice. Although the seafarer renders birds' sounds more human and human noises more obscure, none of them is necessarily verbal. They are, however, mostly corporeal. The avian utterances evoke merriment, laughter, and the warmth of a drink, so noise matters to the bodily sensation it induces. But it also matters to the bodily act by which it is produced: notwithstanding the vagary of *gamen*, the laughing, drinking men and the cawing, calling birds perform especially guttural acts. The processes of hearing and listening are similarly corporeal and emotional; although these sounds are bereft of words, they are certainly not without meaning.

---

24   Anlezark, 'From Elegy to Lyric', pp. 75–76.
25   For another perspective on the confluence of avian and human voices, see Ellard, 'Communicating between Species and between Disciplines'.
26   *The Seafarer*, ed. by Gordon, p. 35.
27   Here and below, I have adapted the translation of Magennis, 'The Solitary Journey', p. 305.

The memory of human speech continues to recede from the maritime landscape as the waves of the sea and the birds of the air engage in non-verbal, clamorous conversation. Their shared discourse moves along intermingled elements; waves roar through the air, and ice grips onto wings. Again, the materiality of the environment is crucial to the ways in which sensory experience creates emotional affect. The physical interaction of the sea and the air is elemental: as storms rage at the cliffs, the tern calls in answer 'oncwæð' and the eagle shouts out 'bigeal'.[28] These seem to be uttered in response to one another, but now none of the noises has discernible meaning. Everything from the shouting sea and the sleeting sky reverberates from the cliffs which keep silent the human voice. This echoing soundscape is almost claustrophobic, joining the physical suffering of confinement with the emotional anguish of displacement. Yet like the birds from their frozen feathers, the seafarer speaks out from his gripping sorrow.

An aural approach enriches the ever-vexing passage:

gielleð anfloga,
hweteð on wælweg   hreþer unwearnum,
ofer holma gelagu.   Forþon me hatran sind
Dryhtnes dreamas   þonne þis deade lif,
læne on londe. (62b–66a)[29]

> [the lone-flyer yowls, urges the chest irresistibly onto the whale-road and over the waters of the ocean, for the joys of the Lord are hotter to me than this dead life, transitory on land.][30]

These lines mark the first noise that the sailor makes at sea. The sound comes not from his throat, or his mouth, or even his mind, but from his disembodied spirit. This is especially striking in contrast to the self-containment of the opening line: 'Mæg ic be me sylfum | soðgid wrecan' (I can recite a true tale about myself).[31] He opens with the claim that he can tell his own story, but here, at the exact middle of the poem, the figure of the seafarer disappears from view as soon as he speaks aloud.

This singular speech act reshapes the sonic environment by inverting it. Instead of interpreting birdsong as human noise, he divests his voice of any human association. The *DOE* asserts that *gyllan* is primarily 'of birds: to make a loud cry'. Its secondary and tertiary meanings, 'of a wolf/dog' and 'of an inanimate object', respectively, reaffirm its discordance with human voice.[32] There is nothing at all guttural about this yell — it exists entirely beyond the body, in the space between, or rather beyond, the sea and the sky. The seascape, too, is redrawn: there are no cliffs, or storms, or ice-feathered birds to echo his sound. Even the

---

28 For extensive lexical notes, see *The Seafarer*, ed. by Gordon, p. 36.
29 *The Seafarer*, ed. by Gordon, pp. 41–42.
30 Magennis, 'The Solitary Journey', p. 306.
31 *The Seafarer*, ed. by Gordon, p. 33. The translation is my own.
32 See 'gyllan', in Cameron, Amos, and Healey, *Dictionary of Old English*.

atmosphere is different; the heart surges towards the heat of the Lord instead of being locked in by the cold. In this place over water and ice and air, voice dissipates into noise and obliterates human relevance. Like Ceolfrith, the spirit of the seafarer offers one last, transcendent sound before crossing into a state of pilgrimage. The sea and the river are the final frontiers for familiar voices of the transitory world; silence, more than space, marks the ultimate departure from it.

By inciting the breast to seek out the sea, the yawping spirit may also be heard to mimic the *keleustes* of the classical epic.[33] This naval officer would use his voice or an instrument to keep up the time (and the spirits) of rowers as they reached their oars across the sea. It is to his cry, the *celeuma*, that we now turn.

Classical *Celeumae*

Despite only two distinct survivals in British sources, the pre-Conquest *celeuma* in its original, nautical usage demands a considerable introduction. In his comprehensive survey of the term, Daniel J. Sheerin notes that the ancient Greek *keleusma* originally referred to a 'pattern of shouts to keep time' for oarsmen.[34] This historical practice was later adapted by the Romans, who preferred to use an instrument rather than the human voice to keep rhythm, but whose Latin epics preserved the Greek tradition of vocal rather than instrumental noise.[35] This practice was ubiquitous and flexible enough to accommodate any kind of song, on actual or imagined ships. The benefit of the *celeuma* from the rower's perspective is twofold: it helps the oarsman keep time, and it alleviates the pain and boredom of demanding and repetitive exertions.

The *celeuma*'s fluidity made it adaptable to some of the earliest writings of Christianity. According to Sheerin, Clement's description of 'musical joy in the day-to-day activities of the Christian life' and the 'traditional symbol of the Church as a ship, with Christ or the Holy Spirit at the helm' contributed to conditions in which writers like Paulinus of Nola could use the *celeuma* in Christian context.[36] This had an extraordinary effect on the already versatile term, and helped ensure that the shouts and songs of sailors would resound through antique classical and early Christian traditions alike. In commentaries, treatises, and personal correspondence, the Church Fathers deployed the bosun's

---

33 For an extended investigation of the seafarer's *celeuma*, see Shores, 'Nautical Narratives in Anglo-Latin Hagiographies'.
34 Sheerin, '"Celeuma" in Christian Latin', p. 46.
35 Sheerin, '"Celcuma" in Christian Latin', p. 49.
36 Clement wrote, 'Throughout our entire lives, then, we celebrate a feast, persuaded that God is present everywhere and in all things; we plough the fields while giving praise, we sail the seas while singing hymns, and on every other public occasion we conduct ourselves skillfully'; McKinnon, *Music in Early Christian Literature*, no. 60, p. 36 (*Stromata VII*, vii. 35. PG ix. 452; GCS iii. 27–28). For Paulinus's *Carmen 17*, see Paulinus, *Carmina*, ed. by Dolveck.

beat to sound out their own positions on the interpretation of Scripture, the nature of the soul, and the transformative power of sound.[37]

The literary use of the *celeuma* spread quickly to and through north-west Europe.[38] Its temporal and geographic distribution makes its rarity in British sources somewhat surprising. For the sixth-century monk Gildas, the *celeuma* invokes an idealized past when sailors would sing — presumably cheerful — songs across the sea. But in the place of this musical exuberance, British refugees of the Saxon invasions cry out in despair. Under the 'cum ululatu magno ceu celeumatis vice hoc modo sub velorum sinibus cantantes: "dedisti nos tamquam oves escarum et in gentibus dispersisti nos"' (swelling sails they loudly wailed, singing a psalm (Ps. 43. 12) that took the place of a shanty: 'You have given us like sheep for eating and scattered us among the heathen').[39] Breaking from the expectation of a cheerful song (or even a slightly less despondent verse) heightens the pathos of the scene. Gildas shows these exiles to be not merely removed from the comforts of their homeland, but even beyond the safety of the shore by this verse of the psalm in the middle of the sea. Like everything else in the wake of the Saxon invasions, the *celeuma* has been distorted and misused.

In Stephen's early eighth-century *Vita Wilfridi*, Christianized song warps into a cacophonous storm that shipwrecks the saint on the pagan shores of Sussex. Sailing across the English Channel, Wilfrid and his priests are 'canentibus clericis et psallentibus laudem Dei pro celeumate in choro' (praising God with psalms and hymns, giving the time to the oarsmen), when a squall arrives.[40] Their pulsing psalms are muted by the wind, which 'flante namque vento euroaustro dure' (blew hard from the southeast) while 'albescentia undarum cullmina in regionem Australium Saxonum [...] proiecerunt eos' (the foam-crested waves hurled them onto the land of the South Saxons and left the ship and the men high and dry).[41] Sandra Duncan sees the act of singing psalms and hymns as a means of 'ensuring a correct progress through the waves', perfectly suited to the Christian tradition of imagining the vicissitudes of life as a perilous sea from which the ship of the Church offers mankind's only refuge.[42] But *celeuma* could refer to a song of happy sailors, a shout for lunging oarsmen, or a salve for mortal labour; its depth of association animates and emboldens the otherwise disembodied voices on the ship. And given Wilfrid's — and perhaps also Stephen's — interest in music, it is worth considering that the aural elements of this scene are more resonant than the visual metaphor allows.[43]

---

37 Sheerin, '"Celeuma" in Christian Latin'.
38 Sheerin, '"Celeuma" in Christian Latin'.
39 Gildas, *The Ruin of Britain*, ed. and trans. by Winterbottom, pp. 27–28, 97–98.
40 Stephen, *The Life of Bishop Wilfrid*, ed. and trans. by Colgrave, pp. 26–27.
41 Stephen, *The Life of Bishop Wilfrid*, ed. and trans. by Colgrave, pp. 26–27.
42 Duncan, 'Prophets Shining in Dark Places', p. 89.
43 See Billett, 'Wilfrid and Music'.

Having cast Wilfrid's ship ashore, the sea 'terras fugiens, litoraque detegens, et in abyssi matricem recessit' (fled from the land, and, laying the shores bare, withdrew into the depths of the abyss).[44] A pagan army arrives and a battle ensues. When Wilfrid tries to ransom his surviving men, the locals decline his offer, 'dicentes superbe, sua esse omnia quasi propria, quae mare ad terras proiecit' (proudly declaring that they treated as their own possessions all that the sea cast upon the land).[45] Their reaction should not have surprised Wilfrid, since Icelandic, Norse, Faroese, English, and Continental laws made specific provisions for what was washed ashore.[46] In this almost mercantile moment, the law of flotsam marks the heathens as landed citizens and brings Wilfrid's status as exile into sharp relief. He is, as ever, a pilgrim without a country. In this case, though, he is also without a means of finding one. Only slightly better off than the *celeuma*-singers of Gildas, Wilfrid has made it safely to shore, but stands, shakily in this moment, as a stranger in a foreign land.

Wilfrid's exile-as-pilgrimage bubbles up in moments like the one above, but erupts through Stephen's Old Testament parallels. Wilfrid prays just like Moses; and just like the Israelites, 'ita et hic isti pauci christiani feroces et indomitos paganos tribus vicibus in fugam versus strage non modica obruerunt' (this little band of Christians overthrew the fierce and untamed heathen host).[47] After the third battle, Wilfrid 'orante sacerdote magno ad Dominum Deum suum, qui statim iussit ante horam penam, priusquam consuerat, mare venire' (prayed to the Lord his God, who straightaway bade the tide return before its usual hour) and 'tunc mare redundans fluctibus tota litora implevit, elevataque nave, cimba processit in altum' (the sea came flowing back and covered all the shore, so that the ship was floated and made its way into the deep).[48] While these links between Stephen and Moses seem clear enough, the depiction of the tides and the prayer on the ship signal more subtle connections.

Stephen's littoral diction speaks to the 'long-continued medieval belief that tides arise out of the abyss through a spring or spiracle in the sea bottom'.[49] The discrete mechanisms of marine movement and distinct topography of the tidal shore invoke the narrative of the Israelites crossing the Red Sea in the Book of Exodus. Back at sea, Wilfrid and his retinue 'returned thanks to God' before arriving in Sandwich.[50] It is easy to imagine, given the parallels

---

44 Stephen, *The Life of Bishop Wilfrid*, ed. and trans. by Colgrave, pp. 26–27.
45 Stephen, *The Life of Bishop Wilfrid*, ed. and trans. by Colgrave, pp. 28–29.
46 Whatever washed up from the sea, 'from whales to driftwood to goods from wrecked ships', had such potential value 'that most monarchs across Europe took pains to legally claim "drift"'. Szabo, *Monstrous Fishes and the Mead-Dark Sea*, p. 250.
47 Stephen, *The Life of Bishop Wilfrid*, ed. and trans. by Colgrave, pp. 28–29.
48 Stephen, *The Life of Bishop Wilfrid*, ed. and trans. by Colgrave, pp. 28–29.
49 Brown, 'Bede, a Hisperic Etymology, and Early Sea Poetry', p. 424.
50 'Gloriose autem a Deo honorificati, gratias ei agentes, vento flante ab affrico, prospere in portum Sandicae salutis pervenerunt'. Stephen, *The Life of Bishop Wilfrid*, ed. and trans. by Colgrave, pp. 28–29.

cited above, that this prayer of gratitude was inspired by the Canticle of Moses, recited first by Moses (Exodus 15. 1–18) and later by Miriam (Exodus 15. 20–21), to the accompaniment of dancing and tambourines. Wilfrid is written into the exile of Moses and the Israelites by God's tidal rescue from landbound heathens. And as the enveloping *celeuma* and prayer of thanks suggest, he is also incorporated into the earliest biblical attestation of a song sung at sea.[51]

Problematic *Proretae*

The loud and joyful sea-crossings resounding in the hagiographical accounts written by Byrhtferth of Ramsey celebrate the act of sailors keeping and passing time as in Aldhelm's earlier writings — without ever precisely mentioning *celeuma*. What is known of Byrhtferth's biography is ensconced in the history of Ramsey Abbey, where he spent much of his life.[52] Ramsey was founded on an island, and according to Lapidge, Oswald identified its position within a network of navigable waterways as crucial to its success: 'The value of the site — which Bishop Oswald realized as soon as he had first seen it (*VSO* iii. 16) — is that it lay only a mile's distance from both Ramsey Mere and Ugg Mere, and less than five miles from Wittlesey Mere', through which the old course of the River Nene would reach 'upstream to the quarries of Barnack [...] and downstream to the Wash and the North Sea'.[53] We do not know how much of Byrhtferth's childhood and adolescence had been lived out on this island, but we can surmise that he was a young man by the time Abbo of Fleury arrived there in 995 and 'provided a focus for young Byrhtferth's energies'.[54] These energies were indeed well directed: they generated one of the most prolific writing careers in pre-Norman England. To his initial work, a 'vast compendium of texts of a computistical and astronomical interest', Byrhtferth quickly added three hagiographies, the third of which was Oswald's; by the age of thirty, he had likewise been ordained and appointed instructor to Ramsey's brethren.[55] It feels ironic to know so little about the life of such a productive figure, especially when the life of Oswald is so comparatively rich in detail.[56]

---

51 Similarly, the poet of the Old English *Exodus* emphasizes the sound of the celebratory song after the crossing of the Red Sea; see *Exodus*, ed. by Lucas, ll. 574–79. The Canticle of Exodus is still referred to as 'The Song of the Sea' in Jewish tradition.
52 Byrhtferth of Ramsey, *Life of St Oswald*, ed. and trans. by Lapidge, p. xv.
53 Byrhtferth of Ramsey, *Life of St Oswald*, ed. and trans. by Lapidge, p. xvi.
54 Byrhtferth of Ramsey, *Life of St Oswald*, ed. and trans. by Lapidge, p. xxvii.
55 Byrhtferth of Ramsey, *Life of St Oswald*, ed. and trans. by Lapidge, pp. xxviii, xxix.
56 'St Oswald, bishop of Worcester from 961 and archbishop of York from 971 until his death on 29 February 992, is one of the best-documented figures in tenth-century English history, even though there are unfortunate lacunas in our knowledge of him', Byrhtferth of Ramsey, *Life of St Oswald*, ed. and trans. by Lapidge, p. lxv.

Unlike the *Vita Ceolfridi*, the *Vita Oswaldi* offers up joyful sounds of nautical endeavours of travelling monks. Having been summoned to Fleury by Oda, Oswald 'concito pergit cursu undisonos fluctus maris cernere; intrans liburnam cum sociis suis celerrime, hortante deinde proreta et sonante nauclero, nauigant cum mentis gaudio' (set off on a speedy course to see the rolling waves of the sea; getting swiftly into a ship with his companions, they set sail joyfully with the pilot barking commands and the skipper beating time (for the rowers)).[57] Byrhtferth's enlivened description of this highly derivative passage exemplifies his intentionally 'elevated register of expression'.[58] Within this higher register, Byrhtferth's choice to replace the Aldhelmian 'crepitante naucleru' with 'sonante nauclero' is intriguing: he has changed the non-human strike into an embodied voice.

Perhaps the most peculiar moment of nautical noise occurs after a procession to the Church of St Mary. Oswald was a guest at Ramsey during Rogation Days and participated in the communal custom of walking barefoot in an act of penitence.[59] After the Mass, the priests wish to return home by boat, claiming that walking across the bridge would take more time. Lapidge tries to make sense of the priests' dubious choice by suggesting that 'perhaps the best solution [...] is to suppose that some words have fallen out after *exiuimus*, explaining why it was quicker to return home by boat'.[60] It may be more likely that Byrhtferth is stretching the truth a bit, and that sore feet, rather than missing text, best explain this imprudent decision.

Straining under the weight of its cargo, the boat 'perducta in meditullio profundi stagni' (drifted out into the middle of the deep lake) and began to sink.[61] The priests began to shout for help and Oswald, safely on shore, 'audi uociferantium uoces "sancte Benedicte, succurre nobis!"' (heard our voices shouting, 'Save us, St Benedict!').[62] Byrhtferth uses asyndeton to convey the causal relationship between hearing and understanding: 'Quorum uoces audiens, interrogat causam; quam agnoscens, eleuauit sanctam manum sue dextere et in Domino fiducialiter confidens dixit, "Celitus aduenit nobis benedictio Christi!"' (Hearing these cries he asked the reason; learning the reason he raised his holy right hand, and, trusting confidently in the Lord, said, 'Christ's blessing shall come quickly to us from on high!').[63] Such a connection reflects the writings of early Christian authors like Nicetas and Jerome, who ruminated on the sonic implications of 'I will sing with the spirit,

---

57 Byrhtferth of Ramsey, *Life of St Oswald*, ed. and trans. by Lapidge, pp. 38–40, 39–41.
58 Byrhtferth of Ramsey, *Life of St Oswald*, ed. and trans. by Lapidge, p. xliv. Lapidge suggests Jerome, Sedulius, Aldhelm, and Cassiodorus as sources for *pergit cursu, undisonos fluctus maris, hortante [...] nauclero*, and *mentis gaudio*, respectively, pp. 39–40.
59 Byrhtferth of Ramsey, *Life of St Oswald*, ed. and trans. by Lapidge, p. 133.
60 Byrhtferth of Ramsey, *Life of St Oswald*, ed. and trans. by Lapidge, p. 134, n. 150.
61 Byrhtferth of Ramsey, *Life of St Oswald*, ed. and trans. by Lapidge, p. 134.
62 Byrhtferth of Ramsey, *Life of St Oswald*, ed. and trans. by Lapidge, p. 134.
63 Byrhtferth of Ramsey, *Life of St Oswald*, ed. and trans. by Lapidge, p. 135.

I will sing with the mind also' (1 Corinthians 14. 15), 'singing in your hearts' (Ephesians 5. 19), and 'sing with understanding' (Psalm 46. 8).[64]

With this aurally induced understanding (modelled perhaps on the passage from Ephesians), Oswald's voice takes on a new shape to cut through the air at a miraculous speed. The power of his prayer also annotates the deficiency of those which had been directed to Benedict rather than Christ. To my knowledge, this is the only nautical rescue of any Anglo-Latin hagiography in which a ship is not saved by a divine meteorological act such as calming a storm, redirecting the winds, or returning the tide. Instead, the boat is returned by the swift rescue of the 'nauclero et proreta' (skipper and helmsman), whose appearance is as miraculous as the boat's buoyancy.[65]

The skills of Byrhtferth's *proreta* originated in Aldhelm's late seventh-century *De Virginitate* three centuries earlier than the *Vita Oswaldi*. Aldhelm compares the mental effort of his readers to the physical trials of athletes, including rowers. In each of five examples, a motivational mentor (Christ) supports the athlete's remarkable corporeal strength:

> alius clasicis nautarum cohortibus stipatus et densis remigantum agminibus circumsaeptus per uitreos oceani gurgites celerrimam agens liburnam aut lintrem instanter hortante proreta et crepitante nauclerii portisculo spumosis algosisque remorum tractibus trudit.[66]

> [another one, surrounded by the naval companies of sailors and encircled by dense throngs of rowers, driving his swift galley or skiff through the glassy waters of the ocean, with the steersman urgently inciting (them) and the master-rower beating time with his truncheon, presses on with foamy and sea-weedy strokes of the oars.][67]

Here, the work is loud, strenuous, crowded, wet, and, most importantly, led by a keen and encouraging coach.[68] Winterbottom notes that this final example is marked by curious diction:

> The *proreta*, who has not appeared in Latin literature since Plautus's *Rudens*, is egging the rowers on; and the *portisculus*, an old friend from the same author's *Asinaria*, is making its customary noise in the hands of the

---

64 'psallam spiritu, psallam et mente et', 'cantantes in cordibus uestris', 'psalite sapienter': *Sancti Eusebii Hieronymi Epistulae*, ed. by Hilberg, p. 134 (ll. 7–10). Jerome cites all three in his letter to Rusticus, Epistle CXXV, trans. by McKinnon, *Music in Early Christian Literature*, no. 328, p. 143.
65 Byrhtferth of Ramsey, *Life of St Oswald*, ed. and trans. by Lapidge, pp. 134, 135. The editors acknowledge that this is 'an Aldhelmian phrase' in n. 153.
66 Aldhelm, *Prosa de Virginitate*, ed. by Gwara, II. ll. 33–37.
67 Aldhelm, *Prosa de virginitate*, in Aldhelm, *The Prose Works*, trans. by Lapidge and Herren, p. 60.
68 For the peculiar diction throughout the 'five athletes' passage, see Griffith, 'Old English Poetic Diction'.

skipper. Isidore cites *portisculus* from Plautus, and no doubt Aldhelm has been glossary-hunting. As for *nauclerus*, it is not an unknown word, for it was available in the Vulgate Acts (XXVII.11) [and] appears in Isidore.[69]

Yet *proreta* does not appear in Isidore, and Winterbottom 'doubt[s] that Aldhelm knew the word at first hand'.[70] There is no debating that the *proreta* belonged on the *prora* — the bow (or prow) of a boat — but the word's sense, and its sensory register, have vexed copyists, glossators, and even modern editors for centuries.[71] Katrin Thier finds the following Old English glosses for the term: *plihtere* (a hapax legomenon), *ancerman*, and *stēora*. While *stēora* suggests that the *proreta* might have 'also had steering duties, perhaps with a bow rudder', she believes that he was, first and foremost, a lookout.[72] In any event, with the *proreta* cheering and the *nauclerus* striking a beat, both the vocal and instrumental versions of the presumed *celeuma* are retained. And thanks to enthusiastic borrowing by Byrhtferth, each remains gainfully employed throughout the tenth century.

Reflecting on the power of Oswald's prayer to rescue the torpid boat, Byrhtferth connects the tenth-century saint with his seventh-century predecessor, Cuthbert. Byrhtferth recalls the antiphon for Cuthbert's feast day with only some forgetfulness: 'Dum iactantur puppes salo, sanctus solo mox orat uentorum uis motata, naues uertit ad littora' (When boats were being tossed about on the sea, the saint, stationed on the shore, fell to prayer at once; the wind changed direction and turned the boats towards the shore).[73] Setting this scene in relation to the Bedan 'rafts episode' connects the two saints in the act of littoral prayer.[74] Their spiritual voices each reach God when others' supplications do not, and while Cuthbert's prayer is answered by the changing of the wind, Oswald's is answered by voices from the ship. We end here, as we began: on the shores of the Northumbrian monastery, amid the sounds of salvation.

Navigable waterways offered early medieval authors spaces in which to consider the terrestrial nature of earthly existence. Sounds made underway manifested the embodied experiences of the human trial adrift in the temporal

---

69 Winterbottom, 'Aldhelm's Prose Style and its Origins', p. 40.
70 Winterbottom, 'Aldhelm's Prose Style and its Origins', p. 42, n. 2.
71 *Proreta* is occasionally miscopied as *propheta*, for instance.
72 Thier, *Old English Sea Terms*, p. 91.
73 Byrhtferth of Ramsey, *Life of St Oswald*, ed. and trans. by Lapidge, p. 135. Lapidge (*Lives of St Oswald and Ecgwin*) and Gretsch identify that the citation is 'somewhat garbled' (Gretsch, *Ælfric and the Cult of Saints*, p. 99). Gretsch provides her own translation alongside the Latin text of the hymn, from Hohler and Hughes, 'The Durham Services in Honour of St Cuthbert', pp. 156–57: 'Dum iactantur puppes salo | sanctus orans heret solo | mox uentorum uis mutata | naues uertit ad litora' (While the ships [...] are tossed by the sea, the saint in prayer clings to the earth; the power of the winds having swiftly changed, the ship returns to the shore), *Ælfric and the Cult of Saints*, p. 99.
74 See ch. 3 of Bede, *VCP* and Chapter 2 of Shores, 'Nautical Narratives in Anglo-Latin Hagiographies'.

world. Yet far from being a superficial metaphor for abstract visions of isolation and despair, being at sea realizes the physicality of mortal flesh, the materiality of the world around it, and the power of sound to dissolve them both.

Works Cited

Primary Sources

Aldhelm, *Prosa de Virginitate cum glosa Latina atque anglosaxonica*, ed. by Scott Gwara, CCSL, 124, 2 vols (Turnhout: Brepols, 2001)
——, *Aldhelm: The Prose Works*, trans. by Michael Lapidge and Michael Herren (Cambridge: Brewer, 1979)
Ambrose, *Exposito Psalmi CXVIII*, ed. by Michael Petschenig and Michaela Zelzer, CSEL, 62 (Vienna: Verlag der Österreichischen Akademie der Wissenschaften, 1999)
Augustine of Hippo, *Confessiones*, ed. by Luc Verheijen, CCSL, 27 (Turnhout: Brepols, 1990)
——, *Enarrationes in Psalmos*, ed. by Eligius Dekkers and Iohannes Fraipont, CCSL, 40 (Turnhout: Brepols, 1956)
Bede, *Vita S. Cuthberti prosaica*, in *Two Lives of Saint Cuthbert: A Life by an Anonymous Monk of Lindisfarne and Bede's Prose Life*, ed. and trans. by Bertram Colgrave (Cambridge: Cambridge University Press, 1940), pp. 141–307
Byrhtferth of Ramsey, *The Lives of St Oswald and St Ecgwine*, ed. and trans. by Michael Lapidge (Oxford: Clarendon Press, 2009)
Exodus, ed. by Peter J. Lucas (Exeter: Exeter University Press, 1994)
Gildas, *The Ruin of Britain, and Other Works: History from the Sources*, ed. and trans. by Michael Winterbottom (London: Phillimore, 1978)
Paulinus of Nola, *Carmina*, ed. by Franz Dolveck, CCSL, 21 (Turnhout: Brepols, 2015)
Sancti Eusebii Hieronymi Epistulae, Pars III, Epistulae cxxi–cliv, ed. I Hilberg, CSEL, 56 (Vienna: Tempsky, 1918)
The Seafarer, ed. by Ida Gordon, Exeter Medieval English Texts and Studies, 2nd edn (Exeter: Exeter University Press, 1997)
Stephen, *The Life of Bishop Wilfrid by Eddius Stephanus*, ed. and trans. by Bertram Colgrave, rev. edn (Cambridge: Cambridge University Press, 1985)
Vita Ceolfridi, in *Abbots of Wearmouth and Jarrow*, ed. and trans. by Christopher W. Grocock and Ian N. Wood, Oxford Medieval Texts (Oxford: Clarendon Press, 2013), pp. 77–121

Secondary Works

Allen, Valerie J., 'Broken Air', *Exemplaria*, 16 (2004), 305–22
Anlezark, Daniel, 'From Elegy to Lyric: Changing Emotion in Early English Poetry', in *Understanding Emotions in Early Europe*, ed. by Michael Champion and Andrew Lynch (Turnhout: Brepols, 2015), pp. 73–98

Billett, Jesse D., 'Wilfrid and Music', in *Wilfrid: Abbot, Bishop, Saint. Papers from the 1300th Anniversary Conferences*, ed. by N. J. Higham (Donington: Shaun Tyas, 2013), pp. 163–85

Brown, Alan K., 'Bede, a Hisperic Etymology, and Early Sea Poetry', *Mediaeval Studies*, 37 (1975), 419–32

Cameron, Angus, Ashley Crandell Amos, and Antonnette diPaolo Healey, eds, *Dictionary of Old English: A to H Online* (Toronto: Dictionary of Old English Project, 2016), <https://www.doe.utoronto.ca/pages/index.html> [accessed 15 February 2020]

Cramp, Rosemary, *Wearmouth and Jarrow Monastic Sites*, vol. 1, English Heritage Archaeological Monographs (Swindon: English Heritage, 2005)

Duncan, Sandra, 'Prophets Shining in Dark Places: Biblical Themes and Theological Motifs in the *Vita Sancti Wilfridi*', in *Wilfrid: Abbot, Bishop, Saint. Papers from the 1300th Anniversary Conferences*, ed. by N. J. Higham (Donington: Shaun Tyas, 2013), pp. 80–93

Ellard, Donna Beth, 'Communicating between Species and between Disciplines — Lessons from the Old English Seafarer', *Exemplaria*, 30 (2018), 293–315

Foys, Martin K., 'A Sensual Philology for Anglo-Saxon England', *Postmedieval: A Journal of Medieval Cultural Studies*, 5 (2014), 456–72

Gretsch, Mechthild, *Ælfric and the Cult of Saints in Late Anglo-Saxon England*, Cambridge Studies in Anglo-Saxon England, 34 (Cambridge: Cambridge University Press, 2006)

Griffith, Mark, 'Old English Poetic Diction Not in Old English Verse or Prose – and the Curious Case of Aldhelm's Five Athletes', *Anglo-Saxon England*, 43 (2014), 99–131

Haines, John, 'A Musical Fragment from Anglo-Saxon England', *Early Music*, 36 (2008), 219–29

Hohler, Christopher, and Anselm Hughes, 'The Durham Services in Honour of St Cuthbert', in *The Relics of Saint Cuthbert*, ed. by C. F. Battiscombe (Oxford: Oxford University Press, 1956), pp. 155–92

Klein, Stacey S., 'Navigating the Anglo-Saxon Seas', in *The Maritime World of the Anglo-Saxons*, ed. by Stacey S. Klein, William Schipper, and Shannon Lewis-Simpson, Essays in Anglo-Saxon Studies, 5 (Tempe: ACMRS, 2014), pp. 1–20

Magennis, Hugh, 'The Solitary Journey: Aloneness and Community in *The Seafarer*', in *Text, Image, Interpretation: Studies in Anglo-Saxon Literature and its Insular Context in Honour of Éamonn Ó Carragáin*, ed. by Alastair Minnis and Jane Roberts (Turnhout: Brepols, 2007), pp. 303–18

McKinnon, James, ed., *Music in Early Christian Literature*, Cambridge Readings in the Literature of Music (Cambridge: Cambridge University Press, 1987)

Ó Carragáin, Éamonn, *The City of Rome and the World of Bede*, Jarrow Lecture (Newcastle: Parish Church Council of St. Paul's Church, Jarrow, 1994)

Sheerin, Daniel J., '"Celeuma" in Christian Latin: Lexical and Literary Notes', *Traditio*, 38 (1982), 45–73

Shores, A. R. P., 'Nautical Narratives in Anglo-Latin Hagiographies, *ca.* 700–1100' (unpublished doctoral thesis, University of North Carolina, 2017)

Szabo, Vicki, *Monstrous Fishes and the Mead-Dark Sea: Whaling in the Medieval North Atlantic* (Leiden: Brill, 2008)

Thier, Katrin, *Old English Sea Terms* (Ely: Anglo-Saxon Books, 2014)

Turner, Sam, Sarah Semple, and Alex Turner, *Wearmouth and Jarrow: Northumbrian Monasteries in an Historic Landscape* (Hatfield: University of Hertfordshire Press, 2013)

Winterbottom, Michael, 'Aldhelm's Prose Style and its Origins', *Anglo-Saxon England*, 6 (1977), 39–76

ELIZABETH A. ALEXANDER

The Sailors, the Sea Monster, and the Saviour: Depicting Jonah and the *Ketos* in Anglo-Saxon England

The prophet Jonah is the most frequently depicted Old Testament figure in early Christian art;[1] however, depictions of Jonah seem to have been less popular in the early medieval period than they were in the early Christian world. It is possible that this early medieval under-representation of Jonah could be attributed to poor rates of survival, though it is difficult to see how such artwork would have been disproportionately affected by the ravages of time. A more likely explanation is that changes in the contexts in which Jonah initially featured led to his relegation in the visual imagination. The majority of the early Christian Jonah scenes survive in considerable numbers in funerary contexts, either on sarcophagi or as frescos in catacombs, but only two known examples of Anglo-Saxon Jonah scenes survive from the pre-Viking period, and neither is preserved in a funerary context. Rather, they are preserved as copies of Anglo-Saxon miniatures found in the Antwerp Sedulius,[2] which depict Jonah Lowered from the Boat (fol. 9ᵛ) and Jonah Regurgitated (fol. 10ʳ). While they are preserved as manuscript miniatures, these nevertheless form part of a greater pictorial cycle that replicates a high proportion of the Jonah scenes featured on early Christian sarcophagi and catacomb frescos. These images, besides the two scenes of the prophet on the boat and being regurgitated, often included occasional depictions of him Embarking on the Boat at Joppa and, more frequently, Relaxing under the Gourd. Clearly the theme of Jonah and the sea monster in early Christian and early medieval art would seem to relate, unequivocally, to ideas relating to the waters of the ocean. However, as will become clear, the ways in which the sea monster

---

1   Narkiss, 'The Sign of Jonah', p. 63.
2   Antwerp, Museum Plantin-Moretus, MS M. 17. 4.

Elizabeth A. Alexander • (ea502@york.ac.uk) is a Research Affiliate in the History of Art Department at the University of York.

was articulated in Anglo-Saxon art indicates that it denoted much more than this, in fact, deriving much of its inspiration from visualizations that associated it with 'the deep': aquatic, but more importantly in terms of its symbolic significances, hell. Here, it functions multivalently to signify the crucial nature of salvation.

As far as the extant images of the sea monster are concerned, there is only one further visual reference to Jonah's ordeal in the deep, dating from the latter part of the Anglo-Saxon period: contained within the illustration of Psalm 103 (104) in the Harley Psalter.[3] This early eleventh-century psalter is an Anglo-Saxon 'copy' of the Utrecht Psalter,[4] a ninth-century Carolingian manuscript.[5] While not strictly an image of Jonah and the sea monster — rather the creature and boats depicted in Psalm 103 (104). 26[6] — the iconography of the Harley Psalter's illustration is deeply rooted in the episode and likely recalls Christ's allusion to it at Matthew 12. 40.

The most prominent character of the Jonah story is, of course, the sea monster that swallows the prophet; and there are two further representations of this beast in Anglo-Saxon England, which both illustrate the constellation of the *cetus* and clearly demonstrate that the Anglo-Saxons had access to images of the creature, even if these were in the form of astronomical drawings rather than tied to depictions of the Jonah narrative. The first of these astronomical drawings is contained within a ninth-century Carolingian manuscript of Cicero's *Aratea*, known to have been circulating in Anglo-Saxon England during the eleventh century.[7] The second is a drawing of the *cetus* constellation by the Anglo-Saxon artist named the 'Master of Ramsey' found in a late tenth-century Continental manuscript of the same text.[8] Both drawings show the sea monster as a long-snouted creature that appears to have the head and front paws of a canine and the long twisted body and tail of a fish or other sea-dwelling animal. These astronomical drawings, the Antwerp Sedulius miniatures, and the Harley Psalter's sea creature all point towards the fact that, at the very least and despite the dearth of surviving examples, the Anglo-Saxons had access to and engaged with artistic representations of Jonah and his sea monster.

---

3   BL, MS Harley 603, fol. 51^v.
4   Utrecht, Universiteitsbibliotheek, MS Bibl. Rhenotraiectinae I Nr 32.
5   Panofsky, 'The Textual Basis of the Utrecht Psalter Illustrations', p. 50. For discussion regarding the copying process, see Gameson, 'The Anglo-Saxon Artists of the Harley (603) Psalter'; van der Horst, Noel, and Wüstefeld, *The Utrecht Psalter in Medieval Art*, pp. 6–9.
6   'illic naves pertransibunt draco iste quem formasti ad inludendum ei' (There the ships shall go. This sea dragon which thou hast formed to play therein). In this chapter all English biblical citations are from the Douay-Rheims and all Latin citations are from the Vulgate.
7   BL, MS Harley 647, fol. 10^r; Gneuss and Lapidge, *Anglo-Saxon Manuscripts*, p. 346.
8   BL, MS Harley 2506, fol. 42^r; Gneuss and Lapidge, *Anglo-Saxon Manuscripts*, p. 350; Mostert, 'Relations between Fleury and England', p. 203.

The *Ketos* in Early Christian Art

All of the Anglo-Saxon representations of Jonah and the sea monster continue and refigure the tradition found in early Christian examples. The formulaic layout of Jonah imagery on the late antique sarcophagi of the fourth and fifth centuries shows that even at this early stage of Christianity, the prophet had developed a fixed iconography, with the three most frequently depicted scenes being Jonah Lowered from the Boat (Jonah 1. 15); Jonah Regurgitated by the Whale (Jonah 2. 11); and Jonah Relaxing under the Gourd (Jonah 4. 6). In all of these scenes the whale is depicted as a sea monster classified as a *ketos*.

The first of these — Jonah Lowered from the Boat — usually depicts the prophet shown in the act of being lowered from the boat by the shipmen; he has his arms outstretched towards the mouth of a *ketos*, emphasizing his desire to sacrifice himself to calm the seas. The boat nearly always has a sail, which billows in the wind, and the sea is shown in a manner suggesting turbulence, highlighting the storm sent by God because of Jonah's refusal to carry out his command (Jonah 1. 4). Usually one or more of the sailors hold oars, referencing the moment in the narrative where, before sacrificing Jonah to the sea, the shipmen attempt to row back to shore (Jonah 1. 13). Jonah Regurgitated by the *Ketos* depicts him exiting the *ketos* head first and arms outstretched. The creature is shown with either its head emerging from the surface of the water or is depicted in full, its head and body on top of the water line, so the prophet emerges skywards and often towards land. There is a variant where the sea monster regurgitates the prophet onto land, where he is shown reclining and conflates the Jonah Regurgitated scene with Jonah Relaxing under the Gourd. This final scene in the Jonah cycle — Jonah Relaxing under the Gourd — shows the prophet relaxing naked under an arching plant with hanging gourds.[9]

The term *ketos* refers to a very specific type of sea creature whose visual origins lie in Greek art. Depictions of the monster tend to present it as canine headed, with razor-sharp teeth, pointed ears, two front paws, and an elongated and twisted body that loops round on itself before ending in a tail. Its iconographic roots lie in depictions of the Greek myth of Andromeda being sacrificed to the *ketos*, where she is shown fettered to a rock, while the heroic figure Perseus saves her from imminent death by attacking the beast.[10] Pliny the Elder in his *Natural History* describes this event as having taken place in Joppa,[11] which is the same location from which Jonah set sail

---

9 The pose of Jonah in this scene is closely related to that of the sleeping Endymion, who, in Greek mythology, was a beautiful youth that spent much of his life in an endless sleep. See Schultze, *Archäologische Studien über altchristliche Monumente*, p. 81; Narkiss, 'The Sign of Jonah', p. 67.
10 Papadopoulos and Ruscillo, 'A *Ketos* in Early Athens', p. 207.
11 Pliny, *Naturalis Historia*, ed. and trans. by Rackham, pp. 316–17.

in the biblical account of his sea voyage.[12] It is probable, therefore, that the evolution of the whale that swallowed Jonah came out of the association of these two events having taken place in the same locality.[13] This understanding is further supported by the reference in the Greek Septuagint version of the Old Testament to the creature that swallows Jonah as a *mega ketos* (big fish),[14] which is a translation from the Hebrew text where this creature is also a *dag gadol* (great fish). In Jerome's Latin Vulgate translation of the biblical book the creature is again referred to as a *piscis grandis* (great fish). However, when transcribing the passage of Matthew 12. 40, Jerome chooses not to use the term *piscis grandis* for the creature, preferring to Latinize the Greek *ketos* as *cetus*:

> sicut enim fuit Ionas in ventre *ceti* tribus diebus et tribus noctibus sic erit Filius hominis in corde terrae tribus diebus et tribus noctibus. (emphasis added)
>
> > [For as Jonah was in the whale's belly three days and three nights: so shall the Son of man be in the heart of the earth three days and three nights.][15]

As we shall see, this passage would create a direct association between the story of Jonah and the whale (specifically the *ketos*) and the Harrowing of Hell, which is said to have taken place between Christ's death on the cross and his resurrection three days later (1 Peter 3. 19). The lexical choice on Jerome's part may not only represent the combined momentum of a fully developed tradition of the Harrowing in the late fourth century when he was making the Vulgate translation, but also reflect the fact that Jerome, who was then living in Palestine where both events were believed to have begun, wished to associate the whale of the Jonah story with Andromeda's sea monster in this particular verse.

It is evident, then, that the iconography of the *ketos* represents a specific link between a perceived 'real' deep-sea creature, which becomes synonymous with a whale or other large fish (forming the etymological root of the present taxonomic classification of cetacean),[16] and the biblical creature that swallowed Jonah. Furthermore, as a result of the Old Testament event being seen as a prefiguration of Christ's death, descent, and resurrection (Matthew 12. 40), the sea creature also developed in the early medieval visual imagination as an embodiment of hell.

---

12 Jonah 1. 3.
13 Jerome, *Commentariorum in Jonam Prophetam*, ed. by Adriaen, 1. 3, p. 383; English translation found in Jerome, *Ancient Commentaries in English*, ed. by Litteral, trans. by MacGregor, p. 13.
14 Jonah 1. 17.
15 Matthew 12. 40.
16 Liddell, Scott, and Drisler, *A Greek-English Lexicon*, pp. 949–50; Ranneft, Eaker, and Davis, 'A Guide to the Pronunciation and Meaning of Cetacean Taxonomic Names', p. 185.

Copying the Anglo-Saxons: The Antwerp Sedulius

Lamentably, there are no representations of Jonah and the *ketos* which survive directly from Anglo-Saxon England, though this does not mean that they did not at one time exist. Nevertheless, we have some indication that Anglo-Saxons were familiar with the subject if the manuscript sources are taken into consideration. Through the accidents of survival, there is clear evidence for at least one manuscript known in Anglo-Saxon England that contained a depiction of Jonah and the *ketos*. The original manuscript is now lost, but there is evidence to suggest that at some point during the early ninth century images were copied from an Anglo-Saxon exemplar by Carolingian artists into a book that does survive. The scribes of the Antwerp Sedulius carefully replicated the colophon 'FINIT · FINES · FINES · CUDVVINI'[17] identifying it as a copy of an Insular exemplar, though this lost book may have been a foreign import, or a copy of one.[18] The Anglo-Saxon Cuthwine mentioned in the colophon has been identified as an eighth-century East Anglian bishop, who, according to Bede, brought back an illuminated manuscript from Rome. It is tempting to think this might perhaps be the lost manuscript, but it could easily have been another Italian book from Cuthwine's library.[19]

The Carolingian manuscript that now contains the two copies of depictions of Jonah and the *ketos*, copied from Cuthwine's book, is an illuminated version of the *Pascale carmen* by Sedulius, which was, for over a century, one of the most popular poems in western Europe.[20] Likely written between AD 425 and 450, Sedulius's biblical epic details the life of Christ based on the four Gospel books of the Bible and provides an overview of a select few episodes

---

17 Antwerp, Museum Plantin-Moretus, MS M. 17. 4, fol. 68^v.
18 Levison, *England and the Continent in the Eighth Century*, pp. 133–34; see more recently Hawkes, 'Anglo-Saxon *Romanitas*', pp. 21–22.
19 'Quod ita intelligendum, ab antiquis ita intellectum, testator etiam picture eiusdem libri, quam reverendissimus ac doctissimus Cudum Orientalium Anglorum antistes, veniens a Roma secum in Britanniam detulit, in quo videlicet libro omnes pene ipsius apostolic passions sive labores per loca opportune errant depictæ' (That it is to be understood in this way and was understood in this way by the ancients is also attested by the picture of the Apostle in the book which the most reverend and most learned Cuthwine, bishop of the East Angles, brought with him when he came from Rome to Britain for in that book all of his sufferings and labours were fully depicted in relation to the appropriate passages); Bede, *De octo quaestionibus*, 2, ed. by Migne, p. 456; Bede, *On Eight Questions*, trans. by Foley and Holder, p. 151; Lapidge argues the manuscript referenced by Bede in this passage is an illuminated copy of some *Passio S. Pauli* which has disappeared without an identifiable copy and that Cuthwine must have had another illuminated manuscript which served as the exemplar to the Antwerp Sedulius. See Lapidge, *The Anglo-Saxon Library*, pp. 26–27.
20 Springer, *The Gospel as Epic in Late Antiquity*, p. ix.

from the Old Testament, including the story of Jonah.[21] However, the Jonah narrative has been radically condensed into just five lines:

> Ionas puppe cadens, coeto sorbente uoratus
> In pelago non sensit aquas, uitale sepulchrum
> Ne moreretur habens, tutusque in uentre ferino
> Depositum, non praeda fuit, uastumque per aequor
> Venit ad ignotas inimico remige terras.
>
>> [Now, Jonah, swallowed by a whale when fallen from a ship, escaped the sea: this life-sustaining tomb kept him from death, within the animal he was an offering, not just some prey, and through the sea he came to foreign lands: by enmity of oarsmen, in the oarsmen's enemy.][22]

The two Jonah images are adjacent to these five lines and serve to illustrate the text. The first — Jonah Lowered from the Boat — appears at the bottom of fol. 9^v, with the next scene in the cycle — Jonah Regurgitated by the *Ketos* — appearing at the top of fol. 10^r. Both of them provide key insight to our understanding of the complex iconography of water and 'the deep'.

Jonah Lowered from the Boat (Figure 5.1) includes many of the details found in early Christian art — the two sailors lower Jonah from the boat, one of whom holds an oar under his arms — but does not include the rough seas or billowing sails; nor does Jonah embrace his sacrifice, and there is no creature awaiting him. The shipmen and the boat are isolated with no background, water, or sail, while Jonah is carefully lowered from the vessel. The overall result of this iconographic adaptation is to emphasize the act of lowering Jonah from the boat; it is an emphasis highlighted by the sentence immediately above the scene, 'Ionas puppe cadens, coeto sorbente uorantus' (Jonah falling from the ship, he was devoured).[23] Here text and image work together with the verbal statement spelling out Jonah's act of self-sacrifice and 'death'.

It is almost certain that the viewers of this scene would have been aware of its significance as an allegory of Christ's passion. The passage from Matthew, as well as exegesis on the Jonah story by the early Church Fathers, such as Jerome, all explicitly link Jonah being thrown to the sea with the Crucifixion.[24] Jerome takes the comparison a step further, explaining:

> Si consideremus ante passionem Christi, errores mundi, et diversorum dogmatum flatus contrarios,

---

21 Springer, *The Gospel as Epic in Late Antiquity*, p. 23.
22 Antwerp, Museum Plantin-Moretus, MS M. 17. 4, fol. 9^v; Swanson, 'Easter Poem', p. 293.
23 Translation author's own. Antwerp, Museum Plantin-Moretus, MS M. 17. 4, fol. 9^v.
24 Jerome, *Commentariorum in Jonam Prophetam*, ed. by Adriaen, 2. 1b, p. 393; Jerome, *Ancient Commentaries in English*, ed. by Litteral, trans. by MacGregor, p. 23.

THE SAILORS, THE SEA MONSTER, AND THE SAVIOUR   133

Figure 5.1. 'Jonah Lowered from the Boat', Antwerp, Museum Plantin-Moretus, MS M. 17. 4, fol. 9ᵛ. Early ninth century. Image courtesy of Museum Plantin-Moretus, Antwerp – UNESCO, World Heritage.

et nauiculam totumque humanum genus, id est creaturam Domini periclitantem, et post passionem eius tranquilitatem fidei, et orbis pacem, et secura omnia, et conuersionem ad Deum, et uidebimus quomodo post praecipitationem Ionae steterit mare a furore suo.

> [If we consider before the suffering of Christ, the confessions of the world, the contrary winds of different opinions, the ship and all human kind, that is all creation to be in danger, then, after the suffering of Christ there is a calm of faith, the peace of the world, universal safety, conversion to God, and we will see how after Jonah has been thrown overboard the sea ceases from its raging.][25]

Here the boat becomes humanity surrounded by the storm of sin, which is calmed by Christ's crucifixion; the implied sea into which Jonah is being lowered becomes the embodiment of the Devil destroyed through Christ's sacrifice. It is more than likely that those responsible for the careful design of the Antwerp Sedulius were aware of this interpretation;[26] if this is indeed the case, then the lack of rough seas and billowing sails could reflect deliberate choices, highlighting the 'calm of faith' (*tranquilitatem fidei*) brought about by Christ's crucifixion and resurrection. Sacrifice and its salvific effect are

---

25   Jerome, *Commentariorum in Jonam Prophetam*, ed. by Adriaen, 1. 15, p. 392; Jerome, *Ancient Commentaries in English*, ed. by Litteral, trans. by MacGregor, pp. 20–21.
26   There are four surviving manuscripts containing Jerome's commentary on Jonah known to have been circulating in Anglo-Saxon England, with a high probability that there would have been more copies that have been lost to the ravages of time. See Gneuss and Lapidge, *Anglo-Saxon Manuscripts*, pp. 25, 144, 185, 475.

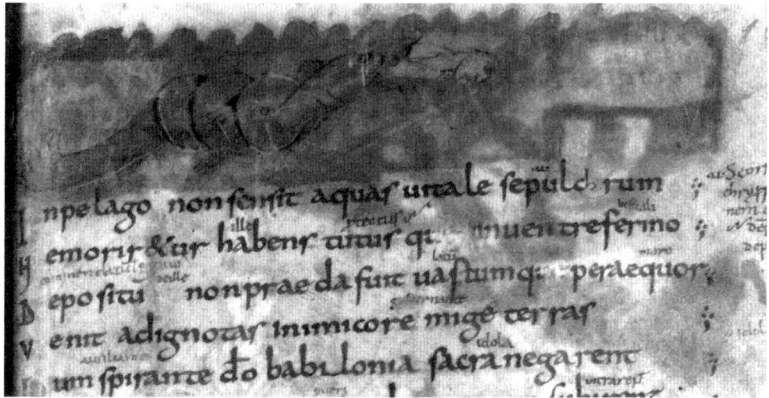

Figure 5.2. 'Jonah Regurgitated', Antwerp, Museum Plantin-Moretus, MS M. 17. 4, fol. 10ʳ. Early ninth century. Image courtesy of Museum Plantin-Moretus, Antwerp – UNESCO, World Heritage.

presented as one and the same moment: as Jonah is lowered in sacrifice to the rough sea brought about by God's wrath, so is Christ offered in sacrifice to end the suffering of humanity brought about by the original sin.

This interpretation is further supported by the removal of any obvious water from the scene; the sea is not just calmed by Jonah's sacrifice, it is absent altogether. A possible explanation for this absence could be that those responsible for the design of the Antwerp Sedulius scene intended to reference Revelation 21. 1: 'Et vidi cælum novum et terram novam. Primum enim cælum, et prima terra abiit, et mare jam non est' (And I saw a new heaven and a new earth. For the first heaven and the first earth was gone, and the sea is now no more). With the lack of water signifying the end of days, the scene can be seen to depict not only the Old Testament event of Jonah thrown from the boat, prefiguring Christ's salvific death, but also the Second Coming and the establishment of the New Jerusalem.

At the top of the page immediately following the scene of Jonah Lowered from the Boat, is an illustration of the next part of the Jonah cycle: Jonah Regurgitated (Figure 5.2). In the biblical account of Jonah, after the shipmen throw the prophet from the boat, he is swallowed by a large sea creature that carries him in its belly for three days and nights before regurgitating Jonah onto the land. Unlike early Christian examples in which Jonah exits the *ketos* head first and with arms outstretched, the Antwerp Sedulius's Jonah is not illustrated being immediately deposited under the gourd; his exit from the creature is more passive: he does not appear to have his arms outstretched, but slips out with his arms tucked in to his sides. Conversely, the depiction of the *ketos* itself is similar to many late antique exemplars, with only slight variations in the smaller size of its front paws, streamlined body, and more fish-like than canine head. Although the Antwerp Sedulius scene recalls the late antique examples in many ways, it thus remains iconographically distinct. Once again Jerome offers an explanation for these changes in his *Commentary on Jonah*:

> Morti et inferno praecepit Dominus, ut prophetam suscipiat [...]. Ventrem autem inferi, aluum ceti intellegamus, quae tantae fuit magnitudinis, ut instar obtineret inferni. Sed melius ad personam Christi referri potest, qui sub nominee Dauid cantat in psalmos: 'Non derelinques animam meam in inferno, nec dabis sanctum tuum uidere corruptionem.' Qui fuit in inferno uiuens, inter mortuos liber [...]. Praecipitur ergo huic magno ceto, et abyssus et inferno, ut terries restituant Saluatirem, et qui mortuus fuerat, ut liberaret eos qui mortis uinculis tenebantur, secum plurimos educat ad uitam.

> [The Lord commanded death and the underworld to receive the prophet [...]. By the 'belly of hell' we understand the stomach of a whale of such a great size that it took the place of hell. But this can better be referred to the person of Christ, who under the name of David sings in the psalm: 'you will not leave my spirit in hell, and you will not allow your saints to see putrefaction' (Psalm 15. 10), living in hell free among the dead [...]. The great whale, the deep and hell are then ordered to give back the Lord to dry earth; thus he who had died to free those detained by the chains of death, can lead with him many others towards life.][27]

Here, Jerome describes the deep-sea creature as both embodying hell and belonging to hell, referring to the underworld in relation to both the prophet Jonah and Christ. This concept of the *ketos* as a hell beast does find its way visually into contemporary Carolingian manuscripts: so much so that in a depiction of a soul being dragged into hell in the ninth-century Stuttgart Psalter (fol. 10^v) a creature with a canine head, pointed ears, two front paws, and an elongated body that loops and ends in a tail is portrayed within hell itself.[28] This closely resembles not only late antique models of the *ketos*, but also the *ketos* of the two Jonah scenes found elsewhere within the Psalter.[29]

One possible explanation for the small divergences from the established *ketos* iconography in the Antwerp Sedulius is that the representation of the creature that swallowed Jonah in this manuscript was intended to recall both the *ketos* of Late Antiquity and, simultaneously, the serpent of the Adam and Eve narrative. This would explain why the creature, while closely resembling its late antique counterparts, also looks more serpentine than the traditional representations of the *ketos*. If this is the case then the figure exiting the mouth

---

27 Jerome, *Commentariorum in Jonam Prophetam*, ed. by Adriaen, 2. 1a–11, pp. 393–403; Jerome, *Ancient Commentaries in English*, ed. by Litteral, trans. by MacGregor, pp. 22–31.
28 Stuttgart, Württembergische Landesbibliothek, Bibl. fol. 23, fol. 10^v.
29 Stuttgart, Württembergische Landesbibliothek, Bibl. fol. 23, fols 90^v and 147^v.

of the creature could be intended to portray both Jonah and Christ, who through his willing sacrifice, death, descent, and resurrection absolved mankind of the original sin which was brought about by the temptation of Adam and Eve by the serpent. This interpretation is further emphasized by the trifold twist in the creature's body, which may refer to the sacred number three, specifically, the three days spent in hell by Christ, after sacrificing himself to reverse the original sin, alongside the three days spent in the belly of the beast by Jonah.

In an undoubted Anglo-Saxon evocation of this rich interpretative tradition, Bede's commentary *On Ezra and Nehemiah* debates the role of the serpent under the waves of the sea as representative of vices and wicked works. While discussing the watery transport of the logs for building the temple acquired from Joppa (the city from which Jonah sets sail), a passage which must be endured to reach the city (representing virtue), Bede states that the Devil, 'qui dictus est a propheta draco rex omnium quae sunt in aquis, hoc est impiorum; quorum conuersatio non est in caelis, sed in mari est pertubationibus saeculi fluentuantis' (whom the prophet calls a 'serpent, the king of everything in the waters', which is to say, of the ungodly, whose dwelling is not in the heavens but in the sea (that is, the fluctuations of this restless world)).[30] Here Bede, referring to Revelation 12. 9,[31] calls the Devil a serpent, but unlike the biblical passage, he does not describe the fallen angel thrown down to earth, but presents him as a sea dweller. It seems, in this light, that the sea, specifically the sea surrounding the city of Joppa, was understood by commentators to be a place where evil dwelt, making it the particular home of the serpentine Devil.[32] This may further explain why those responsible for the design of the Antwerp Sedulius's depiction of Jonah Regurgitated chose to carefully adapt their inherited iconography of the *ketos* and to show it as more serpentine and so more devil-like.

Water plays a key part here, as it does in the preceding scene of Jonah Lowered from the Boat. In this case, however, rather than a lack of water, there is an abundance of it. The waves surround Jonah and the *ketos* as the creature regurgitates the prophet; this again differs from early Christian examples of the scene, where the prophet is thrown up by the beast above the water, rather than being surrounded by it. The emphasis on water in the scene clearly places

---

30 Bede, *In Ezram et Neemiam*, ed. by Hurst, 1. 3. 7, p. 275; Bede, *On Ezra and Nehemiah*, trans. by DeGregorio, p. 56.

31 'Et projectus est draco ille magnus, serpens antiquus, qui vocatur diabolus, et Satanas, qui seducit universum orbem: et projectus est in terram, et angeli ejus cum illo missi sunt' (And the great dragon was cast out, that old serpent, called the Devil, and Satan, which deceiveth the whole world: he was cast out into the earth, and his angels were cast out with him).

32 The dangerous and evil nature of the sea/ocean is referenced on several occasions in the Bible: it was the primeval abyss (Genesis 1. 2); the home of the Leviathan (see, for example, Psalm 103 (104). 26), Satan (Revelation 12. 9, 20. 1–3), and demons (Luke 8. 31); and it would be the place where Apocalyptic beasts would emerge from at the end of days (Revelation 11. 7). Bede's evocation of the Devil dwelling off the coast of Joppa, therefore, continues in this rich tradition. See further, O'Loughlin, 'Living in the Ocean', p. 13; Pickles, 'Anglo-Saxon Monasteries as Sacred Places', pp. 40–41.

the creature and the prophet in the deep, surrounded by waves, which are represented by the undulating line of the surface of the water. Again, Jerome provides a context for explaining this in his *Commentary on Jonah*, as:

> Quantum ad personam Jonae non est difficilis interpretatio, quod ceti clausus aluo in profundissimo et in medio maris fuerit, fluminibusque uallatus sit. Quantum ad Dominum Saluatorem sexagesimi octaui psalmi sumamus exemplum in quo loquitur: 'Infixus sum in limo profundi, et non est substantia: Veni in profundum maris, et tempestas demersit me' [...] omnis terrena habitatio plena est fluctibus, plena tempestatibus. Porro cor maris significatur infernus, pro quo in euangelio legimus: '*In corde terrae*'.
>
> > [the interpretation of the person of Jonah is not difficult: from the moment when he was closed in the stomach of the whale and found himself at the deepest and middle of the sea, he was surrounded by waves. For the Lord, the Saviour, prefiguring Psalm 68 in which he says, 'I am enshrouded in the deep mud where there is no ground. I have come of the deepest part of the sea and the storm engulfs me' (Psalm 68. 3) [...] all habitation on earth is full of waves, full of storms. And the 'heart of the sea' means hell, for which we read in the Gospel, 'in the heart of the earth' (Matthew 12. 40).][33]

Not only is the *ketos* a representation of hell, but it resides within another representation of hell, the sea, just as the creature is seen as both hell and belonging to hell in the depictions of the *ketos* in the ninth-century Stuttgart Psalter. The use of water in the Antwerp Sedulius is clearly key to understanding the complex iconography of the scene. It visually articulates the complex relationship between the abyss, the *ketos*, and hell, and the salvific deaths of both Jonah and Christ. As through sacrificing himself to calm the storm, Jonah saved the lives of the sailors, while Christ saved mankind from death through his crucifixion. Together, the two Antwerp Sedulius Jonah scenes present a complex set of iconographic references, which closely follow the account of Matthew 12. 40, illustrating how the events of the Old Testament story of Jonah prefigure Christ's death (through Jonah being lowered from the boat), descent (the three days spent in the body of the *ketos*, further emphasized by the triple loop and the use of water to highlight the links between the creature in the abyss and the Devil in hell), and resurrection (Jonah being regurgitated).

One final, important aspect of the images must be noted: the nakedness of Jonah throughout the two scenes. This is not an unusual feature; the majority of the late antique examples also depict Jonah in this unclothed state, despite the absence from the biblical story explaining the motives for showing him in this state. The tradition and ongoing artistic decision to depict Jonah in

---

33 Jerome, *Commentariorum in Jonam Prophetam*, ed. by Adriaen, 2. 4, p. 396; Jerome, *Ancient Commentaries in English*, ed. by Litteral, trans. by MacGregor, p. 24.

Figure 5.3. 'Illustration of Psalm 104 (103)', London, British Library, MS Harley 603, fol. 51ᵛ. Early eleventh century. Image © British Library Board.

his nakedness is thus more than likely due to the inherited understanding of the symbolic significance of Christ within the Jonah narrative. Proposed by Plato and subsequently discussed by St Augustine in his *De civitate Dei*, it was argued that the human soul returned to God pure, that is, naked.[34] It is thus possible that by depicting Jonah as naked, the viewer was prompted to view Jonah's naked body as Christ's soul entering hell. Hell is then represented symbolically as the *ketos* on the proceeding page, which exegetes such as Jerome believed was a creature of the abyss, a creature of hell, and the embodiment of hell simultaneously. In this respect, even the placement of the images on the pages can be seen to have symbolic significance. The boat is positioned at the bottom of the page and has the shipmen lowering Jonah/Christ down into hell, while Jonah/Christ is regurgitated by the *ketos*/hell at the top of the following page, completing the cycle of death and redemption of sin brought about by Christ's passion and crucifixion.

Copying the Carolingians: The Harley Psalter

The early eleventh-century Harley Psalter (Figure 5.3) presents more allusions to the *ketos* and its association with hell. As noted, the psalter is an almost direct copy of the ninth-century Utrecht Psalter, a Carolingian manuscript which is thought to have reached Anglo-Saxon England by AD 1000.[35] The

---

34 Plato, *Cratylus*, ed. by Duke and others, p. 403B; Plato, *Gorgias*, ed. by Burnet. pp. 523E–524D; Augustine, *De civitate Dei*, ed. by Dombart and Kalb, 13. 16, p. 397.
35 Benson, 'New Light on the Origins of the Utrecht Psalter', p. 14.

reference to the Jonah narrative appears in the illustration for Psalm 103 (104), on fol. 51ᵛ. Directly below a depiction of Christ standing on the 'wings of the winds',[36] and the psalmist at a feast (illustrating verse 15: 'et vinum laetificat cor hominis ut exhilaret faciem in oleo et panis cor hominis confirmat' (And that wine may cheer the heart of man. That he may make the face cheerful with oil: and that bread may strengthen man's heart)), there are two empty ships on a body of water. One has a sail billowing in the wind, and the turbulent water contains a sea creature with a long, twisted body reminiscent of a *ketos*. The contents of the Psalm identify this creature as the Leviathan, another biblical sea monster who according to Augustine is synonymous with the Devil;[37] however, the iconography of the scene is very closely related to depictions of Jonah and the *ketos*. The adaptation of the iconography of Jonah to represent Psalm 103 (104). 26 was a deliberate attempt by those responsible for the design of the illustration to recall the Jonah narrative and Matthew 12. 40.

The reasoning behind this adaptation lies in the symbolic significance of the story of Jonah and the *ketos* and its associations with Christ. The positioning of the boats and sea monster beneath an image of Christ in paradise, standing on personifications of wind and flanked by four angels, was surely intended to prompt the viewer to link these two elements. Through recalling the iconography of Jonah and the *ketos* in the illustration of Psalm 103 (104). 26 those responsible for the design of the scene invite the viewer to contemplate Matthew 12. 40 and the prophet Jonah as a prefiguration of Christ's death, descent, and resurrection. Furthermore, the choice to depict the Leviathan as a *ketos* visually links the Leviathan with hell. The *ketos* was a creature of the deep, the symbolic representation of hell and a creature belonging to hell. By using and adapting its iconography to represent the Leviathan, the artist similarly aligns this sea monster with hell. The Leviathan would later become the visual embodiment of the hell mouth in the full-page miniature of the Fall of the Rebel Angels in the mid-eleventh-century *Old English Hexateuch*,[38] where its long, twisting body and a gaping open jaw, complete with razor-shape teeth, holds a mandorla-shaped object containing the fettered Satan.

As the Harley Psalter's illustration of Psalm 103 (104) is a very close copy of the Utrecht Psalter, all these considerations are present in the earlier ninth-century manuscript, with those responsible for the design of the Anglo-Saxon Psalter accurately reproducing these contemplations for their copy. However, the Harley Psalter does diverge in one major way from its source model and that is through its use of colour. The Utrecht Psalter's illustrations are largely monochrome, whereas the Harley Psalter uses colour, predominately blue, green, red, yellow, and mauve. Yet, for the illustration to

---

36 'qui tegis in aquis superiora eius qui ponis nubem ascensum tuum qui ambulas super pinnas ventorum'. Psalm 103. 3.
37 Augustine, *Enarrationes in Psalmos*, ed. by Dekkers and Fraipont, *sermo* 4, 103. 36, p. 1525.
38 BL, MS Cotton Claudius B.iv, fol. 2ʳ.

Psalm 103 (104) the artist principally uses just two colours: red and blue. The overall effect of this creates a vivid scene, which stands out when compared to those that surround it: the blue pigment colours the sky and the water, with red being used predominately for the landscape. The decision to use such large quantities of blue and red clearly serves a symbolic purpose.

Bede in his commentary *On Genesis*, following in the footsteps of Gregory the Great and Isidore of Seville,[39] discusses the rainbow as a continued sign of the covenant made between God and Noah, going on to describe how the earth will not be destroyed again by flood, but by fire:

> Neque enim frustra ceruleo simul et rubicundo colore resplendet, nisi quia ceruleo colore aquarum quae praeterierunt, rubicundo flammarum quae uenturæ sunt nobis testimonium perhibet.
>
> [For not without reason does it gleam blue and red at the same time, since by the colour blue it bears witness to us of the waters that have gone past and by the colour red, of the flames that are to come.][40]

This sets out clearly that the colours blue and red were understood to signify Noah's Flood and the Last Judgement respectively, and so their usage can be seen to symbolize the Second Coming of Christ. Therefore, it is likely that the choice of blue and red for the illustration to the psalm in the Harley Psalter also conveyed a similar message. The dominance of blue for the water and sky, and the use of red for the landscape, alongside some of the outlines of the people and creatures depicted on the page, could have been a deliberate addition to the illustration to visually depict the impending Last Judgement through the use of colour,[41] adding yet another layer of symbolic significance to the miniature.

The Harley Psalter's depiction of Psalm 103 (104) calls upon a complex set of iconographic and exegetical references, building on those already present in the manuscript's source model, the Utrecht Psalter. The Leviathan, in the guise of a *ketos*, is situated in a sea where 'illic reptilia quorum non est numerus animalia pusilla cum magnis' (there are creeping things without number: Creatures little and great)[42] and can be viewed, like the sea monster of the Jonah story, as a representation of hell, while simultaneously being a creature of hell. The sea, filled with creeping creatures, can likewise be seen as hell and the dwelling place of the Devil, as referred to by Bede in his commentary *On Ezra and Nehemiah*.[43]

---

39 Gregory the Great, *Homiliae in Hiezechihelem*, ed. by Adriaen, 1. 8. 29; Isidore, *De natura rerum*, ed. by Migne, 31. 2. 19–22, pp. 1003–04.

40 Bede, *Libri quatuor in principium Genesis*, ed. by Jones, 2. 9. 13–15, p. 410; Bede, *On Genesis*, trans. by Kendall, p. 208.

41 It has been argued that the years surrounding the millennium witnessed an increase in anxiety due to the perceived impending apocalypse. See Prideaux-Collins, '"Satan's Bonds Are Extremely Loose"', p. 290.

42 Psalm 103 (104). 25.

43 Bede, *In Ezram et Neemiam*, ed. by Hurst, 1. 3. 7, p. 275; Bede, *On Ezra and Nehemiah*, trans. by DeGregorio, p. 56.

Christ holding a book, appearing in the sky above, surrounded by angels, and standing on personifications of wind recalls Christ in Majesty scenes, such as that of the eighth-century Codex Amiatinus, where an enthroned Christ holding a book and making a blessing sign is flanked by two angels. Although not shown seated, Christ in the Harley Psalter references early Christian and early medieval depictions of the *maiestas*, and it is possible that this depiction of the Saviour contributes to the eschatological imagery of the scene as a whole.

While these elements combined — the Leviathan as a representation of hell and Satan and the allusion to Christ in Majesty — clearly reference the overcoming of the Devil by Christ allowing humanity to experience life everlasting once again, the addition of colour and the symbolic implication of the colours blue and red as signifiers of the first and second covenants and the impending Last Judgement add yet another layer to this incredibly rich iconographic programme. Demonstrating, therefore, that the Anglo-Saxon copiers of the Utrecht Psalter deeply considered the symbolism of their source model and sought to enrich it further in their copy.

Conclusion

While there is no direct evidence for the depiction of Jonah and the *ketos* in Anglo-Saxon England, this small body of evidence suggests that those responsible for their designs engaged with and visually depicted the Old Testament prophet and his imprisoner. From what survives it is possible to infer that the Anglo-Saxons were aware of and had access to early Christian depictions of Jonah and the *ketos*, and they were eager to fit in with the well-established iconographic tradition surrounding the pair. However, they also desired to expand and improve on the visual symbolism of the Jonah cycle to reflect their deep understanding of the exegesis surrounding Jonah and his prefiguration of Christ, explicitly stated in Matthew 12. 40.

The almost exclusive use of blue and red in the Harley Psalter's illustration to Psalm 103 (104), for example, was a deliberate and well-considered choice to highlight associations between these two colours and Noah's Flood and the Last Judgement. When considered alongside the contents of the illustration the use of colour further emphasizes the rich eschatological programme of the scene. Although not a true depiction of the Jonah narrative, the evocation of the *ketos* for the Leviathan, in a turbulent body of water near ships with billowing sails, with Christ in Majesty situated above, was surely intended to call to mind Matthew 12. 40 and thus the Harrowing of Hell. The transformation of the *ketos* into a more serpentine creature in the Antwerp Sedulius was likely intended to visually connect this creature with the Devil in his guise as a serpent, again drawing associations between the Jonah narrative and the Harrowing of Hell. It seems clear, therefore, that in Anglo-Saxon England the *ketos* was understood to be not only a creature of the deep, but a representation of the Devil and the embodiment of hell.

Furthermore, the use (or absence) of water plays a substantial role in the symbolic significance in all three scenes. The abundance of turbulent water surrounding the *ketos* in the Antwerp Sedulius was a conscious decision by those responsible for its design to associate the dwelling place of the sea monster in the Jonah narrative — the abyss — with the abode of the Devil: hell. Likewise, the absence of the sea as the prophet is thrown from the boat in the same manuscript emphasizes the tranquillity of faith attained through Christ's sacrificial death and Second Coming, and through referencing the heavenly Jerusalem and its lack of sea as alluded to in Revelation 21. 1. The manipulation of water, therefore, in the Antwerp Sedulius clearly functions to visually represent complex exegetical understandings of the significance of water, its inhabitants, and the role they will play in the Last Judgement: 'et cum finierint testimonium suum bestia quae ascendit de abysso faciet adversus illos bellum et vincet eos et occidet illos' (And when they shall have finished their testimony, the beast, that ascendeth out of the abyss, shall make war against them, and shall overcome them, and kill them).[44] Similar associations are invoked in folio 51^v of the Harley Psalter, with the colours blue and red being used to signify the Flood and End of Days, further highlighting the links between the *ketos*, hell, and the salvific nature of Christ's death, descent, and resurrection. Clearly the ways in which the sea monster was articulated in Anglo-Saxon art indicates that it denoted much more than simply a creature that dwelt in the ocean: in fact, deriving much of its inspiration from early Christian visualizations which associated the aquatic beast of the Jonah narrative with the *ketos* of the Andromeda story, this creature was more significant for its symbolic implications in early medieval England — as the embodiment and inhabitant of hell.

Works Cited

Manuscripts

Antwerp, Museum Plantin-Moretus, MS M. 17. 4
London, British Library, MS Cotton Claudius B.iv
London, British Library, MS Harley 603
London, British Library, MS Harley 647
London, British Library, MS Harley 2506
Stuttgart, Württembergische Landesbibliothek, Bibl. fol. 23
Utrecht, Universiteitsbibliotheek, MS Bibl. Rhenotraiectinae I Nr 32

---

44 Revelation 11. 7.

Primary Sources

Augustine, *De civitate Dei*, ed. by Bernard Dombart and Alphonso Kalb, CCSL, 48 (Turnhout: Brepols, 1955)

——, *Enarrationes in Psalmos*, ed. by D. E Dekkers and John Fraipont, CCSL, 40 (Turnhout: Brepols, 1956)

Bede, *De octo quaestionibus*, in *Patrologia cursus completus, series Latina*, ed. by Jacques-Paul Migne, 221 vols (Paris: Migne, 1841–65), XCIII (1862)

——, *In Ezram et Neemiam*, ed. by David Hurst, CCSL, 119A (Turnhout: Brepols, 1969)

——, *Libri quatuor in principium Genesis usque ad nativitatem Isaac et eiectionem Ismahelis adnotationum*, ed. by Charles W. Jones, vol. I of *Bedae Venerabilis Opera: Opera exegetica*, CCSL, 118A (Turnhout: Brepols, 1967)

——, *On Eight Questions*, in *Bede: A Biblical Miscellany*, trans. by W. Trent Foley and Arthur G. Holder, Translated Texts for Historians, 28 (Liverpool: Liverpool University Press, 1999), pp. 149–66

——, *Bede: On Ezra and Nehemiah*, trans. by Scott DeGregorio, Translated Texts for Historians, 47 (Liverpool: Liverpool University Press, 2006)

——, *Bede: On Genesis*, trans. by Calvin B. Kendall, Translated Texts for Historians, 48 (Liverpool: Liverpool University Press, 2008)

Gregory the Great, *Homiliae in Hiezechihelem*, ed. by Markus Adriaen, CCSL, 142 (Turnhout: Brepols, 1971)

Holy Bible Douay-Rheims Version, with Challoner Revisions 1749–52 (Baltimore, MD: John Murphy Company, 1899)

Isidore, *De natura rerum*, in *Patrologia cursus completus, series Latina*, ed. by Jacques-Paul Migne, 221 vols (Paris: Migne, 1841–65), LXXXIII

Jerome, *Ancient Commentaries in English: Commentary on Jonah by Saint Jerome*, ed. by John Litteral, trans. by Robin MacGregor (Ashland, KY: Litteral's Christian Library Publications, 2014)

——, *Commentariorum in Jonam Prophetam*, ed. by Markus Adriaen, CCSL, 76 (Turnhout: Brepols, 1969)

Plato, *Cratylus*, in *Platonis Opera*, vol. I, ed. by E. A. Duke, W. F. Hickens, W. S. M. Nicoll, D. B. Robinson, and J. C. G. Strachan (Oxford: Oxford University Press, 1995), pp. 188–275

——, *Gorgias*, in *Platonis Opera*, vol. III, ed. by John Burnet (Oxford: Oxford University Press, 1922), pp. 447–527

Pliny, *Naturalis Historia*, ed. and trans. by H. Rackham, vol. II, Loeb Classical Library, 352 (Cambridge, MA: Harvard University Press, 1942)

Secondary Works

Benson, Gertrude R., 'New Light on the Origins of the Utrecht Psalter', *Art Bulletin*, 13 (1931), 13–79

Gameson, Richard, 'The Anglo-Saxon Artists of the Harley (603) Psalter', *Journal of the British Archaeological Association*, 143 (1990), 29–48

Gneuss, Helmut, and Michael Lapidge, *Anglo-Saxon Manuscripts: A Bibliographical Handlist of Manuscripts and Manuscript Fragments Written or Owned in England up to 1100*, Toronto Anglo-Saxon Series, 15 (Toronto: University of Toronto Press, 2014)

Hawkes, Jane, 'Anglo-Saxon *Romanitas*: The Transmission and Use of Early Christian Art in Anglo-Saxon England' in *Freedom of Movement in the Middle Ages: Proceedings of the 2003 Harlaxton Symposium*, ed. by Peregrine Horden (Donnington: Shaun Tyas, 2007), pp. 19–36

Horst, K. van der, William Noel, and Wilhelmina C. M. Wüstefeld, *The Utrecht Psalter in Medieval Art: Picturing the Psalms of David* (Utrecht: Hes & De Graaf, 1996)

Lapidge, Michael, *The Anglo-Saxon Library* (Oxford: Oxford University Press, 2006)

Levison, Wilhelm, *England and the Continent in the Eighth Century* (Oxford: Clarendon Press, 1946)

Liddell, Henry G., Robert Scott, and Henry Drisler, *A Greek-English Lexicon* (Oxford: Harper and Brothers, 1894)

Mostert, Marco, 'Relations between Fleury and England', in *England and the Continent in the Tenth Century: Studies in Honour of Wilhelm Levison (1876–1947)*, ed. by David Rollason, Conrad Leyser, and Hannah Williams (Turnhout: Brepols, 2010), pp. 185–236

Narkiss, Bezalel, 'The Sign of Jonah', *Gesta*, 18 (1979), 63–76

O'Loughlin, Thomas, 'Living in the Ocean', in *Studies in the Cult of St Columba*, ed. by Cormac Bourke (Dublin: Four Courts Press, 1997), pp. 11–23

Panofsky, Dora, 'The Textual Basis of the Utrecht Psalter Illustrations', *Art Bulletin*, 25 (1943), 50–61

Papadopoulos, John K., and D. Ruscillo, 'A *Ketos* in Early Athens: An Archaeology of Whales and Sea Monsters in the Greek World', *American Journal of Archaeology*, 106 (2002), 187–227

Pickles, Thomas, 'Anglo-Saxon Monasteries as Sacred Places: Topography, Exegesis and Vocation' in *Sacred Text, Sacred Space: Architectural, Spiritual and Literary Convergences in England and Wales*, ed. by Joseph Sterrett and Peter Thomas (Leiden: Brill, 2011), pp. 35–56

Prideaux-Collins, William, '"Satan's Bonds Are Extremely Loose": Apocalyptic Expectation in Anglo-Saxon England during the Millennial Era', in *The Apocalyptic Year 1000: Religious Expectation and Social Change, 950–1050*, ed. by Richard Landes, Andrew Gow, and David C. Van Meter (New York: Oxford University Press, 2003), pp. 289–310

Ranneft, D. M., H. Eaker, and R. W. Davis, 'A Guide to the Pronunciation and Meaning of Cetacean Taxonomic Names', *Aquatic Mammals*, 27 (2001), 183–95

Schultze, Victor, *Archäologische Studien über altchristliche Monumente* (Vienna: W. Braumüller, 1880)

Springer, Carl P. E., *The Gospel as Epic in Late Antiquity: The Paschale Carmen of Sedulius*, vol. II (Leiden: Brill, 1988)

Swanson, Roy A., 'Easter Poem', *Classical Journal*, 52 (1957), 289–97

MEG BOULTON

Pearls before Paradise: Considering the Material Associations of Heavenly Waters, Precious Stones, and Liminality in the Art of the Medieval West

Water, as recognized across the papers that make up this volume, is a substance that is ubiquitous, indispensable, and life-giving, possessed of both sacred and secular understandings in the medieval world and beyond. Within the medieval period, water was an integral part of both physical and metaphysical experience, deeply linked to societal systems and spiritual significances. Although fluid and mutable in both its nature and associations, it was also a constant presence on the medieval ecclesiastical landscape, whether understood to be a symbolic fluid marking entry into the Church; a substance that had been transformed and blessed by saintly relics, thus becoming a portable sacred essence; an image of Paradisal waters that evoked spiritual significances, histories, or geographies for its viewers; the pragmatic medium for the thriving waterscapes of trade and transport inscribing the landscape of early medieval England and the Continent; the potential, drama, and threat of the wide open ocean beyond the mapped edges of the world; or, indeed, performing some or all of these fluid and fluidic identities simultaneously. In all such instances, through its various depictions and identities, water was an integral part of the natural, religious, and cultural understandings of the early medieval world.

This contribution to the discussion of medieval water will examine its symbolic significance as a motif in the repertoire of Christian art. Specifically, it considers the use and representation of water as a visual metaphor, particularly where it has been employed as an image for something understood to be made of water but that also represents something 'other',[1] and/or to conceptually contain the transformative properties of water as understood in the Christian tradition. This consideration of symbolic water and its material objects moves from a

---

[1] Isidore of Seville, *The Etymologies*, trans. by Barney and others.

Meg Boulton • (meg.boulton@york.ac.uk) is a Teaching Fellow in Art History at the Edinburgh College of Art, University of Edinburgh.

Meanings of Water in Early Medieval England, ed. by Carolyn Twomey and Daniel Anlezark, Studies in the Early Middle Ages, 47 (Turnhout: Brepols, 2021), pp. 145–165
BREPOLS PUBLISHERS 10.1484/M.SEM-EB.5.122145

discussion of selected examples of early Christian art — from the monumental mosaic programmes of the early Christian world, to subsequent and related visual expressions in the art of early medieval England which evoke similar symbolic resonances or experiences for their viewers. These representations and objects will be examined in order to consider the manner in which water was depicted (both literally and figuratively) in a medieval Christian context. Further, this chapter suggests that these aquatic and quasi-aquatic material forms and iconographies were systematically employed in order to express a symbolic understanding of sacred space, place, and materialities within the Church as an institution — along with the associated metamorphic ideas of transformation offered by the sacred waters flowing through the Church, its history, and its art.

This implicit transformation of and through water is expressed through the visual portrayals of the four rivers of Paradise in early Christian mosaics and the more allusive uses of water seen on later medieval monuments in England that were the inheritors of this artistic tradition. Alongside the more expected, naturalistic depictions of aqueous bodies that flow across and through these art works, and the literal watery landscapes which may have framed and bounded them, images of things and objects such as pearls will be discussed here as articulations of heavenly water, following the reading of Isidore of Seville given in his *Etymologies*.[2] In all of these articulations and presences, it is important to note that water is a potent visual signifier that at once evokes the sacred through its liminal presence, and recalls and actualizes the historic associations and narratives it presents, as well as being the active agent of transformation upon entering the faith through the sacrament of baptism for catechumens.[3]

It is the celestial and sacred connections between the various (im) material forms of water that form the focus of this discussion. Water is emblazoned across the monumental iconographic schemes of the early Church, presenting sacred narratives that blur and cross time and space, forming a metanarrative of salvation that connects architectural church to institutional Church, and thus also connects place to space. Water, in its various riverine guises, quite literally frames and emphasizes the edges of several mosaic programmes adorning the ecclesiastical spaces of the early Church, serving to underscore the rituals and places of the church as a potent

---

2   Isidore of Seville, *The Etymologies*, trans. by Barney and others. For further discussion of the possible symbolic significance of jewels in the borders of early Christian art, see also Boulton, 'The Conceptualisation of Sacred Space in Anglo-Saxon Northumbria'; Boulton, '"The End of the World as We Know It"'; Boulton, '(Re-)Viewing "Iuxta Morem Romanorum"'; Boulton, 'Bejewelling Jerusalem'; and Boulton, 'Art History in the Dark Ages'.

3   For the most recent discussion of baptism in the early medieval world, see Twomey, 'Living Water, Living Stone' and 'Rivers and Rituals: Baptism in the Early English Landscape' in this volume.

Figure 6.1. Detail of the apse mosaic at San Vitale, Ravenna, consecrated 547. Photo by Heidi Stoner.

iconographic device. The waters that flow across the edges of these images perform a twofold function: First, they emphasize the sacred thresholds at play, both between the architectonic spaces of the church (choir and apse, altar and dome), but also emphasize the metaphysical connections understood to resonate between the apse, the ecclesiastical dome, and the heavens beyond — serving to bring paradisal waters flooding into the space of the Church (Figure 6.1).[4] Second, these liminal depictions of water, understood to simultaneously evoke the waters of Paradise and those of baptism throughout the institution of the Church, serve to construct and emphasize a sense of universality within the Church, as has recently been argued more widely for the mosaic programmes employed across Rome by Eric Thunø.[5]

---

4 For further discussion of water in mosaics, see Maguire, *Earth and Ocean*.
5 Thunø, *The Apse Mosaic in Early Medieval Rome*. See also Doig, *Liturgy and Architecture* and Brenk, *The Apse, the Image and the Icon*.

Ideas of viewer, viewing, and perception are an integral part of the study of imagery. In the specific context of the medieval Christian world, a central aspect of visual culture and expression is the pervasive and permeable relationship between the earthly world and the heavenly kingdom. Arguably, one of the most interesting aspects of studying these visual forms is considering how these images, composed of recognizable earthly forms and motifs, call into being the abstract, imagined, and envisioned form, space, and presence of heaven — something which was understood to exist far beyond earthly experience. These two disparate sites of physical and metaphysical experience and imagining, the earthly and the heavenly, are nonetheless linked by iconographical constants. These were (re)imagined, (re)interpreted, and (re)presented across the earthly Christian landscape and the material/ities of the artworks which inscribe it — one of which is water, which flows between and within these spaces, linking them through its visual constant and its various and simultaneous symbolic significances.

On the Edge: Water, Pearls, and Material Meaning in Mosaics

As in the Church more widely, water is a constant, indeed ubiquitous presence in the mosaics and images of the early Christian world, most often depicting the historic waters of the Flood or the River Jordan, or the imagined, envisioned waters of Paradise. Such watery imagery is perhaps most loaded and significant when seen in early Christian baptisteries such as those surviving in Rome, Ravenna, and Naples, where, in conjunction with the font, such riverine depictions are understood to make viscerally present the River Jordan — the space of Christ's own baptism — within these spaces (sometimes literally through the embodied presence of the river god as in the Ravennate baptisteries). In these instances, such references are often accompanied by iconographic gestures to the atemporal salvific water of the sacrament, sometimes conjured through the visual motif of the spiritually slaked hart,[6] sometimes through the Tree of Life motif, as at the Lateran in Rome, where the fifth-century Chapel of Santa Rufina preserves an elaborate mosaic, and always presented through the physical presence of the baptismal waters themselves. These baptismal programmes present defined and discrete iconographies of sacrament-based salvation, which exist within wider ecclesiastical iconographies found outside of baptistery contexts. These often centre on powerful eschatological schema showing Christ in Majesty at the end of time — such as those at San Vitale, San Lorenzo, San Marco, or Sancti Cosmas and Damian,[7] which juxtapose

---

6   Psalm 41 (42). 2.
7   Several of which also have paradisal waters making up part of their iconographical programmes.

the aspatial, liminal, cosmological Majestus with the geographically located waters of Paradise, mapped onto the known cartographies of the world, and frequently depicted below Christ in these eschatological programmes. While several of these mosaics contain portrayals of the waters of Paradise as part of their visual programme, emphasizing how salvation through Christ will return humanity to a paradisal state at the end of time, in this discussion I focus on a less literal depiction of paradisal waters than the flowing rivers seen under the feet of Christ — namely, the pearls which so often surround him.

The link between pearls and water emphasized here is neither astonishing nor nonsensical, according to Isidore of Seville in his *Etymologies*: 'People say it is called a pearl (*margarita*) because this kind of stone is found in shell-fish from the sea [...]. It is made from Celestial dew, which shellfish absorb at a certain season of the year'.[8] For Isidore, then, pearls are *literally* made from the waters of heaven — making them a more than fitting subject for consideration here, as a refracted material from heaven, formed through its waters.[9] The association of pearls and water was clear in the medieval world, as shown in the image of Catching of Pearls in the Bern Physiologus.[10] Moreover, the connection between pearls and heavenly waters was both pervasive and long-lasting in its usage, as these understandings are still clearly demonstrated in the later bestiary tradition from the twelfth century, as shown in the image of pearls and oyster shells in Paris, Bibliothèque nationale de France, MS lat. 14429, folio 117^v (Figure 6.2). In this depiction, the oyster shells are open, receiving the waters and light of heaven which are shown as wavy lines, reaching into the shell to form pearls. This later bestiary tradition further reinforces the identity of the pearl as a material form/ed of heaven, through its depiction of celestial bodies as well as celestial waters.

The celestial and moreover *heavenly* connection with the material form of pearls is also given in Scripture, where pearls are described in a passage from Revelation. The Bible presents an eschatological vision of the heavenly city of Jerusalem as it would be seen at the end of time: here, the heavenly city is described as 'foursquare [... with] a wall great and high, having twelve gates [...]' and the foundations of the wall of the city were adorned with all manner of precious stones', including pearls. These gems are 'Jasper, Sapphire, Chalcedony, Emeralds, Sardonyx, Sardius, Chrysolite, Beryl, Topaz, Chrysoprasus, Jacinth,

---

8   Isidore of Seville, *The Etymologies*, trans. by Barney and others, XVI. x. 1, p. 324.
9   This connection, noted by Isidore, goes back to Pliny the Elder (*Natural History*, IX. 54–59), who states: 'Pearls are the offspring of shells similar to oysters; at the breeding season, the shells gape open and become filled with a dew from the sky that makes them pregnant. The quality of the pearl depends on the quality of the dew received, and on the state of the sky: a clear sky produces a clear pearl, but a cloudy sky produces a pale pearl'.
10 Bern, Burgerbibliothek, Codex Bongarsianus 318, is a ninth-century illuminated copy of the Latin translation of the Physiologus. It was probably produced at Reims about 825–50. It is believed to be a copy of a fifth-century manuscript. See Calkins, *Illuminated Books of the Middle Ages*, p. 21.

Figure 6.2. Oysters in Bestiary, Paris, Bibliothèque nationale de France, lat. 14429, fol. 117ᵛ, twelfth century. Image by permission of the Bibliothèque nationale de France.

Amethyst, Pearl, Gold', a list that indicates the sheer number of precious stones associated with the space of heaven.[11] These gems present a bewildering and glittering space to the dazzled imagination of those encountering heaven textually (and so also visually), through either the textual description or the visions the passage conjures. But, more importantly, in presenting this profusion of precious stones — the most valuable 'things' the earth could offer — these stones provide some indication of the significance these decorative materials held within the wider Church when it came to constructing a visual identity portraying the unknowable space of the heavenly city. The precious stones set into the envisioned architecture of heaven provide an appropriate iconography by which to portray the heavenly city, which is later reflected in the reimagining of the highly ornate space/s of the earthly church as presenting or foreshadowing heaven. Indeed, within the passage describing the walls of heaven, pearls are given specific weight and emphasis: 'and the twelve gates are twelve pearls, one to each: and every several gate was one of several pearl'.[12] Here the liminality associated with the object narrative of the pearl as discussed by Isidore, formed on earth from the fallen waters of the heavens above, is also emphasized: pearls are often depicted hanging in the threshold spaces of the gates in the celestial

---

11 Revelation 21. 16–21. These stones are analysed further by Bede in an Insular context, in his treatise on the Apocalypse, Bede, *Commentary on Revelation*, trans. by Wallis, p. 278, a discussion which includes pearls, emphasizing their associations with light and enlightenment. Interestingly, these do not appear alongside the list of heavenly gems in the eleventh-century *Old English Lapidary*, in Evans and Serjeantson, *English Medieval Lapidaries*. This connection, of pearls and heaven, is seen metaphorically in a New Testament parable (Matthew 13. 45–46), wherein Jesus compared the Kingdom of Heaven to a 'pearl of great price', a connection also emphasized by Bede.

12 Revelation 21. 21.

wall in Christian art, each gate one of several pearls, as in the textual description of heaven itself. Just as pearls were understood to be formed from a liminal, fluid substance that had traversed the distance between the heavens and the earth, so too were pearls used to articulate physical spatial transitions by materially bridging these spaces. These symbolic, shining, gemmed iconographies as described in the Book of Revelation and depicted in mosaics make present both historical time and eschatological space within the early Church, through a series of allusive connections between place, space, memory, and medieval exegetical viewing practices.

Like the watery forms of flood and paradisal rivers that flow through these spaces , and the pearls that hang in the gates of heaven, the peripheral iconography of bejewelled edges is seen throughout these artworks, forming a series of precious frames and mutable, permeable thresholds that coexist with the more elaborate narrative features of the iconographies they accompany. For the viewer, both water and pearls (and by extension, the wider programme of precious stones that frequently accompany these pearl-strewn borders) present shifting, shining, mutable surfaces, through which greater ecclesiastical mysteries might be revealed. This is accomplished both through individual iconographic elements as well as the meditative, contemplative act of viewing the rhythmic repetition of chromatic, stylized stones and flowing water dividing one scene from the next within mosaic programmes. With both water and pearls, their material presence adds to the rich symbolic allusions of these visual schemes, forming transitional spaces that may well be suggested to echo the spiritual growth and transformation of the Christian soul as it comes closer to the ultimate threshold of judgement. These borders do not function merely as decoration, and are not a passive part of these iconographic schemes, but rather serve to make the sacred present on earth, foreshadowing and actualizing the space of the heavenly city within the church via their symbolic significances and material identities.[13]

Over and over in these artworks, pearls frame more central iconographic narratives as quasi-aquatic material presences, like the flowing rivers at the bottom of scenes. Most often the extant scholarship on mosaics is conducted through an examination of their central spaces and iconographies, rather than the marginal, flowing bodies of water that encircle the scenes and the fixed, architectural jewelled borders that surround them. However, these often overlooked framing elements exist as a set of abbreviated iconographies in their own right. These iconographies are emphasized by the wider architectural symbolism of the monumental triumphal arches set before apsidal spaces, intended to reference traditional imperial spaces of victory, and thus also emphasizing an implied entrance into the heavenly city when used in an ecclesiastical context. What is important here is not simply the appropriation

---

13   This trait is embodied most fully in the fourteenth-century poetic work *Pearl*, ed. by Anderson.

of imperial architectural forms within Christian contexts, but also the eschatological spaces of victory and judgement referred to by the triumphal arch recast in an ecclesiastical setting, and the implied entrance into the heavenly city created by these structures. The implied entrance to Jerusalem is often demarcated, recognized, and understood through the physical presence of gems, as I have discussed elsewhere, and the liminal boundary of depicted water, both of which are bound up with the eschatology of the Church, and the ultimate entrance, or lack thereof, into the institution, and so also into heaven.

As well as paradisal waters inhabiting the liminal edges of the ecclesiastical narratives shown in the mosaics, there are other, notable instances of water that are perhaps less immediately recognizable than the four rivers, especially the water-formed pearl. The forms and motifs of the carved and painted borders occupying the liminal spaces of artworks found across the medieval Christian world can be used to explore how the imagined space of the heavenly kingdom was made viscerally present for the earthbound viewer. In Ravenna, for instance, the sixth-century mosaic programme at Sant'Apollinare in Classe emphasizes the presence of the sacred in its Transfiguration scene, which plays out against a very recognizable and hyper-saturated local landscape of indigenous flora and fauna recast in a paradisal setting. The physical presence of the Divine in Sinai, now Classe, is made manifest through the presence of the *manus dei* (the Hand of God), which emerges from a bank of red and blue (apocalyptic) clouds, with a pearl frieze containing the whole (Figure 6.3). The portrait bust of Christ, the central actor of the Transfiguration in its biblical context,[14] is placed on the *crux gemmata* — the jewelled cross in the heavens,[15] that is itself heavily studded with pearls. The use of pearls in this context makes connections for the viewer between the multivalent salvific space of the jewelled cross and the eventual eschatological entrance into the bejewelled heavenly city of Revelation, made present and possible by Christ's sacrifice on the rood — a connection that is visually emphasized by the presence of the pearls on and around the cross, and the scriptural description of pearls in heaven in Revelation 21. The programme thus refers to the space of heaven in multiple forms — not least articulated through the material presence of the pearl, scripturally present in the textual description of the heavenly Jerusalem, but also understood to be made of a material *from* heaven as given in Isidore, here given spatial referent and material presence through the pearls 'hanging' from the top of the frame of the mosaic; the

---

14 Matthew 17. 1–5, 7. For further discussion of the jewelled border motif found in this mosaic, see Boulton, 'Art History in the Dark Ages' and Boulton, 'Bejewelling Jerusalem'. For discussion of the symbolism of jewels and heaven in this period, see Janes, *God and Gold in Late Antiquity*, pp. 80–96, 105, 128–29. See also Schapiro, *The Language of Forms*. For discussion of visual matter and medieval relationships, see Kessler, *Seeing Medieval Art*. See also Bynum, *Christian Materiality*, pp. 53–123.

15 For recent discussion of the True Cross and the *crux gemmata*, as it may have influenced early medieval artwork in England, see Hawkes, 'The Body in the Box'.

Figure 6.3. Detail of the apse mosaic at Sant'Apollinare in Classe, consecrated 549. Photo by Heidi Stoner.

whole again surrounded by a bejewelled border. These pearls make present the liminal threshold of heaven — notionally at the top of the mosaic — but also, through their shared aqueous and celestial materiality, powerfully evoke the material presence of heaven within the space of the church.

Additionally, the jewelled space of the gemmed borders so often seen in parallel with pearls in these mosaics may also contain some of the material and symbolic resonances of water (and so of heaven), through an understanding of the materialities they present. Earthly precious stones are recast as reflective, enlightening gems,[16] with the implicit heavenly waters contained within the body of the pearl. These liminal, marginal aspects reflect and refract symbolic ideas greater than the sum of their material parts. There are a series of concrete textual, temporal, and material motifs within the iconographic repertoire of these mosaics that allowed for the imagined construction and actualization of heaven and the sacred for theologically informed viewers. Further, it is possible to suggest these motifs developed in an anachronic idiom outside of earthly time. This timelessness is performed through the material presence of the pearl itself — whose body is *made* of the material of heaven, bringing those materials to the earth. Thus, through an implicit material synchronicity of pearls on earth and pearls in heaven, material forms

---

16 For discussion of the chromatic understandings these gems may elicit for a viewer, although in different material contexts, see Pulliam, 'Eyes of Light' and Pulliam, 'Color'.

that were understood to exist in both spaces, but moreover, to be comprised of the waters of heaven itself, depictions of pearls in an ecclesiastical setting are understood to present a performative actualization of the atemporal locus of the kingdom of heaven. In other words, pearls, and associative depictions of pearls, bring the timeless space of heaven into the church through their material presence. Either as they are or as they are depicted, pearls embody the condensed waters of heaven, presenting some of the (meta)physical qualities of that space through their own materiality. This is made all the more potent when these are presented alongside images of the paradisal waters which flow across both time and topography, and when they are placed on liminal thresholds, such as the gate of the heavenly city (symbolically presented within the church as the 'triumphal' architectural space in front of the apse, or as a bejewelled depiction of the heavenly city itself).

Mosaics and the tesserae that make up these artworks — much like the shifting, tensile, reflective surface of water — are material carriers particularly suited to acting as signifiers of the sacred and perform this role in several ways. The glittering mural surface seems otherworldly to those viewing these vast expanses of shining colour, peopled with sacred figures and narratives, presenting other spaces, other materials, other forms, and other times to their viewers, through their shifting, glittering presence. Peter Brown in *The Cult of the Saints* discusses the potent materiality of such mosaics in creating sacred space (and evoking the presence of the sacred) within ecclesiastical architecture, describing them as bringing the light of heaven itself into the space of the church or shrine they adorned.[17]

However, like the water so often captured in their glass tesserae, these works also present an artistic reality that is comprised of both presence and absent presence, the substance and the insubstantial, as understood through their complex minutiae of coloured, gilded glass. These fractured, vitreous forms recall the scriptural passage from Revelation 21. 18 describing heaven as a shining place, 'the city itself pure gold, like to clear glass', through the shifting play of light on their surface/s. Likewise, the jewel-like mural surface of the mosaics is understood to present the sacred space of heaven, and so also the presence of the divine for their viewer/s (evinced, in part, through the scriptural narratives and historicizing scenes displayed in them as well as the myriad associations of God and light). That said, it is important to recognize that the actualized vision of the space of heaven that is implied (and fleetingly experienced) through these visual forms is endemic to the material properties of mosaics, but is also seen more specifically, indeed, one might say more literally, in the systematic employment of the motif of jewelled borders and watery thresholds across these artworks.

The early medieval understanding of highly ornate churches symbolizing the heavenly Jerusalem, and ecclesiastical artworks presenting rich and nuanced

---

17 Brown, *The Cult of the Saints*, p. 4.

iconographies foreshadowing the heavenly future or illuminating the biblical past is exceptionally widespread. Such images are complex, and the viewing of such images was understood to be an act which both engaged and altered the body — a transformative process that involved an exchange between eyes, mind, and soul evoking a metaphysical shift, completed by a performative act on the behalf of the viewer. Such viewing experiences are not limited to the Continental mosaics that have been the focus of this discussion thus far, and can be found across the Christian world, including (but not limited to) the monumental carved stone forms occupying the ecclesiastical landscape of early medieval England. As well as the consistent employment of bejewelled borders across the mosaic programmes of the Continental Church there are other, decidedly non-mosaic examples found throughout the art of the Church, including that of England, that involve analogous understandings of sacred materiality and require similar acts of performative viewing to those monumental mosaics.

Water, Pearls, and Liminality in Insular Sculpture

The glittering, shining jewelled borders which perform such a pivotal function in creating spatial significance and symbolism in Continental churches translate indirectly into early medieval England. There is no extant material evidence for the use of mosaics in the ecclesiastical architecture of the early English Church; however, there is plenty of evidence in the contemporary textual accounts and the material record for an interest in, and a marked preference for, jewel-toned and chromatic adornment. The popular practice of decorating the structures and objects of the early medieval Church with precious stones and shining metal created a sophisticated interrelation of surface, object, and subject which, as with the Continental mosaics, also powerfully acts on our contemporary gaze, as well as the period eye. Notably, as with the Continental mosaics, the best preserved examples of such significant performative spaces and iconographies in early medieval England occur on borders, threshold spaces, or on the edges of things, maintaining familiar and potent liminal associations of crossing from one place to another, moving between the secular and the sacred, and of a shifting recognition of forms and features, both actual and actualized, as seen within the wider Church.

The bejewelling of Jerusalem was a visual mainstay in the ecclesiastical structures of early Christian England, as it was in the Mediterranean Church. Rather than mosaic, medieval churchmen in the Insular Church used a multivalent iconographic vocabulary of carved and painted surfaces and objects to demonstrate a universal understanding of the symbolism of the bejewelled Jerusalem. Churches and their accompanying ecclesiastical ornament filled the Christian landscape of the conversion and post-conversion societies of England, and so too did water. Water was a particularly vital force shaping the Insular landscape, with its riverways and sea routes, as well as being a symbolic force that retained and inscribed potent significance for the Christian Church throughout the conversion period. However, despite recent scholarly attention paid to the manner in which physical water was employed in Insular iconographies,[18] it remains true that water itself is a rare aspect of early Insular iconographies — to the point of being almost invisible. Of course, the lack of presence of natural features on these carved stone objects is by no means limited to the depiction of water: across this corpora of carved objects, it is notable that there is a lack of trees, rocks, mountains, etc., even in the iconographic scenes which most usually contain them. Limited exceptions can be found to this rule, such as the clouds of the Ascension scene on the fragmented base of the Rothbury Cross, and the living wood of the cross of Christ abundantly recalled both through Crucifixion scenes, the monumental presence of the stone crosses themselves, and the seemingly living and growing vine scrolls on their faces. The reasons for this lack of naturalism might well be twofold: some details might have been painted on, as is frequently suggested for such sculptures,[19] and, more importantly, the natural environment existing alongside and read against and around the crosses might have negated the need for their particular depiction on the stone.[20]

Indeed, it has been suggested that there is a specific connection between these carved stone sculptures and water as a medium for the Christianization of early medieval England — and further, that these standing stone crosses should be considered in the various discussions of baptismal places in conversion-period England.[21] Like the architectural spaces of churches built in wood and stone on the Insular landscapes, crosses (also constructed in wood and stone) are understood to have marked significant liturgical and sacred places in the landscape. As most recently noted by Carolyn Twomey in her discussion of baptism, water, and stone in early medieval England, James

---

18 Pulliam, 'Blood, Water and Stone'. See also Ó Carragáin and Ó Carragáin, 'Singing in the Rain on Hinba?'.
19 See Lang, 'Survival and Revival in Insular Art' and Rodwell and others, 'The Litchfield Angel'.
20 I would like to thank my colleague Nicholas G. Baker for discussion of crosses inhabiting the landscape in this manner, following research presented in an unpublished conference paper 'Climbing a Stairway to Heaven', where he drew attention to Cassiodorus, *Explanation of the Psalms*, particularly Psalm 21 (22) (trans. by Walsh, pp. 224–25): 'They have dug my hands and feet. Before coming to the beginning of the passion itself, we must examine why He chose for himself such a death, whereas He said: I have power to lay down my life, and I have power to take it up again. A first reason is that the setting of the cross is such that its top points to the heavens yet its base does not quit the earth. When implanted it touches the depths of the realm below, and its breadth, with arms so to say extended, stretches towards the regions of the whole world; when flat it marks out the four points of the earth'. This discussion reveals an extant reading of the cross, as embedded in the material landscape of the earth and connected to the material of heaven, that functioned very much as with the substance and symbolism of the pearl, to join heaven and earth through a material substance.
21 Lang, 'The Apostles in Anglo-Saxon Sculpture', pp. 280–82; see also Twomey, 'Living Water, Living Stone'.

Lang and Éamonn Ó Carragáin have interpreted stone crosses specifically as places of baptism, through their formal associations with the pillars of the Temple, their iconographic narratives, and their physical proximity to water,[22] although this reading is not universally accepted as the sole explanation (or role) of these monuments in wider scholarly discussion, as they are usually accepted to have had, in all probability, multivalent functions and associations. Indeed, these crosses are variously interpreted as having been constructed for a variety of reasons, including demarcating sacred boundaries, commemorating individuals and holy sites, but perhaps more significantly, to have the primary purpose of standing physically in the landscape, planted in the soil as atemporal beacons for prayer, preaching, and perhaps even the celebration of the sacraments,[23] as well as an allusion to the victorious narratives of the triumph of the Church and the salvation and resurrection to come. Furthering the possible connection between the stone crosses and water, Lang suggested that the common geographic placement of stone crosses near rivers and other water sources may have indicated a baptismal function, and an explicit apostolic association in the case of the eighth- and ninth-century sculptural group he identified from Northern Yorkshire and Eastern Mercia, including the Easby Cross (Figures 6.4a and b).[24] Certainly this is not an entirely implausible suggestion, as such striking monuments would make natural gathering points on the landscape, and their proximity to water would make a baptismal function a practical addition to their possible range of functions. As recently noted by Twomey, such potent signifiers in the landscape may well have recalled the crosses standing beside the River Jordan (known to an Insular audience through their description in Adomnán's *De locis sanctis*, later also written on by Bede) which marked the location of the baptism of Christ, lending further credence to Lang's suggestion.[25]

In a Continental context, this association may find further support in the slightly later treatment of water and stone found in the ninth-century apse mosaic in the private oratory of Germigny-des-Prés — built by Theodulf,

---

22 Lang, 'The Apostles in Anglo-Saxon Sculpture', pp. 280–82, at p. 280. Lang also discussed possible Roman and Ravennate exemplars for these stone sculptures, particularly at pp. 279–82.
23 It has been suggested that standing crosses 'could have a variety of functions: as a devotional focus inside a church or as part of a line of churches; as a grave-marker; as a liturgical station; as outlying preaching-point; as marker on the boundary of precinct, parish or sanctuary zone'; see Blair, 'Churches in the Early English Landscape', p. 11; see also Bailey, 'Crosses, Stone', pp. 132–33.
24 Lang, 'The Apostles in Anglo-Saxon Sculpture', pp. 280–82. The slight issue with this argument is that the monuments in the group addressed by Lang all originate in a monastic setting, which renders such a deliberate baptismal connection uncertain.
25 Twomey, 'Living Water, Living Stone', pp. 91–97; Lang, 'The Apostles in Anglo-Saxon Sculpture', pp. 280–82. See also Ó Carragáin, 'The Ruthwell Cross and Irish High Crosses', pp. 120–22, and Ó Carragáin, *Ritual and the Rood*, pp. 54–56 and 296–67; Adomnán, *De Locis Sanctis*, ed. by Meehan, II. 16.

Figure 6.4a. The Easby Cross (N. Yorks), Victoria & Albert Museum, c. 800. Photo by author, with permission of the Victoria & Albert Museum.

the Visigothic advisor to Charlemagne.[26] This is a unique survival of mosaic art — initially made and assembled in 805.[27] The scheme is an interesting one, centring on the Ark of the Covenant and its golden angels in the centre of

---

26 For background into the Carolingian attitudes to images, iconography, and iconoclasm, see Noble, *Images, Iconoclasm, and the Carolingians*. For discussion of Theodulf and Germigny-des-Prés, see McClendon, *The Origins of Medieval Architecture*, pp. 128–36. See also Meyvaert, *The Art of Words*, ch. VIII, which presents Meyvaert and Freeman, 'The Meaning of Theodulf's Apse Mosaic'. See also Grabar, 'Les Mosaïques de Germigny-des-Prés', and McClendon, *The Origins of Medieval Architecture*, pp. 129–30.

27 Freeman, 'Theodulf of Orléans: A Visigoth at Charlemagne's Court' and Freeman, 'Theodulf of Orléans and the *Libri Carolini*'. See also McClendon, *The Origins of Medieval Architecture*, pp. 128–36, and Meyvaert, *The Art of Words*, ch. VIII, pp. 11–23.

Figure 6.4b. Detail of the Easby Cross (N. Yorks), Victoria & Albert Museum, c. 800. Photo by author, with permission of the Victoria & Albert Museum.

the apse.[28] However, despite the foregrounding of the ark, there is a specific feature at Germigny-des-Prés that might be more specifically relatable to the function and perception of the stone crosses of early medieval England. Found in the linear mosaic detail running along and under the ark are parallel, waving lines of dark blue and brown bisecting the gold ground from the blue cosmos above. These shifting bands of colour have been notably discussed by Meyvaert and Freeman, as well as by several scholars, including Paul Clemen's encyclopaedic study and Jean-François Bradu's guide to the site;

---

28 See the cherubim of Exodus 25. 18–21 and 1 Kings 6. 22–28. For the direct exegetical account of the four cherubim of the Ark as given by Bede (four, not two), as identified by Meyvaert and Freeman, 'The Meaning of Theodulf's Apse Mosaic'; see Bede, *On the Temple*, trans. by Connolly, pp. 49–52. See also O'Reilly, 'Introduction'.

nonetheless it is a detail that repays close attention.²⁹ The chromatic lines are visible from the floor of the oratory, but compared to the iconographic detail of the mosaic as a whole, are read by scholars as chromatic shifts bisecting the apse mosaic — neither seemingly particularly significant within the overall scheme. However, these lines have been identified as relating to the biblical narrative of Joshua (3. 17–4. 9) which tells how the waters of the Jordan parted to allow the ark to be carried across the river to enter the promised land and how, subsequently, twelve stones were set up to monumentalize this event. If one counts the raised bumps on the brown line, there are twelve spaced across the apse, grouped in four sets of three, as has been noted by Meyvaert and Freeman. The addition of this motif, understood to represent standing stones memorializing this iconic crossing, shows the ark (and thus also the Temple and the Church), caught between moment and monument, past and future, on the threshold of the promised land in a powerful demonstration of eschatological imagery, speaking to the future salvation of the present Church, alongside the marking of the safe passage of the Old Testament object in the past. These lithic memorials, set up in the wilderness at the moment of crossing the Jordan, may well find a resonance in the visual practices of early medieval Christian England, as noted by Lang, Nicholas Howe, and Flora Spiegel, who discuss the possible connections between Jewish history and early English material practices and societal identity, of which the prominent stone crosses standing in the landscape, often in proximity to water, may be another instance.³⁰

This being so, it is possible to add a memorial function to the stone crosses of early Christian England that looks to the past as well as to the eschatological future to their suggested meanings and readings as they stand in the landscape. As outlined, the polychrome, mixed media stone sculptures of early medieval England were often inset with metal and paste glass, and so share a symbolic materiality with the aforementioned mosaics. They thus may have functioned in an analogous manner to the mosaics of the Continent — bringing the sacred into the ecclesiastical landscape they inhabit. As noted by scholars such as Jane Hawkes and Catherine Karkov, among others including myself, such sculptures engage their viewer, allowing for an active, cognitive repositioning of surface and subject, producing a multivalent viewing encounter that transcends space and plane between viewer and viewed object. They encourage those looking at such surfaces to

---

29 For further, see Clemen, *Die Romanische Monumentalmalere in den Rheinlanden*; and Bradu, 'Une mosaïque unique', 'Description de la mosaïque', and 'Histoire de la mosaïque', in Jacques and Bradu, *La Mosaïque de Germigny alliance et eucharistie*. Also of interest are Jacques's essays in the same volume: 'La Mosaïque de Germigny' and 'L'Art de la mosaïque'.

30 See Howe, *Migration and Mythmaking in Anglo-Saxon England*; Howe, *Writing the Map of Anglo-Saxon England*; and Spiegel, 'The *Tabernacula* of Gregory the Great'. I would also like to thank a former student, Emily Maskas, for sharing her perceptions on this topic while completing her MA at the University of York in 2016.

look beyond the frame, to look through it, to experience spaces that recede from the frame or penetrate out from it — spaces beyond the represented and the real, presenting visions of the heavenly eschaton, here recognized through the glowing beacon of the cross.

However, like the Continental mosaics discussed above, when viewed in a more literal, iconographical sense, alongside the myriad associations conjured by their (glowing) surfaces, material presentations, and iconographic narratives, Insular crosses may have a further element on their carved surfaces which evoked the bejewelled space of heaven. Along the periphery of various panels of such crosses — particularly notable on the delicate carving of the Easby Cross (Figures 6.4a and b), and the stone fragment of the cross head displayed at Jarrow — are decorative borders most usually read in the scholarship in terms of evoking skeuomorphic metalwork: large-scale replications of the beading and bosses found on the decorative metal crosses of the era. Indeed, following Lang (although his discussion makes no mention of pearls), if this reading is followed, then the connection between (pearls), water, the sacred, and salvation may be especially notable at Easby, where the cross stood 'close to the shallows of the Swale', explicitly furthering the connection between the carved stone and a flowing, watery environment.[31]

While I am not suggesting that all examples of this motif must be read in terms of jewels, I would nonetheless argue that some of these carved details ought to be read as pearls — particularly if they were painted white, instead of the gold or silver of metalwork — becoming an evocation of heaven, as well as a form of monumental *crux gemmata*: speaking of salvation, water, and the entrance to the heavenly kingdom.[32] In addition to their other symbolic functions and messages, these crosses — if understood as decorated with pearls — may gain a further dialogic, symbolic significance that speaks to the salvific power of water, and to the eschatological sacred. Placed on the edge of these powerful and emotive pieces of sculpture, a border motif of pearls would function much the same as the gemmed borders or aquatic thresholds of other ecclesiastical artworks, demarcating the sacred and bringing heaven to earth through their material presences. Water, as both a physical and metaphysical substance, was deeply linked to social systems and spiritual significances of the early medieval Church; the images which embodied or depicted water marked entry into the Church in the same manner as its baptismal sacrament, joining the earth with the heavens beyond, and reinforcing the sacraments and sacred mysteries of the Church. Such aqueous art was capable of evoking multivalent and complex spiritual significances, histories, and geographies for its viewers through the reflective, shining surfaces it presented — be they in the guise

---

[31] Lang, 'The Apostles in Anglo-Saxon Sculpture', p. 280.
[32] If so, then this border motif may well function in an analogous manner to that suggested for marginal motifs in Insular artworks. See Pulliam, *Word and Image in the Book of Kells*, pp. 49–95, following Françoise Henry, Carol Farr, and Martin Werner.

of flowing water or fixed pearl. Such depictions, both literal and figurative, were employed within ecclesiastical art throughout the Christian world, in a shared artistic inheritance that stretched from Rome to Ripon, all expressing a symbolic understanding and actualization of sacred space — bringing some of the envisioned and imagined qualities of heaven to these earthly places, along with associated metamorphic ideas of transformation offered by the sacred and salvific waters flowing through the Church, its history, and its art.

Works Cited

Manuscripts

Bern, Burgerbibliothek, Codex Bongarsianus 318
Paris, Bibliothèque nationale de France, MS lat. 14429

Primary Sources

Adomnán, *De Locis Sanctis*, ed. by Denis Meehan, Scriptores Latini Hiberniae, 3 (Dublin: Dublin Institute for Advanced Studies, 1958)
Bede, *Bede: Commentary on Revelation*, trans. by Faith Wallis, Translated Texts for Historians, 58 (Liverpool: Liverpool University Press, 2013)
——, *Bede: On the Temple*, trans. by Seán Connolly, Translated Texts for Historians, 21 (Liverpool: Liverpool University Press, 1995)
Cassiodorus, *Explanation of the Psalms*, vol. I, trans. by P. G. Walsh, Ancient Christian Writers: The Works of the Fathers in Translation, 51 (Mahwah, NJ: Paulist Press, 1990)
Isidore of Seville, *The Etymologies of Isidore of Seville*, trans. by Stephen A. Barney, W. J. Lewis, J. A. Beach, and Oliver Berghof (Cambridge: Cambridge University Press, 2011)
Pearl, in *Sir Gawain and the Green Knight; Pearl; Cleanness: Patience*, ed. by J. J. Anderson (London: J. M. Dent, 2005), pp. 1–46

Secondary Works

Bailey, Richard, 'Crosses, Stone', in *The Blackwell Encyclopaedia of Anglo-Saxon England*, ed. by Michael Lapidge, John Blair, Simon Keynes, and Donald Scragg, 2nd edn (Oxford: Blackwell, 2014), pp. 132–33
Baker, Nicholas G., 'Climbing a Stairway to Heaven: Early Insular Images and the Contemplative's Journey to the Divine' (unpublished conference paper, Subterranean in the Medieval World, University of York, 2014)
Blair, John, 'Churches in the Early English Landscape: Social and Economic Contexts', in *Church Archaeology: Research Directions for the Future*, ed. by John Blair and Carol Pyrah (Walmgate: Council for British Archaeology, 1996), pp. 6–19

Boulton, Meg, 'Art History in the Dark Ages: (Re)Considering Space, Stasis and Modern Viewing Practices in Relation to Anglo-Saxon Imagery', in *Stasis in the Medieval West? Questioning Change and Continuity*, ed. by Michael D. J. Bintley, Martin Locker, Victoria Symons, and Mary Wellesley (New York: Palgrave Macmillan, 2017), pp. 69–86

——, 'Bejewelling Jerusalem: Architectural Adornment and Symbolic Significance in the Early Church in the Christian West', in *Islands in a Global Context: Proceedings from the Seventh International Insular Arts Conference*, ed. by Connor Newman, Mags Mannion, and Fionia Gavin (Dublin: Four Courts Press, 2017), pp. 15–23

——, 'The Conceptualisation of Sacred Space in Anglo-Saxon Northumbria in the Sixth to Ninth Centuries' (unpublished doctoral thesis, University of York, 2013)

——, '"The End of the World as We Know It": The Eschatology of Symbolic Space/s in Insular Art', in *Making Histories: Proceedings of the Sixth International Insular Arts Conference*, ed. by Jane Hawkes (Donington: Shaun Tyas, 2013), pp. 279–90

——, '(Re-)Viewing "Iuxta Morem Romanorum": Considering Perception, Phenomenology, and Anglo-Saxon Art and Architecture', in *Sensory Perception in the Medieval West*, ed. by Simon C. Thompson and Michael D. J. Bintley (Turnhout: Brepols, 2016), pp. 207–26

Bradu, J. F., 'Description de la mosaïque', in *La Mosaïque de Germigny alliance et eucharistie*, ed. by P. Jacques and J. F. Bradu (Gien: Renaissance de Fleury: la revue du monastère, 2005), pp. 5–13

——, 'Histoire de la mosaïque', in *La Mosaïque de Germigny alliance et eucharistie*, ed. by P. Jacques and J. F. Bradu (Gien: Renaissance de Fleury: la revue du monastère, 2005), pp. 14–24

——, 'Une mosaïque unique', in *La Mosaïque de Germigny alliance et eucharistie*, ed. by P. Jacques and J. F. Bradu (Gien: Renaissance de Fleury: la revue du monastère, 2005), pp. 3–4

Brenk, Beat, *The Apse, the Image and the Icon: An Historical Perspective of the Apse as a Space for Images* (Wiesbaden: Reichert, 2010)

Brown, Peter, *The Cult of the Saints: Its Rise and Function in Latin Christianity* (Chicago: Chicago University Press, 1982)

Bynum, Caroline Walker, *Christian Materiality: An Essay on Religion in Late Medieval Europe* (New York: Zone Books, 2011)

Calkins, Robert G., *Illuminated Books of the Middle Ages* (Ithaca, NY: Cornell University Press, 1983)

Clemen, Paul, *Die Romanische Monumentalmalere in den Rheinlanden* (Düsseldorf: L. Schwann, 1916)

Doig, Allan, *Liturgy and Architecture from the Early Church to the Middle Ages* (Aldershot: Ashgate, 2008)

Evans, Joan, and Mary Serjeantson, eds, *English Medieval Lapidaries*, EETS, o.s. 190 (London: Oxford University Press, 1933)

Freeman, Ann, 'Theodulf of Orléans: A Visigoth at Charlemagne's Court', in *L'Europe héritière de l'Espagne wisigothique: Colloque international du C. N. R. S. tenu à la Foundation Singer-Polignac, Paris, 14–16 mai 1990*, ed. by Jacques Fontaine and Christine Pellistrandi, Collection de la casa de Velázquez, 35 (Madrid: Casa de Velázquez, 1992), pp. 185–94

——, 'Theodulf of Orléans and the *Libri Carolini*', *Speculum*, 32 (1957), 695–703

Grabar, Andre, 'Les Mosaïques de Germigny-des-Prés', *Cahiers Archéologiques*, 7 (1954), 171–83

Hawkes, Jane, 'The Body in the Box: The Iconography of the Cuthbert Coffin', in *Crossing Boundaries: Interdisciplinary Approaches to the Art, Material Culture, Language and Literature of the Early Medieval World*, ed. by Eric Cambridge and Jane Hawkes (Oxford: Oxbow Books, 2017), pp. 78–89

Howe, Nicholas, *Migration and Mythmaking in Anglo-Saxon England* (New Haven, CT: Yale University Press, 1989)

——, *Writing the Map of Anglo-Saxon England: Essays in Cultural Geography* (New Haven, CT: Yale University Press, 2007)

Jacques, P., 'L'Art de la mosaïque', in *La Mosaïque de Germigny alliance et eucharistie*, ed. by P. Jacques and J. F. Bradu (Gien: Renaissance de Fleury: la revue du monastère, 2005), pp. 29–31

——, 'La Mosaïque de Germigny', in *La Mosaïque de Germigny alliance et eucharistie*, ed. by P. Jacques and J. F. Bradu (Gien: Renaissance de Fleury: la revue du monastère, 2005), pp. 1–3

Jacques, P., and J. F. Bradu, eds, *La Mosaïque de Germigny alliance et eucharistie* (Gien: Renaissance de Fleury: la revue du monastère, 2005)

Janes, Domenic, *God and Gold in Late Antiquity* (New York: Cambridge University Press, 1998)

Kessler, Herbert, *Seeing Medieval Art*, Rethinking the Middle Ages, 1 (New York: Broadview Press, 2004)

Lang, James, 'The Apostles in Anglo-Saxon Sculpture in the Age of Alcuin', *Early Medieval Europe*, 8 (1999), 271–82

——, 'Survival and Revival in Insular Art: Northumbrian Sculptures of the Eighth- to Tenth-Centuries', in *The Age of Migrating Ideas*, ed. by Michael Spearman and John Higgitt (Edinburgh: National Museums of Scotland; Stroud: Alan Sutton, 1993), pp. 261–67

Maguire, Henry, *Earth and Ocean: The Terrestrial World in Early Byzantine Art*, College Art Association, Monographs on the Fine Arts, 43 (University Park: Pennsylvania State University Press, 1987)

McClendon, Charles B., *The Origins of Medieval Architecture* (New Haven, CT: Yale University Press, 2005)

Meyvaert, Paul, *The Art of Words: Bede and Theodulf* (Aldershot: Ashgate, 2008)

Meyvaert, Paul, and Ann Freeman, 'The Meaning of Theodulf's Apse Mosaic at Germigny-des-Prés', *Gesta*, 40 (2001), 125–39

Noble, T. F. X. *Images, Iconoclasm, and the Carolingians* (Philadelphia: University of Pennsylvania Press, 2009)

Ó Carragáin, Éamonn, *Ritual and the Rood: Liturgical Images and the Old English Poems of the Dream of the Rood Tradition* (London: British Library, 2005)
——, 'The Ruthwell Cross and Irish High Crosses: Some Points of Comparison and Contrast', in *Ireland and Insular Art, A.D. 500–1200*, ed. by Michael Ryan (Dublin: Royal Irish Academy, 1987), pp. 118–28
Ó Carragáin, Éamonn, and Tomás Ó Carragáin, 'Singing in the Rain on Hinba? Archaeology and Liturgical Fictions, Ancient and Modern (Adomnán, *Vita Columbae* 3.17)', in *Listen, O Isles, Unto Me: Studies in Medieval Word and Image in Honour of Jennifer O'Reilly*, ed. by Elizabeth Mullins and Diarmuid Scully (Cork: Cork University Press, 2011), pp. 204–18
O'Reilly, Jennifer, 'Introduction', in Bede, *Bede: On the Temple*, trans. by Seán Connolly, Translated Texts for Historians, 21 (Liverpool: Liverpool University Press, 1995), pp. xvii–lv
Pulliam, Heather, 'Blood, Water and Stone: The Performative Cross', in *Making Histories: Proceedings of the Sixth International Insular Arts Conference*, ed. by Jane Hawkes (Donington: Shaun Tyas, 2013), pp. 262–78
——, 'Color', in 'Medieval Art History Today — Critical Terms', special issue, *Studies in Iconography*, 33 (2012), 3–14
——, 'Eyes of Light: Colour in the Lindisfarne Gospels', in *Newcastle and Northumberland: Roman and Medieval Architecture and Art*, ed. by Jeremy Ashbee and Julian M. Luxford, British Archaeological Association Conference Transactions, 36 (Leeds: Maney, 2013), pp. 54–72
——, *Word and Image in the Book of Kells* (Dublin: Four Courts Press, 2006)
Rodwell, Warwick, Jane Hawkes, Emily Howe, and Rosemary Cramp, 'The Litchfield Angel: A Spectacular Anglo-Saxon Painted Sculpture', *Antiquaries Journal*, 88 (2008), 48–108
Schapiro, Meyer, *The Language of Forms: Lectures on Insular Manuscript Art* (New York: Pierpont Morgan Library, 2005)
Spiegel, Flora, 'The *Tabernacula* of Gregory the Great and the Conversion of Anglo-Saxon England', *Anglo-Saxon England*, 36 (2007), 1–13
Thunø, Eric, *The Apse Mosaic in Early Medieval Rome: Time, Network, and Repetition* (Cambridge: Cambridge University Press, 2015)
Twomey, Carolyn, 'Living Water, Living Stone: The History and Material Culture of Baptism in Early Medieval England, *c.* 600–*c.* 1200' (unpublished doctoral Thesis, Boston College, 2017)

JOHN J. GALLAGHER

'Streams of Wholesome Learning': The Waters of Genesis in Early Anglo-Saxon Exegesis

The earliest biblical exegesis to survive from Anglo-Saxon England is a body of biblical commentaries associated with the late seventh-century Canterbury School of Theodore of Tarsus, Archbishop of Canterbury, and Hadrian the African, Abbot of St Augustine's. Prior to the discovery of the texts in the Ambrosiana by Bernhard Bischoff in 1936 and their publication (in part) in 1994, relatively little was known about the Canterbury School except the high regard in which its scholarship was held by near-contemporary sources. The Venerable Bede speaks highly of the School in his *Historia Ecclesiastica* when he says:

> Et quia litteris sacris simul et saecularibus, ut diximus, abundanter ambo erant instructi, congregata discipulorum caterua scientiae salutaris cotidie flumina inrigandis eorum cordibus emanabant.
>
> [And because both of them were extremely learned in sacred and secular literature, they attracted a crowd of students into whose minds they daily poured the streams of wholesome learning.][1]

The imagery which Bede uses to describe this golden age of biblical learning in the early Anglo-Saxon Church is not too far removed from the reality of

---

My thanks are due to Dr Jodi-Anne George at the University of Dundee for reading and commenting upon this chapter in its early stages. It was her Bible as Literature reading group at the University of Dundee that inspired me to first begin thinking about the question of rain in Genesis. Many thanks are due to my supervisor Dr Christine Rauer at the University of St Andrews and to Dr Kees Dekker at the University of Groningen for reading early drafts of this chapter and for providing many helpful comments.

1 Bede, *HE*, IV. 2, pp. 332–33.

John J. Gallagher • (john.gallagher@ucd.ie) gained a PhD in 2019 from the University of St Andrews where he taught in the School of English and the School of Divinity. He is currently a Government of Ireland Postdoctoral Research Fellow in the School of English, Drama, and Film at University College Dublin.

the curriculum at Canterbury. Streams of wholesome learning did, indeed, flow into the hearts of students as attested in the biblical commentaries, which address, among other topics, the subject of water in the Book of Genesis. The Canterbury Commentaries are thought to represent records of the viva voce exposition of the biblical text by the School's two illustrious masters.[2] Exegesis is the exposition of difficult aspects of the biblical text, and water is one such difficulty expounded in the Commentaries. This chapter will explore how the Canterbury Commentaries understand three key passages of Genesis that relate to water: the primordial abyss of waters described at Genesis 1. 2; the creation of the firmament amidst the waters at 1. 6–7; and the absence of rain at 2. 5. The first part of this chapter will explore how and why the Canterbury Commentator understands the waters of Creation and the waters of the Flood to be related. The second part of this chapter will examine the absence of rain before the Flood and how the Commentator attempts to bring the Bible into accord with a scientific understanding of the hydrological cycle. It is this tension between the world as early medieval science understood it and the biblical picture of it that is the central concern of these biblical water glosses. These two cases prove that far from simply recapitulating patristic interpretations, Anglo-Saxon biblical scholars such as the Canterbury Commentator represented the active continuation of the exegetical tradition in Anglo-Saxon England.

Primordial Water and Diluvial Water

'In principio creavit Deus caelum et terram. Terra autem erat inanis et vacua, et tenebrae erant super faciem abyssi: et spiritus Dei ferebatur super aquas' (In the beginning God created heaven and earth. And the earth was void and empty, and darkness was upon the face of the deep; and the spirit of God moved over the waters).[3] The Commentaries explicate the cosmology and makeup of the material world at this earliest point in Creation. The Commentator follows traditional exegesis in explaining that 'caelum et terram' refer to the entirety of the physical and material world, both celestial and terrestrial; that the earth was 'inanis' (void), not of matter, but of human, animal, and plant life; and that there was 'tenebrae' (darkness), since the luminaries had not yet been created.[4] Although Genesis does not say so explicitly, water was part of this initial phase of Creation. God then created a firmament amidst the primordial waters that separated the waters, trapping a body of water above

---

2 *Biblical Commentaries*, ed. by Bischoff and Lapidge, p. 267.
3 Genesis 1. 1–2. All citations from the Vulgate are taken from *Biblia sacra iuxta vulgatam versionem*, ed. by Weber. All English translations of the Vulgate are taken from the Douay-Rheims version (1582–1609).
4 *PentI* 17–20, in *Biblical Commentaries*, ed. by Bischoff and Lapidge, pp. 304–05. This study will follow the sigla and numbering used in this edition.

the firmament.⁵ The waters below the firmament were then parted, thus allowing dry land to emerge. At 2. 5, it is stated that God had not yet caused rain to fall upon the earth. At 7. 4, God decided to destroy life on earth through a great Flood that will be brought about by rain. The Flood then begins with the opening of the 'floodgates' of heaven at 7. 11.

According to Genesis, at the earliest stage of Creation, the world was entirely covered by water. The Canterbury Commentaries connect the earliest reference to water in Genesis with the Flood, arguably the most prominent account of water in the Bible. The Commentator's explanation of 'the deep', literally 'the abyss', draws a comparison between the waters of Creation and those of the Flood.⁶ In discussing the darkness that was upon the face of the deep, the Commentator explains the composition of the physical world at this stage: 'Super faciem abissi .i. aquas dicit abyssos quia sic erant super materiali terra sicut in diluuio' (Upon the face of the deep, i.e. he calls the abyss 'waters', since they were then lying upon the material earth, just as in the Flood).⁷

The 'abyss' refers to the water that is subsequently mentioned in the same verse. From the text of Genesis, it is clear that water is the characterizing element of this early stage of Creation. The material world at this stage was indistinct, comprising only a great depth of water. The world did not exist in a chaotic state awaiting composition and ordering, as in Mesopotamian creation narratives on which Genesis draws. Early medieval Christian exegetes understood the world to be fully formed as part of this initial creation of the heavens and earth, awaiting revelation by the parting of the waters at Genesis 1. 9.⁸ This primordial abyss of water is sufficiently deep to warrant comparison with the floodwaters. The Flood submerged the earth and obliterated all life outside of the ark, barring the creatures of the water. The Commentator clarifies that the 'abyss' comprises a great body of water that lay upon the earth in the same manner as the Flood. The primordial waters and the Flood's waters are alike as both cover the material world entirely.

Connecting these two periods of water was a logical conclusion for biblical scholars explaining water in the Book of Genesis to reach. This section will be concerned with unravelling the threads of biblical and patristic precedents that possibly undergird this association. The edition of the Commentaries does not discuss this aspect of the *interpretamentum* and does not suggest the

---

5   Genesis 1. 6–7.
6   The Douay-Rheims and other translations of the Latin Bible translate 'abyssi' as 'deep'. 'Abyss', however, better captures the idea of a depth of water and will be used throughout this chapter.
7   PentI 21, in *Biblical Commentaries*, ed. by Bischoff and Lapidge, pp. 304–05.
8   Bede highlights that other elements besides water existed during this primordial period, but stipulates that these were not mixed formlessly. The earth was fully ordered as in its current form, just like the bottom of the sea. Bede here is drawing on Pseudo-Eustathius's *Commentaria in Hexameron*, which in turn draws heavily on Basil's *Hexaemeron*, a series of homilies that greatly influenced the development of hexameral literature. Bede, *Libri quatuor in principium Genesis*, ed. by Jones, I. 89–106. Bede, *On Genesis*, trans. by Kendall, 1. 2a/b, pp. 70–71.

possible sources behind this connection. Water is our primary concern here, thus possible biblical and patristic sources and early medieval analogues will be surveyed in order to determine how the Commentaries interpret these two bodies of water and how this reading fits into the exegetical tradition.

There are references in the Petrine Epistles to the water of Genesis that provide some biblical basis for understanding a relationship between these two periods of inundation. In I Peter 3. 20–21, the Flood is associated with baptism. It is clear that for the author of II Peter, the event of the Flood carries significant typological and mystical significance. In II Peter, explicit associations are made in his discussion of the Second Coming and the dissolution of the world that link the waters of Creation with those of the Flood: 'Latet enim eos hoc volentes, quod caeli erant prius, et terra de aqua, et per aquam consistens Dei verbo; per quae, ille tunc mundus aqua inundatus, periit' (For this they are wilfully ignorant of, that the heavens were before, and the earth out of water and through water, consisting by the word of God; whereby the world that then was, being overflowed with water, perished).[9] The author of Peter is not necessarily the source for this detail in the Commentaries, but he does articulate an association that might naturally be made by any biblical commentator. Furthermore, for biblical scholars familiar with II Peter, it provides a biblical precedent and certain authority for linking the two waters.

A connection between Creation and the Flood is apparent in the writing of Ambrose who directly connects the 'spiritus' upon the waters at these two stages. Genesis 8. 1–14 recounts the diminution of the Flood's waters, a phenomenon that is brought about by the 'spiritus' (spirit or wind) which God sends forth over the waters.[10] In patristic thought, the nature of the 'spiritus Dei' was understood as the Holy Spirit by allegorical (figurative) exegetes and as wind or air by exegetes of the historical (literal) persuasion. The nature of the 'spiritus' in Genesis was a topic of considerable debate in early Christian exegesis,[11] and continued to be a subject of interest throughout the Anglo-Saxon period. The Venerable Bede, commenting on 1. 2–3, suggests that the 'spiritus' is the force that arranges the world, bringing light onto the abyss of waters, and gathers the waters into one place, thus allowing dry land to emerge.[12] Bede concludes that the force of God is designated by 'spiritus'

---

9   II Peter 3. 7.
10 'Recordatus autem Deus Noe, cunctorumque animantium, et omnium jumentorum, quae erant cum eo in arca, adduxit spiritum super terram, et imminutae sunt aquae' (And God remembered Noe, and all the living creatures, and all the cattle which were with him in the ark, and brought a wind upon the earth, and the waters were abated), Genesis 8. 1.
11 'Spiritus Dei' (the spirit of God) had been understood as the Holy Spirit by Alexandrian exegetes, an idea that was refuted by Antiochene exegetes who understood this as wind or air. Evidence of how patristic authors understood this phrase is laid out in *Biblical Commentaries*, ed. by Bischoff and Lapidge, pp. 434–35.
12 Light is created and brought onto the abyss of waters in Genesis 1. 2–5, which is described as 'the Deep', an ancient cosmological concept of vast primordial waters that is initially devoid of light. The earth is created on the third day by God who gathers and separates the waters to

at 1. 3 in order to illustrate that creation was the work of the Trinitarian God.[13] Ælfric in his *Preface to Genesis* understands this as the Holy Spirit, which is apparent throughout Genesis, hence, the plural 'faciamus' (let us) at 1. 26.[14] In a subsequent comment on the 'spiritus', the Commentator includes both the Alexandrian and Antiochene interpretations of 'spiritus',[15] a most interesting feature given the literal character of the Commentaries as a whole. Adam became a living soul at 2. 7 through the breath of God. The Commentator explains this as the reception of the gift of prophesy lost through the Fall, compared to the Apostles' twofold reception: first, 'soul and breath', that is, life, and secondly, the Holy Spirit which is breathed into them at Pentecost.[16] The Commentaries display a certain interest in the nature of the breath and spirit of God in this world.

In *De Noe et arca*, Ambrose touches upon the abating of the waters and the nature of the 'spiritus' that inaugurated the end of the Deluge.[17] As a biblical scholar of Alexandrian orientation, Ambrose is at pains to refute the idea that the 'spiritus' is akin to wind (surely, today, the seas would evaporate because of wind and storms if this were the case, he states); rather, Ambrose believes that the 'spiritus' represents the presence and working of God in this world.[18] Ambrose's discussion of the 'spiritus' as the force of God in the world, which regenerates the earth following the Flood, strongly echoes the 'spiritus Dei' upon the face of the waters at Creation. To emphasize his point, Ambrose quotes three verses from the Scriptures that demonstrate the creative and regenerative power of God as represented by 'spiritus'.[19] Ambrose sees the

---

reveal dry land in Genesis 1. 9–10. The earth is described as 'void and empty' in the biblical account until this rearrangement of the expansive cosmic waters. Bede understands the description of 'void and empty' as referring to the absence of light, the creation of which by the 'spiritus Dei' permits the form of the material world to be appreciated.

13 Bede, *Libri quatuor in principium Genesis*, ed. by Jones, I. 159–60; Bede, *On Genesis*, trans. by Kendall, I. 2c, p. 73.
14 Ælfric, *Praefatio Genesis anglice*, ll. 58–61: *The Old English Heptateuch*, ed. by Marsden, p. 5.
15 *PentI* 22, in *Biblical Commentaries*, ed. by Bischoff and Lapidge, pp. 304–05. In this comment, the Holy Spirit is issued to breathe life into the waters. This is most interesting since marine life is not brought forward until the Fifth Day of the Hexameron, where Genesis tells us it is generated by God, not the 'spiritus' per se, Genesis 1. 20–23. *PentI* 22 compares the lifeless primordial waters to the Dead Sea. It has been suggested that this represents first hand experience of the Dead Sea, though nowhere do the Commentaries betray that Theodore or Hadrian made it to the Holy Land.
16 *PentI* 32, in *Biblical Commentaries*, ed. by Bischoff and Lapidge, pp. 308–09.
17 Ambrose, *De Noe et arca*, ed. by Schenkl, XVI. 58.
18 Ambrose, *De Noe et arca*, ed. by Schenkl, XVI. 58.
19 'Emittes spiritum tuum, et creabuntur; et renovabis faciem terrae' (Thou shalt send forth thy spirit, and they shall be created: and thou shalt renew the face of the earth), Psalm 103. 30; 'Verbo Domini caeli firmati sunt; et spiritu oris ejus omnis virtus eorum' (By the word of the Lord the heavens were established; and all the power of them by the spirit of his mouth), Psalm 32. 4; 'Spiritus Dei fecit me, et spiraculum Omnipotentis vivificavit me' (The spirit of God made me, and the breath of the Almighty gave me life), Job 33. 4.

'spiritus' that travels over and dissipates the Flood's waters at 8. 1 as identical to the 'spiritus' that is upon the water at 1. 2, which represents the presence of God in primordial Creation. Through this exploration of the 'spiritus', Ambrose explicitly links these two periods and, by extension, the waters that define them. In *De mysteriis*, a treatise on the antiquity of baptism to which water is central, Ambrose makes similar points about the 'spiritus Dei' upon the waters at Creation as representing the moving and working of God.[20] Ambrose also connects the 'spiritus' that works upon the primordial and diluvial waters with that same 'spiritus' that parts the Red Sea.[21] Ambrose's exploration creates firm and direct links between these two periods in biblical history that might allow for an understanding that these waters were related.

While Ambrose could be the source of this association here, the dearth of allegorical and Western Latin exegesis attested in the Commentaries, which are, by and large, literal in character, make this unlikely. The Alexandrian interpretation of 'spiritus' could have been garnered from any of the Greek Fathers. Ambrose and other Latin Fathers, in fact, inherited this idea from Greek exegesis, which is the most likely avenue for its appearance here, given Theodore's Greek education and outlook. Furthermore, there is little correspondence between this comment and Ambrose's treatise. A reading of Ambrose's exploration of 'spiritus' might lead one to connect the waters, but Ambrose himself does not explicitly do so. That this conclusion might be drawn from Ambrose suggests it is orthodox, but, ultimately, it must be derived from elsewhere.

Isidore of Seville dedicated an entire chapter of his scientific treatise *De natura rerum* to the existence of supercelestial water.[22] Isidore's treatise was undoubtedly one of the most influential scientific works in circulation that discusses these waters and is likely to have been known to the Commentator.[23]

---

20 Ambrose, *De mysteriis*, ed. by Faller, III. 9.
21 'Flavit spiritus tuus, et operuit eos mare: submersi sunt quasi plumbum in aquis vehementibus' (Thy wind blew and the sea covered them: they sunk as lead in the mighty waters), Exodus 15. 10.
22 For the Latin text, see Isidore, *De natura rerum*, ed. and trans. by Fontaine, XIV. 1–2, pp. 224–27. For an English translation, see Isidore, *On the Nature of Things*, trans. by Kendall and Wallis, pp. 136–37.
23 *De natura rerum* is not cited in the Commentaries from Canterbury in their present form. Lapidge observes that it is not clear whether or not the Commentator had works of Isidore other than the *Etymologiae* available to him. *De natura rerum* is used in the Leiden Glossary, which is associated with Canterbury. *Biblical Commentaries*, ed. by Bischoff and Lapidge, pp. 204–05. Di Sciacca points out that the Canterbury Commentaries, the Leiden Glossary, and Aldhelm's writings bear witness to a number of works by Isidore, including *De natura rerum*, suggesting that it formed part of the curriculum at the Canterbury School. Di Sciacca, 'Isidore of Seville in Anglo-Saxon England', p. 133. According to Lapidge, the work is cited with certainty in the Leiden Glossary and Aldhelm's *Carmen rhythmicum*. Lapidge, *The Anglo-Saxon Library*, pp. 176 and 181. The work also appears to have influenced Aldhelm's *De metris* and a number of his *Enigmata*, as discussed in Orchard, *The Poetic Art of Aldhelm*, pp. 212–13. The work is quoted by a number of other known Anglo-Saxon authors such as

Isidore does not connect the waters above to the floodwaters, but he establishes their existence. The arguments he puts forward will be relevant to the discussion of rain in the second part of this chapter and are worth outlining in some detail. In establishing the scientific possibility of the biblical notion of supercelestial water, Isidore foregoes the use the biblical term *firmamentum* (firmament) that is present in both the Old Latin and Vulgate versions of Genesis, in favour of the more secular, scientific, and general term *caelum* (sky). Citing Ambrose, Isidore relates that classical philosophers reject the possible existence of water above as it would be lost in the course of the spherical earth's rotation.[24] The existence of supercelestial water also goes against the classical theory of elemental weight and order.[25] In general, *De natura rerum* upholds the classical schema where the four elements of fire, air, water, and earth form a descending hierarchy, but Isidore goes on to state that it would not be beyond the power of God to fix water above the heavens in the form of ice, a point taken from Augustine.[26]

Water may exist above the earth in a solidified form, but the biblically derived notion of a reserve of cosmic water above the firmament is not fully endorsed and no further reference to Genesis is made. In his earlier chapter on heaven, Isidore relates how the classical philosophers believed the earth to have been placed in water in order to rotate in it and be tempered by it.[27] Isidore also relays Ambrose's idea that starlight is the glittering reflection of ice, an idea that is also at odds with classical physics, which proposes that the superterrestrial regions are comprised of a fifth element, αἰθήρ or *aether* (ether), rather than water.[28] The biblical idea of supercelestial waters was clearly an abiding concern for Christian cosmographers, but one which had to be squared with natural science. This idea derives from the Bible and is not organic to classical thought. By including the idea of frozen supercelestial water from Ambrose, Isidore makes some allowance for biblical cosmology by providing a solution that is based on, yet at some variance with, classical science. While *De natura rerum* is not quoted directly in the Commentaries,

---

Bede, Ælfric, and Byrhtferth, giving some indication as to its circulation in this context. Lapidge, *The Anglo-Saxon Library*, p. 310. The appearance of batches of *lemmata* from *De natura rerum* in the Leiden Glossary strongly indicates that the work was available at some stage at the Canterbury School and is likely to have been known to our Commentator, although it is not cited directly.

24 Isidore, *De natura rerum*, ed. and trans. by Fontaine, XIV. 1, pp. 224–27; Isidore, *On the Nature of Things*, trans. by Kendall and Wallis, p. 136.
25 Isidore, *De natura rerum*, ed. and trans. by Fontaine, XIV. 1, pp. 224–27; Isidore, *On the Nature of Things*, trans. by Kendall and Wallis, p. 136.
26 Isidore, *De natura rerum*, ed. and trans. by Fontaine, XIV. 2, pp. 226–27; Isidore, *On the Nature of Things*, trans. by Kendall and Wallis, pp. 136–37.
27 Isidore, *De natura rerum*, ed. and trans. by Fontaine, XII. 4, pp. 218–21; Isidore, *On the Nature of Things*, trans. by Kendall and Wallis, p. 134.
28 Isidore, *De natura rerum*, ed. and trans. by Fontaine, XIV. 2, pp. 226–27; Isidore, *On the Nature of Things*, trans. by Kendall and Wallis, pp. 136–37.

it is likely to have significantly shaped ideas about the firmament and the problematic nature of the biblical text. Isidore incorporated the biblical idea of supercelestial water into his scientific work; it was the task of the medieval scholars who followed him to incorporate his science into their biblical exegesis.

De natura rerum is principally a scientific and secular work that seeks to explain time, the cosmos, and weather according to the principles of classical science and philosophy, as opposed to the Hexameron, the standard model for Christian scholars. Various allegorical, anagogical, and mystical asides are included that give a Christian tone to a work that eschews a purely biblical cosmology in favour of science.[29] A scientific interest in understanding the workings of the natural world and the elements motivated Isidore's patron, the Visigothic king Sisebut, to commission this work.[30] Contemporary events might also have played into the composition of this work and its thoroughly scientific standpoint. In the year of its composition, 612, there was an unusually high number of solar and lunar eclipses.[31] These astronomical events triggered popular apocalyptic anxiety. Isidore sought to explain these perceived apocalyptic signs in the natural world scientifically, rather than biblically. It is also possible that *De natura rerum* was a direct response to the flat-earth theory espoused by Isidore's contemporary, Cosmas Indicopleustas, whose *Topographia Christiana* is based purely on the world as it is presented in the Bible.[32] The *Topographia* was possibly known at the Canterbury School and might even have been quoted in a gloss on the Tower of Babel, which states that it was built as an attack on heaven in retaliation for the Flood.[33] With scientific and less scientific works potentially available, the Commentator and Isidore appear to share the same purpose in their scholarly efforts — to correctly understand, explain, and reconcile the Bible with the world around them.

Having established the possibility of the existence of supercelestial waters, let us return to the connection between the primordial and diluvial waters. Connections between the waters of Creation and the Flood crop up in the discussion of some aspects of the Old English *Exodus*. An analysis of these two biblical waters and their relatedness features in criticism on the description of the floodwaters as 'niwe' (new) in *Exodus*. Edward B. Irving suggests that 'niwe flodas' refers to an 'unprecedented aquatic phenomenon'.[34]

---

29 For example, in the chapter on rain, clouds are understood as teachers and apostles and rain as their teaching of the Word. Isidore, *De natura rerum*, ed. and trans. by Fontaine, XXXIII. 3. Bede's redaction of *De natura rerum* interestingly omits many of the Christian references made by Isidore. See Ferrand, 'The Hydrologic Cycle in Bede's *De Natura Rerum*'.
30 Isidore, *On the Nature of Things*, trans. by Kendall and Wallis, p. 12.
31 Isidore, *On the Nature of Things*, trans. by Kendall and Wallis, p. 10, and Isidore, *De natura rerum*, ed. and trans. by Fontaine, pp. 3–6.
32 Isidore, *On the Nature of Things*, trans. by Kendall and Wallis, p. 13.
33 *Biblical Commentaries*, ed. by Bischoff and Lapidge, pp. 208–11; PentI 91, pp. 320–21.
34 Irving, 'New Notes on the Old English *Exodus*', p. 313, as cited by Anlezark, *Water and Fire*, pp. 197–98.

J. R. Hall disagrees with Irving, seeing the water of Creation as an equivalent phenomenon.[35] Hall's observation that *Genesis A* uses 'flod' in reference to both the primordial waters and the floodwaters strongly suggests a direct correspondence in the mind of the *Genesis A* poet at least.[36] Hall cites Ambrose and Bede in support of the understanding that the two waters were related. Quite naturally, the two periods in biblical history during which the material world was submerged in water are seen as related.

As the foremost Anglo-Saxon exegete, Bede's writing addresses similar questions to the Commentaries and synthesizes sources in a manner that allows us to interrogate how the Commentator fits into the exegetical tradition. For the Canterbury Commentaries, Bede's copious writings represent a rich point of comparison when addressing the peculiarities of these early texts. Associations between the waters of Creation and the Flood appear in Bede's exegesis. In his study of the Flood, Daniel Anlezark observes that in Bede's discussion of the six ages of the world in his commentary *On Genesis*, 'the creation of the waters is naturally associated with the age of the Flood'.[37] Bede relates the arrangements of the firmament in the midst of the primordial waters at 1. 6 to the placing of the ark in the midst of the floodwaters at 7. 11.[38] According to Genesis, the material world at Creation is entirely covered by water; the creation of a firmament in the midst of the waters separates waters beneath the sky from the waters above.[39] Bede's commentary goes on to discuss the opening of the 'floodgates of heaven' at the time of the Deluge.[40] In a subsequent comment on the creation of the firmament amidst the waters at 1. 6, the Commentator, too, refers to the 'floodgates of heaven'.[41] Here, it is suggested that God separated the waters above from those below so that it might be said, 'and the flood gates of heaven were opened'. The association between primordial and diluvial waters hinges on a vital correspondence between the waters released at 7. 11 and those trapped above at the creation of the firmament in 1. 6. It is noteworthy that the description of the beginning of

---

35 Hall, '*Niwe Flodas*', p. 243.
36 The use of 'flod' to refer to primordial waters makes the description of the floodwaters as 'niwe flodas' an apt description, Hall argues. See Hall, '*Niwe Flodas*', p. 243.
37 Anlezark, *Water and Fire*, p. 67; Bede, *Libri quatuor in principium Genesis*, ed. by Jones, I. 1114–18; Bede, *On Genesis*, trans. by Kendall, 2. 3a/b, p. 101.
38 As cited by Anlezark, *Water and Fire*, pp. 67–68; Bede, *Libri quatuor in principium Genesis*, ed. by Jones, I. 1114–18; Bede, *On Genesis*, trans. by Kendall, 2. 3a/b, p. 101. Interestingly, as the commentary continues by linking the days of the Hexameron to the ages of the world, Bede follows Augustine and links the separation of the waters and emergence of dry land and vegetation in 1. 9–11 to Abraham and his separation from idolatrous people, rather than Noah who, alongside his family, began the third age.
39 Genesis 1. 6. This aspect of the Creation narrative reflects the ancient Near Eastern notion that the sky is blue because there is water above it.
40 Genesis 7. 11.
41 *PentI* 26, in *Biblical Commentaries*, ed. by Bischoff and Lapidge, pp. 306–07. For a summary of treatments of the waters above the firmament by the Greek Fathers, see p. 438.

the Flood at 7. 11 in the Vulgate refers to underground waters as 'fontes abyssi', the same words used to describe the great deep at 1. 2.[42] This correspondence is carried over into the vernacular translation of Genesis in the *Old English Heptateuch*, where the 'wellspringas þære micclan niwelnisse' (fountains of the deep) are referred to in the same manner as 1. 2: 'Seo eorþe soðlice wæs ydel and æmtig, and þeostru wæron ofer þære niwelnisse bradnisse, and Godes gast wæs geferod ofer wæteru'.[43] Within the schema of the six ages of the world, the first age is bookended by these two bodies of water, which invites a further degree of association between these waters.

The first explicit reference to rain occurring in the Bible is at the time of the Flood; in Genesis 2, it had not yet rained upon the earth, which was at that point irrigated by the fountain of Paradise, which rose up from the earth.[44] The text of Genesis tells us that the Deluge is inaugurated by a combination of the breaking-up of the 'fountains of the deep' and the release of the 'floodgates of heaven'.[45] Exodus refers to these subterranean waters, a central idea in patristic writing on the origin and function of water in the natural world.[46] It cannot be presumed that the Deluge released all of the waters above the firmament, since these are mentioned later in the biblical text.[47] Bede understands that the first rain is brought about by the release of water that was trapped above the firmament at Creation. In Bede's exegesis, the Flood's waters are, in part, the primordial waters from above the firmament. It is not clear how similar the two waters were in terms of their amount, but it is possible to conjecture that they were alike.[48] Bede's commentary suggests

---

42 Genesis 7. 11. *The Old English Heptateuch*, ed. by Marsden, p. 21.
43 *The Old English Heptateuch*, ed. by Marsden, p. 8.
44 Genesis 2. 5–6.
45 Genesis 7. 11.
46 Exodus 20. 4.
47 Psalm 148. 4.
48 Drawing a connection between the waters of the Flood and those covering the primordial earth raises questions about which body was greater in size. Prior to the creation of the firmament, the waters of Creation occupied the material world, that is, the entirety of area that became separated into the land, sea, and firmament. Bede tells us the water was more attenuated at Creation. Bede, *Libri quatuor in principium Genesis*, ed. by Jones, I. 189–94; Bede, *On Genesis*, trans. by Kendall, 1. 3, p. 74. Both the Commentator and Bede imply that the Flood comprised the waters from above the firmament and the waters that sprung forth from the deep, but it is unclear whether the Commentator believes the Flood to correspond in volume to the primordial waters. The primordial waters are thought to have occupied the area of the material world, but, presumably, with some distance between the uppermost extremity of the material world for the 'spiritus Dei' to move over the abyss. The Flood did not occupy the entirety of the area of the material world, but submerged the earth and could not have reached as high as the waters of primordial Creation due to the firmament. The precise size, volume, and area covered by both bodies of water is not fully clarified. London, British Library, MS Harley 3271, fols 90^r–90^v, includes an Old English note explaining that the floodwaters returned to the subterranean abyss, in the same way that the rivers return to their origin so that they might flow again. This note sees the floodwaters as part of the

that the two periods of inundation are not merely alike, but actually constitute the very same body of water.

Irish biblical scholars address similar difficulties in the Genesis narrative as their Anglo-Saxon counterparts. Concerns with biblical water in these two regions of the Insular world underscore the lacunae that were present in the exegetical tradition of Genesis that early medieval scholars were compelled to resolve. An association between the waters above the firmament and those of the Deluge is made in two seventh-century Irish texts: *De mirabilibus sacrae scripturae*, an anonymous treatise by an author known as the Irish Augustine from c. 655, and Pseudo-Isidore, *De ordine creaturarum* from c. 655–75. Both texts relate that the waters above the firmament were reserved for the Flood.[49] Pseudo-Isidore is indebted to the Irish Augustine for this detail among various others.[50] According to Marina Smyth, 'The belief in the existence of these waters was a novel feature in the cosmology of late antiquity, and a source of much puzzlement and embarrassment to early Christian writers who naturally felt compelled to introduce this biblical data into their view of the world'.[51] Within both texts, the waters serve to regulate the heat of the luminaries, a point that can be traced to Isidore, and to bring about the Flood, a detail for which we have no earlier source. Smyth states, 'I know of no precedent for the belief that the waters above the firmament were intended for the Flood'.[52] I am also unable to identify any commonly known tradition that predates *De mirabilibus*, which links these waters explicitly. As Smyth observes, this association is implied by the biblical account.[53] It seems that the Irish Augustine was the first exegete to make this observation explicitly and that Pseudo-Isidore was the one to popularize it.

A connection between the waters of Creation and those above the firmament and those of the Flood is made in *PentI* 21 and *PentI* 26. Given the rarefied nature of this interpretation within the broader exegetical tradition, I would suggest that the link between these waters in the Canterbury Commentaries is dependent on knowledge of *De ordine*. Exegetes in both of these regions of the early medieval Insular world could have arrived independently at the conclusion that the primordial and diluvial waters were related based on their

---

hydrological cycle, rather than a reserve of cosmic water that was not already present on the earth. Edited and translated in Anlezark, 'Understanding Numbers in London, British Library, Harley 3271', pp. 143–44.

49 *De mirabilibus*, I. 6–7 (ed. by Migne) and *Liber de ordine creaturarum*, III. 1–5 (ed. by Díaz y Díaz). Smyth, 'The Seventh-Century Hiberno-Latin Treatise *Liber de ordine creaturarum*'. See Smyth's chapter on the supercelestial waters in Smyth, *Understanding the Universe in Seventh-Century Ireland*, pp. 94–103.

50 Smyth, 'The Seventh-Century Hiberno-Latin Treatise *Liber de ordine creaturarum*', p. 171, n. 29, and 139–45.

51 Smyth, *Understanding the Universe in Seventh-Century Ireland*, p. 94.

52 Smyth, *Understanding the Universe in Seventh-Century Ireland*, p. 100.

53 Smyth, 'The Seventh-Century Hiberno-Latin Treatise *Liber de ordine creaturarum*', p. 171, n. 29.

interpretation of the 'floodgates of heaven'.⁵⁴ The Irish cosmological texts are the first in the exegetical tradition to link these waters directly. Since both Irish texts predate the Canterbury School, it is plausible to consider them as potential sources for the Commentaries.⁵⁵ While *De mirabilibus* is earlier, there is no evidence for its knowledge in Anglo-Saxon England, unlike *De ordine*, which enjoyed a noteworthy career in this context.⁵⁶ The first known use of *De ordine* outside of Ireland is thought to be in Northumbria where it is drawn upon in Bede's redaction of *De natura rerum*, which contains this detail in its abbreviated chapter on the heavenly waters.⁵⁷ The apparatus to the translation of *De natura rerum* provided by Calvin B. Kendall and Faith Wallis does not identify this detail as Bede's own innovation, but traces it to *De ordine*.⁵⁸ Given the terse nature of the glosses from Canterbury, it is not possible to identify a direct correspondence with *De ordine*. However, the appearance of this rarefied detail in the Canterbury Commentaries suggests reliance on *De ordine*, which is seemingly the first text to make this connection. If this is the case, the work not only was present in Anglo-Saxon England much earlier than scholars have previously thought, but had its entry point in Canterbury, rather than Northumbria, the context with which it is most associated.

The earliest surviving manuscript of *De ordine* is from eighth-century Northumbria.⁵⁹ Channels of influence between Ireland and the Irish-influenced region of Northumbria were, in general, more established than those between Ireland and the Roman-dominated, southern Anglo-Saxon Church. Bede's oeuvre shows a great indebtedness to Hiberno-Latin material, particularly his computus. Dáibhí Ó Cróinín argues that Bede's lack of intimate knowledge of the computistical eighty-four-year cycle, knowledge that was available in texts from Iona, suggests that most of the Irish materials available to Bede reached Jarrow not from Iona, as one might suspect, but from southern or south-western

---

54 We cannot rule out that Irish material could have been inserted into the Canterbury Commentaries at a later stage, perhaps during their life in northern Italy.
55 Smyth observes that *De ordine creaturarum* certainly predates the arrival of Theodore of Tarsus and his introduction of new Roman ideas concerning the fate of the soul after death to the British Isles, which are absent from this work. Smyth, 'The Seventh-Century Hiberno-Latin Treatise *Liber de ordine creaturarum*', p. 138.
56 The work enjoyed some popularity in Anglo-Saxon England, probably because of its mistaken attribution to Isidore. It is used by Bede and the Old English Martyrologist.
57 Bede, *De natura rerum*, ed. by Jones, VIII.
58 Bede, *On the Nature of Things*, trans. by Kendall and Wallis, p. 77, n. 40.
59 While a Northumbrian origin for the text was suggested by Jones, building on the work of Díaz y Díaz, Smyth has shown that reliance on *De mirabilibus*, for which there is no evidence in Northumbria, a variety of Hiberno-Latin phonological features, and Irish themes and details firmly suggest an Irish origin. On the Irish provenance of this text, see Smyth, 'The Date and Origin of *Liber de ordine creaturarum*' and Smyth, 'The Seventh-Century Hiberno-Latin Treatise *Liber de ordine creaturarum*', pp. 149–56.

England.⁶⁰ Aldhelm shows familiarly with some of this Irish material, which Ó Cróinín argues might have come to Anglo-Saxon England from Ireland via Wessex.⁶¹ While Wessex proves to be a closer point of contact with Ireland, we cannot rule out that Irish scholars might have travelled to a renowned centre of learning such as the Canterbury School, bringing materials such as *De ordine* with them. Given Theodore's role in settling the Easter controversy in Anglo-Saxon England, it also seems plausible that he would have had access to computistical material and might have brought this to Northumbria. Given Aldhelm's appearance in this picture, it is wholly plausible that Irish texts such as *De ordine* formed part of the curriculum at Canterbury before making its way to Northumbria for Bede to use a generation later. If this picture that I have sketched is the case, then this transmission opens up fascinating questions about the influence of Irish material on the Canterbury School, the channels of the scholarly interactions between the Canterbury School and Northumbria, and the processes of intellectual exchange between early medieval Ireland and the Canterbury School concerning difficult biblical matters.

Striking differences are apparent, however, between the two waters, such as the area they covered, their mass, and the presence of life in the waters at these two stages. Irish exegesis highlights some further difficulties of associating these waters. For example, does the opening of the 'floodgates' account for the very first rain or simply an unprecedented amount? Where did the waters from above go after the Flood? Did they remain and, so, constitute the cycle of evaporation and rain we know today? Or were they taken back up into the firmament, and, if so, how do we account for rain today? Establishing a clear understanding of these waters is a thorny task, and laying out the lengthy traditions of their interpretation is beyond the scope of this examination. The many difficult questions that arise from a literal and literary reading of Genesis can be circumnavigated, however, since the Commentaries do not suggest that the two periods are exactly similar, but merely alike. These differences are, understandably, not elucidated in the Commentaries given their relatively terse format, which hints at larger, problematic questions and a wealth of influences that are difficult to trace with certainty. In at least one stage of the Canterbury School, Anglo-Saxon students parsing the texts of Genesis would have connected these two periods of biblical history. This correspondence hinges on an intertextual understanding of cosmology and environmental phenomena in 1. 6 and 7. 11 that is made in *De ordine* or which the Commentator determines independently, rather than on any patristic explanation of water in Genesis. The absence of rain prior to the Deluge and the confinement of water above the firmament suggest for early Anglo-Saxon exegetes that these waters were one and the same.

---

60   Ó Cróinín, 'The Irish Provenance of Bede's Computus', p. 242.
61   Ó Cróinín, 'The Irish Provenance of Bede's Computus', p. 244.

Rain before the Flood: The Bible, the Fathers, and the Anglo-Saxons

The history of rain is an intriguing topic that is addressed in the Canterbury Commentaries. The picture of rain given in the Bible does not accord with how this weather is understood in the secular, scientific learning of antiquity. The meteorology of Genesis — the absence of rain, its appearance at the Flood, and its origin beyond the firmament — contrasts to how rain was commonly and scientifically understood to operate in the early Middle Ages. Biblical exegetes were tasked with the responsibility of explicating rain as presented in Scripture and harmonizing this with contemporary scientific explanations of the natural world.

Before delving into the tension between biblical and scientific understandings of weather, it is first necessary to survey what the Bible says about rain. As mentioned above, the Genesis narrative suggests that rain did not exist in Paradise and the postlapsarian world until the Flood.[62] Rain is only mentioned before the Flood to clarify that it did not exist. At 2. 5, it is stated that God had not yet rained upon the earth. Environmental phenomena, at this stage of biblical history, occur solely at the command of God. In the absence of rain it is stated at 2. 6 that the earth was irrigated by a fountain that rose out of the ground to water all of the earth. The precise process of the earth's irrigation is further explained at 10–14 where we read that a river flows out of Paradise, which, presumably, takes as its source the fountain mentioned at 2. 6. This river then divides into four channels, which irrigate a number of lands. It is noteworthy that this is the first mention of the wider world in Genesis up until this point, suggesting the environments and physical conditions of Paradise and the rest of the earth to be intimately linked. That is, there was no rain anywhere on earth at this time. In the beginning the world was covered in water. God created a firmament amidst the waters, thus separating the waters above from those below. God resolves to destroy the world through rain at 7. 4. The 'floodgates of heaven' are then opened at 7. 11, and the waters above the firmament are released. The Flood is the first mention of rain in the biblical text, which seems to come from the waters above.

The absence of rain prior to the Flood is a topic of some debate, as Genesis is not entirely clear whether rain was absent until the Flood or if it simply did not rain until the point of 2. 5–14. Rain is mentioned on numerous occasions throughout the Old Testament. Sometimes, it is simply stated that rain comes from 'heaven' or the 'heavens', which is vague about the precise origin of this rain and the processes of its formation.[63] Elsewhere, we read that rain comes from the clouds, that water is bound up in clouds, and, most intriguingly, that God lifts drops of water upwards, which are then poured out in showers.[64]

---

62 Genesis 7. 4, 7. 6, and 7. 10–13.
63 Deuteronomy 11. 17 and 28. 12.
64 Ecclesiastes 11. 3, Job 26. 8, and Job 36. 27–28, respectively.

Such references show rain to operate in the way it is understood by secular learning. It ought to be noted that a 'scientific' understanding of rain is not confined to scholarly expositions; the connection between rainfall and what modern meteorology calls 'nimbus' clouds would have been observed by the ancient Israelites, as some biblical verses show, as well as the Anglo-Saxons.[65]

The Church Fathers address rain in the Bible and in the natural world in a number of ways. The question of rain is inextricably linked with questions concerning the existence of the waters above the firmament. Processes by which rain was formed are explored in the writings of the Greek Fathers, the evidence for which is laid out by David Sutherland Wallace-Hadrill.[66] Meteorological references in Greek patristic exegesis seem primarily concerned with offering a scientific explanation of rain and its formation, which is shaped by classical Hellenistic cosmology, physics, and astrology. Gregory of Nyssa and Basil, for example, understand that water evaporates, condenses, and returns to earth as rain. Notably, Theophilus of Antioch understands dew to be stored with rain above the firmament, an interpretation that seems more in line with the physical world presented in Genesis than the world as it is understood in Greek natural science.[67]

In the writings of Basil, tensions between both traditions emerge in sharp relief. Basil is, perhaps, the most influential Greek commentator on the Hexameron and is likely to have been known to our Commentator, though his position on rain is by no means unambiguous. Basil describes the necessity of waters above at 1. 6–7, which filter down and regulate the heat of the fiery luminaries below.[68] Drawing on Job's reference to rain sent by God, Basil sees the percolation of water from above as regulated by the divine.[69] A more scientific understanding occurs in his homily on the creation of the luminaries that understands rain to be formed by the 'exhalation' of water from below.[70] As Thomas O'Loughlin has pointed out, Basil elsewhere quotes sources that suggest all rain comes from above; hence the world experiences floods and droughts, phenomena Basil sees as entirely moderated by God.[71] We might note these points of contradiction, but we should also recognize that rain is never systematically expounded by Basil since this detail is absent from the Hexameron of Genesis 1 itself.

In his homily on the creation of the firmament, Basil attempts to integrate and reconcile the commonplace understanding of rain's formation with the

---

65 Judges 5. 4; Isaiah 5. 6.
66 Wallace-Hadrill, *The Greek Patristic View of Nature*, pp. 21–26.
67 Theophilus of Antioch, *Ad Autolycum*, II. 13, as cited by Wallace-Hadrill, *The Greek Patristic View of Nature*, p. 23.
68 O'Loughlin, 'Aquae Super Caelos (Gen 1:6–7)', p. 98.
69 Basil, *The Hexaemeron*, ed. by Schaff and Wace, III. 6.
70 Basil, *The Hexaemeron*, ed. by Schaff and Wace, VI. 8.
71 O'Loughlin, 'Aquae Super Caelos (Gen 1:6–7)', p. 99.

notion of waters above the firmament.[72] Scripture states that rain comes from the heavens; hence, this must refer to the body of water that occupies the higher regions, as per Genesis 1. 6. Basil then goes on to offer an explanation of rain that follows the scientific opinion: 'exhalations' of the air rise and are condensed into clouds by the wind, before returning to earth as rain.[73] While the Bible states that rain comes from the heavens, in Basil's view, this does not preclude its formation as it is classically understood. Basil's interpretation does not entirely address the waters beyond the firmament and the relationship between these waters and rainwater, however. The exegesis proposed by Basil argues that the notion of waters above, contrary to Aristotelian ideas of the proper position of the elements, are possible within a scientific view of meteorology. Basil's exegesis here pushes for a reconciliation of Genesis and the scientific understanding of the atmosphere. Although problems remain, Basil manages, to a certain extent, to marry a scientific view of weather with biblical cosmology and meteorology.

Following on from Basil, Latin exegetes make significant contributions to hexameral exegesis and the understanding of rain, notably Ambrose, Augustine, and Isidore. Generally, Ambrose's position on the Hexameron makes little allowance for Christian apologetics, upholding the inerrancy of the biblical description of the world, the infallibility of the word of God, and the belief in a number of water-related miracles, such as the parting of the Red Sea and the reversal of the Jordan.[74] Although Ambrose seems to come down on the side of a natural science based on the Bible, as opposed to secular learning, his nuanced exploration of water allows for a reconciliation of these opposing perspectives. Ambrose writes that water either comes from the waters placed above the firmament at 1. 6–7 or that it emerges from a high position, presumably the clouds.[75] Ambrose defends the possibility of the suspension of water above the firmament,[76] while also laying out the process by which water evaporates, rises in a vaporized state, and cools, leading to rain, a point that supports the possibility that water can be suspended above the firmament.[77]

As O'Loughlin has observed, Augustine is probably the only Christian exegete with a thorough grasp of classical traditions of natural science.[78] The moot point between biblical cosmology and meteorology and secular traditions is the existence of waters above the firmament. Rather than excusing the natural world presented in Genesis as an unfathomable miracle of God, Augustine seeks to rehabilitate the natural order of the world in the Bible

---

72 Basil, *The Hexaemeron*, ed. by Schaff and Wace, III. 8.
73 Basil, *The Hexaemeron*, ed. by Schaff and Wace, III. 8.
74 Exodus 14. 16–29; Psalm 113. 3.
75 Ambrose, *Exaemeron*, ed. by Schenkl, I. 6.
76 Ambrose, *Exaemeron*, ed. by Schenkl, I. 3.
77 Ambrose, *Exaemeron*, ed. by Schenkl, II. 3.
78 O'Loughlin, 'Aquae Super Caelos (Gen 1:6–7)', p. 103.

with the non-Christian idea of elemental weight in his treatise *De Genesi ad litteram*.[79] O'Loughlin summarizes Augustine's position as follows:

> The other fact noted by this commentator [Augustine] is that water gathers together in clouds and these are near the earth for mountains can be seen to protrude beyond them. So if air is called heaven, reports Augustine, then a firmament is something that separates water in a vaporous state from water in its denser flowing state on earth. The clouds are droplets which apart can stay aloft but when rolled together become heavy and fall — hence rain. So the air between the clouds and sea is the explanation of Genesis![80]

Augustine understands the location of supercelestial waters to be the areas in which clouds condense and the firmament the area between clouds and earth. Augustine manages to uphold the inerrancy of the biblical text, which he shows to be entirely compatible with seemingly opposing cosmological understandings. Interaction between the waters above and those below as Augustine's theory proposes cannot be supported biblically. For even the most accomplished of the Fathers, it seems, harmonizing biblical and natural cosmology and weather is a case of a square peg and a round hole.

Later in *De Genesi ad litteram*, Augustine goes on to address directly the absence of rain at Genesis 2. 5.[81] Augustine does not seem to take any issue with the physical possibility of its absence in Paradise. Genesis 2. 5 describes dormant vegetation, the absence of rain, and the absence of man and his labour. Augustine perceived the absence of rain and mankind as being a demonstration of God's omnipotent power; although vegetation springs forth naturally with rain and through man's labour, ultimately the working of the natural world is ordained by God, rather than his creations.[82] Augustine continues to address other difficulties associated with the fountain of Paradise (was it curtailed as part of the penalty of labour?) and is not concerned with explaining weather. The basic methodology for the application of science to the Bible is laid down by Augustine in his Genesis commentary. As a divinely revealed cosmology, difficult aspects of the Genesis text, such as statements concerning supercelestial waters and the absence of rain in Paradise, must be upheld. It is the job of the exegete to bring these into harmonious accord with scientific understanding and human observation.

Isidore's chapter on rain provides a thoroughly scientific explanation of this phenomenon whereby seawater — and, as other sources of Isidore's propose, water from the land — evaporates, condenses, and is released.[83] To

---

79   Cf. Wisdom 11. 21, 'sed omnia in mensura, et numero et pondere disposuisti' (but thou hast ordered all things in measure, and number, and weight).
80   O'Loughlin, 'Aquae Super Caelos (Gen 1:6–7)', p. 107.
81   Augustine, *De Genesi ad litteram*, ed. by Zycha, v. 6–11.
82   Augustine, *De Genesi ad litteram*, ed. by Zycha, v. 6.
83   Isidore, *De natura rerum*, ed. and trans. by Fontaine, XXXIII. 1–2, pp. 288–91; Isidore, *On the Nature of Things*, trans. by Kendall and Wallis, pp. 160–61.

open this chapter, Isidore includes a quotation from the prophet Amos that reflects this understanding of the hydrological cycle, where seawater is drawn up and poured out upon the earth.[84] A chapter on the nature of the sun also includes this scientific understanding of the hydrological cycle.[85] Questions concerning the absence of rain in Genesis, its first appearance, or the Flood are not addressed in Isidore's treatment of rain. The Bible is drawn upon only to support a scientific view of the hydrological cycle, while the thornier questions of the meteorology of Genesis are omitted. *De natura rerum* is one of the most influential scientific resources that the Anglo-Saxon biblical scholar might have drawn on for a scientific understanding of rain. While it provides an accurate scientific explanation of rain, it offers little resolution to the difficult questions faced by the biblical exegete.

The Anglo-Saxons were similarly tasked with negotiating the clash between biblical and scientific world views, as the biblical commentaries from Canterbury attest. Learned Anglo-Saxons would have possessed a fairly accurate scientific understanding of how weather, water, and rain operated in the natural world. Treatises on natural science and weather prognostics circulated with compustistical material in Anglo-Saxon England, showing an interest in and knowledge of such works. The third day of Creation in the *Old English Martyrology* shows a terse but sophisticated idea of the hydrological cycle.[86] A scientific understanding of water is hinted at in the Exeter Book riddles, a number of which describe water in vaporous and liquid states, some of which draw on Isidore's or Bede's versions of *De natura rerum*.[87] Discussion of the behaviour of water at various stages of Creation in Bede's commentary *On Genesis* similarly reveals a scientific explanation of weather that is incorporated into biblical exegesis. Isidore's *Etymologiae* and *De natura rerum* would have constituted the primary sources for understanding the natural world in the early medieval period. Anglo-Saxon biblical scholars who read such works would have been furnished with an understanding of rain that is not dependent on scriptural accounts of Creation, but which is based on human perceptions of the natural world.

Anglo-Saxon exegetes studying Genesis would have had to bring their understanding of contemporary weather into accord with the operation of the natural world presented in Genesis. Early medieval exegetes were required to include the waters above the firmament in their cosmology, as Smyth has observed.[88] Biblical scholars were also obliged to resolve their scientific understanding of rain with rain as it appears in Genesis. Addressing the difficult question of the existence of rain before the Flood, our Commentator

---

84   Amos 5. 8 or 9. 6.
85   Isidore, *De natura rerum*, ed. and trans. by Fontaine, XV. 1–2.
86   *The Old English Martyrology*, ed. and trans. by Rauer, pp. 66–67 and p. 246.
87   Di Sciacca, *Finding the Right Words*, pp. 133–34. See riddles 1, 2, 3, 33, 74, and 83. See also Jill Frederick's contribution to this volume, '*Modor is monigra mærra wihta*: Watering the World in Exeter Book Riddle 84'.
88   Smyth, *Understanding the Universe in Seventh-Century Ireland*, p. 94.

explains that the absence of rain at 2. 5 refers solely to the first three days prior to the creation of the sun, without which there was no rain.[89] As identified by Bernhard Bischoff and Michael Lapidge, the function of the sun in the hydrological cycle was explained by Aristotle, Basil, and Gregory of Nyssa,[90] in addition to Isidore, one of whom might be the source of this detail here. It seems that this scientific knowledge of evaporation was brought to bear on 2. 5 to show that rain was, indeed, absent, but only until the creation of the luminaries on the third day of the Hexameron. The Canterbury Commentaries thus integrate a scientific understanding of rain with the biblical view.

This interpretation cleverly navigates the challenging details of the biblical text, marrying the peculiar weather conditions prior to the Flood in the biblical narrative with the reality of how weather functions. Such a reading avoids the awkward question of how rain could be spontaneously created after Creation itself had been completed; waters placed above the firmament at Creation, intended for the Flood, and to then remain below the firmament in the form of the water cycle could conveniently solve this issue, but would open up other problematic questions concerning the necessity of two irrigation systems.

The barrenness of the earth, a condition which accompanies the penalty of labour imposed after the Fall, is central to understanding the makeup of the physical world of Genesis. The unwillingness of the earth to produce fruits for the generations up until Noah is thought to be related to a possible absence of rain in the prediluvian world. As the Commentator notes, the earth did not readily produce fruit and the Bible does not say anything in particular concerning rain before the Flood.[91] The Commentator goes on to note that the earth's barrenness is apparent at 2. 5. Rain is attested when it is said at 9. 13 'arcum meum ponam in nubibus' (I will set my bow in the clouds). This is not the first mention of rain, however, which is first referenced in the Bible at 7. 4. For the Commentator, the rainbow is a sign of rain to come, rather than something which occurs after it has rained. Rainbows were understood as vapourized water in the air and would have been known to occur following rain. Despite the imprecision of this comment, the novel phenomenon of the rainbow indicates for the Commentator that rain was also unprecedented until this point. The purification of the world's sin through the Flood seems to lift the punishment of labour imposed first on Adam and Eve for their transgression and also on Cain who is to toil following the first murder. When at 9. 14 it is stated that the rainbow will appear when the sky is covered with clouds, the Commentator explains this as indicating 'the mercy of rain, which was previously sparse, if there was any at all'.[92] He continues, 'And accordingly the earth was infertile, as may be believed, from

---

89   *PentI* 33, in *Biblical Commentaries*, ed. by Bischoff and Lapidge, pp. 308–09.
90   *Biblical Commentaries*, ed. by Bischoff and Lapidge, pp. 439–40.
91   *PentI* 65, in *Biblical Commentaries*, ed. by Bischoff and Lapidge, pp. 316–17.
92   *PentI* 83, in *Biblical Commentaries*, ed. by Bischoff and Lapidge, pp. 318–19.

the aforementioned curse, before it was purified by the Flood'.[93] Addressing the planting of the first vineyard by Noah, the Commentator tells us that not much is said about rain, the vine, or oil before the Flood.[94] The Commentator concludes that the potential for such productivity lay dormant in the earth, 'though their usefulness was not yet evident'.[95] Hence, the productivity of the earth is dependent on external conditions. The absence of these things before the Flood seems to suggest that the required conditions were absent, that is, that there was no rain. The physical conditions of Genesis suggest to the Commentator that rain was, indeed, absent from the prediluvian world (or absent in sufficient quantity so as to be productive).

In his comment on 2. 5, our Commentator is, however, reluctant to conclude that rain was entirely absent from the world. *PentI* 33 goes on to note that the absence of rain here refers specifically to Paradise. The Commentator states that in the absence of rain, Paradise was irrigated by a fountain that is erroneously referred to as the Fountain of Siloam, which rose out of the earth to water Paradise and from there the entire earth.[96] The necessity of a worldwide irrigation system if rain was only absent from Paradise is not clear; perhaps the fountain was necessary to irrigate Paradise, but its watering of the wider world may be superfluous. The Commentator concludes this gloss by stating that it is unclear whether or not there was rain before the Flood. While our Commentator is certain that rain was absent from Paradise at 2. 5, it is unclear if this condition prevailed until the Flood.

In explaining away the suggested absence of rain before the Flood, the Commentator raises questions about the nature of the waters above the firmament and, most importantly, their release in the form of rain at the time of the Flood. The solution arrived at provides a convenient answer to the question of the absence of rain. A conflicting opinion, however, appears in an earlier comment on the creation of the firmament amidst the waters that trapped a body of water above the firmament, which was released when the 'floodgates of heaven' were opened.[97] Reading these comments in light of each other, *PentI* 26 explains the phenomenon of rain at the Flood as the release of primordial waters trapped above the firmament, whereas *PentI* 33 posits that rain existed prior to the Flood. Perhaps *PentI* 26 refers specifically to the sheer volume of the Flood's waters, incorporating both 'ordinary' rain and water from above. There is some tension present in these water glosses explaining the history of rain and its first occurrence. There is evidence to suggest that rain may have been absent from the prediluvian world, but the

---

[93] *PentI* 83, in *Biblical Commentaries*, ed. by Bischoff and Lapidge, pp. 318–19.
[94] *PentI* 84, in *Biblical Commentaries*, ed. by Bischoff and Lapidge, pp. 320–21.
[95] *PentI* 84, in *Biblical Commentaries*, ed. by Bischoff and Lapidge, pp. 320–21.
[96] The conflation of the fountain that rose out of Paradise with the Fountain of Siloam in Jerusalem is related to the gloss on 2. 8 that relates how some scholars understand Jerusalem as the original site of Paradise.
[97] *PentI* 26, in *Biblical Commentaries*, ed. by Bischoff and Lapidge, pp. 306–07.

Commentator entertains the possibility that it may well have been present. The question of the absence of rain is not entirely resolved by the Commentator.

If we read Genesis as a cohesive narrative, the Commentator's conclusion concerning 2. 5 is untenable as there is nothing in the biblical text to suggest that rain was absent only until the creation of the sun on the third day. The understanding presented in the Canterbury Commentaries relies on a particular reading of Genesis 1–2. There is nothing in the biblical text to support the idea that 2. 5 refers only to this period; indeed, a synchronic reading of Genesis in its entirety suggests that the meteorological conditions of 2. 5 prevailed up until the Flood. The Old English *Genesis A* holds to the letter of the Old Testament text, stating that the rains had not yet fallen at this point (2. 5–14): 'Nalles wolcnu ða giet | ofer rumne grund regnas bæron | wann mid winde hwæðre wæstmum stod | folde gefrætwod' (Not then yet had the sky with wind carried the rains over the breadth of the land, nevertheless the earth stood adorned with fruit).[98] The sui generis nature of Creation, which happened only once, might suggest that the world that the Anglo-Saxons knew need not correspond exactly to the environment as presented in the Bible. The conflict between the scriptural account of rain and how human learning has perceived it to operate is problematic for our exegete. In order to uphold the authority of the Bible, a convenient solution is sought, namely that rain was, indeed, absent in the natural world of Genesis, but only until the third day. The Commentator's solution, however, does not stand up to scrutiny, as it cannot be derived from the biblical text itself.

The documentary hypothesis understands the Book of Genesis, and books of the Old Testament more broadly, to consist of various authorial strands that betray a complex textual history of redaction. Within this view, Genesis 1–2 is thought to comprise two narratives: firstly, 1–2. 3, and, secondly, 2. 4–24. The first Creation narrative constitutes a bona fide creation narrative that presents a complete cosmology, outlining the origins of the world, a process which is confined to the Hexameron and which culminates in the creation of man and woman. The second account does not describe the origins of the cosmos, and clearly builds on the cosmology of 1–2. 3, presenting an anthropocentric view of the world at its early stages of formation. The creation of the cosmos does not feature in this second account, and its events are not structured according to days. Thus, the absence of rain at 2. 5, a detail unique to this second account, does not depend on the creation of the sun on a particular day. Of course, medieval biblical scholars would not have been aware of the various sedimentary layers within the biblical text; Genesis 1 and the peculiar recapitulation of events from a slightly different perspective in Genesis 2 would have been viewed as a cohesive unit. From the perspective of modern biblical studies, however, it is most interesting that the resolutions which are sought for certain difficulties in the early medieval period give rise to tensions that reveal the internal contradictions inherent in the biblical text.

---

98 *Genesis A*, ed. by Doane, ll. 212b–215a, p. 155.

Conclusion

Augustine lamented that too much effort had been expended on scrupulously examining the nature of the firmament at the cost of more spiritually edifying and salutary subjects.[99] Nevertheless, throughout the corpus of patristic and medieval exegesis surveyed in this chapter, providing scientific and literal explanations of aspects of the natural world of Genesis, such as water, remained an abiding concern for exegetes. Questions concerning biblical water received various treatments throughout the history of the exegetical tradition, but were by no means conclusively and satisfactorily answered. The Commentaries explored in this chapter demonstrate how sensitive early medieval biblical scholars were to the problematic nature of the biblical text and the need for its correct interpretation by the Church. At the root of these explorations of water was a concern with making sense of the difficult aspects of the Bible and reconciling these with secular learning. Literal biblical exegesis not only was a means of delving deeper into the meaning of the Bible, but was an important location for scientific enquiry and discussion of the watery natural world. The Commentaries discussed here show that biblical water and everyday water were considered one and the same. For the Canterbury Commentator, expounding water in Genesis undoubtedly created intimate links between the distant world of the Bible and daily life in Anglo-Saxon England. There was clearly a cultural fascination with water that was related to the island location, landscape, and climate of the early medieval Atlantic Archipelago.[100] Scholars interpreting the Bible in these contexts would have been all too familiar with high rainfall and must have pondered longingly the prospect of a world without precipitation as they laboured over the biblical text.

Works Cited

Primary Sources

Ambrose, *De mysteriis*, ed. by Otto Faller, in CSEL, 73 (Vienna: Tempsky, 1955), pp. 89–116
——, *De Noe et arca*, ed. by Carl Schenkl, in CSEL, 32.1 (Vienna: Tempsky, 1897), pp. 413–97
——, *Exaemeron*, ed. by Carl Schenkl, in CSEL, 32.1 (Vienna: Tempsky, 1897), pp. 3–261

---

99 Augustine, *De Genesi ad litteram*, ed. by Zycha, IX.
100 Ferrand has concluded that the wet climate in the region of Northumbria today and in the early medieval period was much the same. It is likely that the climate of early medieval Kent and Ireland was as wet as it is today. Ferrand, 'The Hydrologic Cycle in Bede's *De Natura Rerum*', p. 364.

Augustine, *De Genesi ad litteram imperfectus liber*, ed. by Joseph Zycha, CSEL, 28.1 (Vienna: Tempsky, 1894), pp. 459–503

Basil, *The Hexaemeron*, in *Basil: Letters and Select Works*, ed. by Philip Schaff and Henry Wace, Nicene and Post-Nicene Fathers, Second Series, 8 (New York: Christian Literature Company, 1895), pp. 51–108

Bede, *De natura rerum*, ed. by Charles W. Jones, CCSL, 123A (Turnhout, Brepols, 1975)

——, *Bede's Ecclesiastical History of the English People*, ed. and trans. by Bertram Colgrave and R. A. B. Mynors (Oxford: Clarendon Press, 1969)

——, *Libri quatuor in principium Genesis usque ad nativitatem Isaac et eiectionem Ismahelis adnotationum*, ed. by Charles W. Jones, vol. I of *Bedae Venerabilis Opera: Opera exegetica*, CCSL, 118A (Turnhout: Brepols, 1967)

——, *Bede: On Genesis*, trans. by Calvin B. Kendall, Translated Texts for Historians, 48 (Liverpool: Liverpool University Press, 2008)

——, *Bede: On the Nature of Things and On Times*, trans. by Calvin B. Kendall and Faith Wallis, Translated Texts for Historians, 56 (Liverpool: Liverpool University Press, 2010)

Biblia sacra iuxta vulgatam versionem, ed. by Robert Weber, 4th edn (Stuttgart: Deutsche Bibelgesellschaft, 1994)

Biblical Commentaries from the Canterbury School of Theodore and Hadrian, ed. by Bernhard Bischoff and Michael Lapidge, Cambridge Studies in Anglo-Saxon England, 10 (Cambridge: Cambridge University Press, 1994)

De mirabilibus sacrae scripturae: *Sancti Aurelii Augustini, Hipponensis episcopi, opera omnia: Post Lovaniensium theologorum recensionem*, vol. III.2: *Opera exegetica*, in *Patrologia cursus completus, series Latina*, ed. by Jacques-Paul Migne, 221 vols (Paris: Migne, 1841–65), XXXV (1841), 2149–2200

Genesis A: A New Edition, Revised, ed. by A. N. Doane, Medieval and Renaissance Text and Studies, 435 (Tempe: Arizona Centre for Medieval and Renaissance Studies, 2013)

Holy Bible Douay-Rheims Version, with Challoner Revisions 1749–52 (Baltimore, MD: John Murphy Company, 1899)

Isidore, *De natura rerum: Traité de la Nature*, ed. and trans. into French by Jacques Fontaine (Bordeaux: Centre National de la Recherche Scientifique, 1960)

——, *Isidore of Seville: On the Nature of Things*, trans. by Calvin B. Kendall and Faith Wallis, Translated Texts for Historians, 66 (Liverpool: Liverpool University Press, 2016)

Liber de ordine creaturarum: Un anónimo irlandés del siglo VII. Estudio y edición crítica, ed. by Manuel C. Díaz y Díaz, Monografías de la Universidad de Santiago, 10 (Santiago de Compostela: Universidad de Santiago de Compostela, 1972)

The Old English Heptateuch and Ælfric's Libellus de Veteri Testamento et Novo, ed. by Richard Marsden, EETS, o.s. 330 (Oxford: Oxford University Press, 2008)

The Old English Martyrology: Edition, Translation and Commentary, ed. and trans. by Christine Rauer, Anglo-Saxon Texts, 10 (Cambridge: D. S. Brewer, 2013)

Secondary Works

Anlezark, Daniel, 'Understanding Numbers in London, British Library, Harley 3271', *Anglo-Saxon England*, 38 (2009), 137–55

——, *Water and Fire: The Myth of the Flood in Anglo-Saxon England* (Manchester: Manchester University Press, 2006)

Di Sciacca, Claudia, *Finding the Right Words: Isidore's Synonyma in Anglo-Saxon England* (Toronto: University of Toronto Press, 2008)

——, 'Isidore of Seville in Anglo-Saxon England: The *Synonyma* as a Source for Felix's *Vita S. Guthlaci*', in *Isidore of Seville and his Reception in the Early Middle Ages: Transmitting and Transforming Knowledge*, ed. by Andrew Fear and Jamie Wood (Amsterdam: Amsterdam University Press, 2016), pp. 131–58

Ferrand, Lin, 'The Hydrologic Cycle in Bede's *De Natura Rerum*', in *The Nature and Function of Water, Baths, Bathing and Hygiene from Antiquity through the Renaissance*, ed. by Cynthia Kosso and Anne Scott (Leiden: Brill, 2009), pp. 361–80

Hall, J. R.. '*Niwe flodas*: Old English *Exodus* 362', *Notes and Queries*, 22 (1975), 243–44

Irving, Edward B., 'New Notes on the Old English *Exodus*', *Anglia*, 90 (1972), 289–324

Lapidge, Michael, *The Anglo-Saxon Library* (Oxford: Oxford University Press, 2006)

Ó Cróinín, Dáibhí, 'The Irish Provenance of Bede's Computus', *Peritia*, 2 (1983), 229–47

O'Loughlin, Thomas, 'Aquae Super Caelos (Gen 1:6–7): The First Faith-Science Debate?', *Milltown Studies*, 29 (1992), 92–114

Orchard, Andy, *The Poetic Art of Aldhelm*, Cambridge Studies in Anglo-Saxon England, 8 (Cambridge: Cambridge University Press, 1994)

Smyth, Marina, 'The Date and Origin of *Liber de ordine creaturarum*', *Peritia*, 17–18 (2003–04), 1–39

——, 'The Seventh-Century Hiberno-Latin Treatise *Liber de ordine creaturarum*: A Translation', *Journal of Medieval Latin*, 21 (2011), 137–222

——, *Understanding the Universe in Seventh-Century Ireland*, Studies in Celtic History, 15 (Woodbridge: Boydell & Brewer, 1996)

Wallace-Hadrill, David Sutherland, *The Greek Patristic View of Nature* (Manchester: Manchester University Press, 1968)

MICHAEL BINTLEY

Aquas ab Aquis: Aqueous Creation in *Andreas*

Andreas is unusually wet, even for an Old English poem. Water defines the landscape of the city of Mermedonia and the approach to it but does so in a way that, like Grendel's mere in *Beowulf*, continues to frustrate critical interpretation. Dry land is shaped and defined in opposition to water, but Mermedonia confounds our preference for the sharp edges of modern cartography. Water also presents dangerous and uncertain terrain of the kind that swallows mariners in *The Whale* and *Juliana*, and operates in *Andreas* both as a metaphor for the earthly journey through life, and as a means for an initially reluctant, Jonah-like Andrew to demonstrate his heroic commitment to God's plan.[1] Finally, baptismal waters are summoned from the base of a pillar and swell up within the walls of Mermedonia, becoming a flood that prompts its devil-worshipping cannibal inhabitants into a hasty conversion.[2] Though the author's evident interest in the presence and action of water throughout the poem has been a focus of discussion throughout much of its critical history, the coherency of its symbolism has not been fully recognized. These different facets of water's symbolism in *Andreas* are not, as this essay will argue, as disparate as they may first appear, but rather serve to align the author's poetic work with the word of God. The poet creates and orders the watery world of the poem, just as God sets out and orders the parts of the

---

1 Reading, 'Baptism, Conversion, and Selfhood'. Hieatt compares this reluctance with Christ's agony in the Garden of Gethsemane in 'The Harrowing of Mermedonia', pp. 57–58. Breen describes this as 'perhaps, the inverse of a pilgrimage: instead of going willingly, Andrew must be chastised and goaded by God'; see Breen, '"What a Long, Strange Trip It's Been"', p. 71.
2 Essential commentary on the baptismal typology of the flood includes Anlezark, *Water and Fire*, pp. 174–240; also Hill, 'Figural Narrative in *Andreas*'; Hieatt, 'The Harrowing of Mermedonia'; Walsh, 'The Baptismal Flood in the Old English *Andreas*'.

Michael Bintley • (michael.bintley@bbk.ac.uk) is Senior Lecturer in Early Medieval Literature and Culture at Birkbeck College, University of London.

world in Genesis 1. 6, 'aquas ab aquis' (the waters from the waters).³ The poet's work is not merely mimetic; through the poet's authorship, the authority of the divine word flows through the poem, irrigating the barren City of Man with the living waters of heavenly Jerusalem.

The Sea

The sea journey to Mermedonia is a long one, occupying more than a third of the poem's action (ll. 189–828).⁴ Andrew replies to God's instruction to undertake this voyage in obvious terror, which might seem surprising given his professional experience of fishing, but then, as Ælfric's fisherman makes clear when pointing out the difference between hunting whales and river fishing, there is a difference between fishing with nets in the Sea of Galilee (a lake) and crossing vast expanses of open water.⁵ Andrew's words, like those of his father's namesake Jonah,⁶ betray his fear of the 'deop gelad' (deep-sea roads, l. 190), and he suggests that an angel might be better suited to the task:

Ðær mæg engel þin   eað geferan
heah of heofenum,   con him holma begang,
sealte seastreamas   ond swanrade,
waroðfaruða gewinn   ond wæterbrogan,
wegas ofer widland. (ll. 194–98)

> [Your angel can achieve that the more easily
> from on high in heaven, he knows the ocean road,
> the salt sea-currents and the swan's path,
> tumult of shore-tides and terror of the water,
> ways over open country.]

---

3   All references to the Vulgate and translations (modernized by this author) are from the *Holy Bible Douay-Rheims Version*.
4   All references to the Old English text and Modern English translation of *Andreas* are from *Andreas: An Edition*, ed. by North and Bintley, in which the poem is dated to the end of the ninth century.
5   He says that 'gebeorhlicre ys me faran to ea mid scype mynan, þænne faran mid manegum scypum on huntunge hrances' (it is easier for me to go on the river with my boat, than to go with many ships hunting whales); see *Ælfric's Colloquy*, ed. by Garmonsway, pp. 29–30. When we see Andrew and Peter for the first time in Matthew 4. 18–20 and Mark 1. 16–28, they are casting nets close enough to the shore for Jesus to address them. See also further discussion in Momma, 'Ælfric's Fisherman and the *Hronrad*'.
6   Jonah 1. 1–3. Jesus calls Simon Peter *Simon Bar Iona* in Matthew 16. 17. Andrew's address to God in his jail cell also recalls several aspects of Jonah's lament in the belly of the whale in Jonah 2. On Andrew's reluctance, which differs from his response in the analogues, see *The Acts of Andrew in the Country of the Cannibals*, trans. by Boenig, pp. xxiii–xxv. For the iconography and understanding of Jonah in early medieval England, see Elizabeth A. Alexander, 'The Sailors, the Sea Monster, and the Saviour: Depicting Jonah and the *Ketos* in Anglo-Saxon England', in this volume.

Broga is not a word used lightly in Old English poetry, and what Greenfield saw as the 'excess' of these lines indicates the 'enormity of the task from Andrew's point of view': in other words his terror of the deep.[7] To Andrew's mind the sea is its own open country, and crossing it involves confronting all of the dangers that are clearly running riot in his imagination.[8] Jesus, who pilots the ship to Mermedonia in the guise of its helmsman, tests him further before they disembark, saying that 'is se drohtað strang þam þe lagolade lange cunnaþ' (life is tough for the man long exploring ocean paths, ll. 313–14), which Andrew seemingly interprets as pride (or sarcasm). He takes the opportunity to repeat the instruction given to the twelve when Jesus sent them out to spread the good news:

> Farað nu geond ealle   eorðan sceatas
> emne swa wide   swa wæter bebugeð,
> oððe stedewangas   strate gelicgaþ.
> Bodiað æfter burgum   beorhtne geleafan
> ofer foldan fæðm;   ic eow freoðo healde. (ll. 332–36)
>
>> [Go now into all corners of the earth
>> just as far as the water surrounds us
>> or highways lie upon the plains.
>> Preach the bright faith from town to town
>> across the bosom of the land; I will keep you safe.]

The repetition of this instruction, in which Andrew presents the world in terms similar to those found in *Cædmon's Hymn* and the song of creation in *Beowulf* (ll. 90–98), demonstrates Andrew's reaffirmed faith in his mission, and an understanding that however perilous the waters encircling *terra firma* may be, the apostles have been granted protection from them.[9]

Andrew's understanding of the water's significance continues to develop as the sea journey progresses, despite the severity of the conditions they encounter (ll. 372–81) and his concern for his men (ll. 382–95). Christ assures him that all shall be well (ll. 415–26), in ways that begin to reveal the sea's symbolic function, though this is not yet apparent to Andrew, who recounts

---

7 Greenfield, *A Critical History of Old English Literature*, pp. 105–06. *Broga*, as Marilina Cesario writes, 'may also be interpreted as "portent or prodigy", glossing Latin *portentum, ostentum*, or *monstrum*', perhaps anticipating the function of water later on in the poem; see Cesario, 'Fyrenne Dracan in the *Anglo-Saxon Chronicle*', p. 169. My thanks to Dr Cesario for drawing attention to this.

8 The same word in the same form ('wæterbrogan', l. 456) is used once again, this time to refer to the terror of the waters during the calming of the storm. A similar phrase, 'wætres brogan' (terror of the water) is used in line 1395 of *Genesis* to describe the waters from which Noah's ark is protected by God. References to *Genesis* from *The Junius Manuscript*, ed. by Krapp, pp. 3–87.

9 On the development of Andrew and his becoming 'by stages a wiser and holier man', see Hamilton, '*Andreas* and *Beowulf*', p. 95.

the story of the calming of the storm in the Synoptic Gospels (ll. 438–60), further demonstrating his commitment to the voyage.[10] As their conversation continues, the brine beneath them becomes the symbolic sea across which the pilgrim travels on their journey through life.[11] Christ says:

> Oft þæt gesæleð   þæt we on sælade
> scipum under scealcum,   þonne sceor cymeð
> brecað ofer bæðweg   brimhengestum;
> hwilum us on yðum   earfoðlice
> gesæleð on sæwe,   þeh we sið nesan,
> frecna geferan. (ll. 511–16)
>
>> [Often it happens that we on paths at sea
>> in ships under crews, when there comes a storm,
>> plough the bath-way with brim stallions;
>> sometimes on the wave a hardship
>> befalls us at sea, though we survive the journey,
>> pass through the danger.]

Following these lines, Christ goes on to affirm God's power over the oceans, as he who has 'lifes geweald' (power of life), and 'brimu bindeð, brune yða ðyð ond þreateð' (binds sea-brim or crushes and rebukes the dark waves, ll. 518–20). There are strong echoes of the conclusion to Cynewulf's *Christ II* here, which describes the present life in similar terms:

> Nu is þon gelicost   swa we on laguflode
> ofer cald wæter   ceolum liðan
> geond sidne sæ   sundhengestum
> flodwudu fergen.   Is þæt frecne stream
> yða ofermæta   þe we her on lacað
> geond þas wacan woruld,   windge holmas
> ofer deop gelad.[12] (ll. 850–56)
>
>> [Now it is just as though we are carried in water-wood upon the flood,
>> over the cold water sailing in ships,
>> across the wide sea on sea-horses.
>> That is dangerous water
>> with huge waves on which we travel
>> throughout this frail world, and windy are the currents
>> over the deep sea.]

---

10   Matthew 8. 23–27; Mark 4. 35–41; Luke 8. 22–25.
11   Lee, *The Guest-Hall of Eden*, p. 93; also discussed in Hieatt, 'The Harrowing of Mermedonia', p. 58.
12   References to *Christ II* from *The Exeter Anthology of Old English Poetry*, ed. by Muir, pp. 66–81.

Cynewulf then goes on to affirm the protection offered by Christ, in providing a heavenly 'hyþe/hyðe' (port, ll. 859, 864) after stormy seas for those who have endured in fear on the waves.

This metaphorical means of expressing the soul's pilgrimage through life as a journey across perilous seas is found throughout *Andreas* in one form or another.[13] References to travel on ocean paths are at least as important as travel along stone-paved paths, the importance of which was noted by Lisa Kiser, and which she convincingly argued to represent the soul's metaphorical *weg* through life.[14] Karin Olsen, noting that 'the vision of the sea as a tract of land is a common feature in Old English poetry', and that 'ship and sea are associated with horse and land in several poems', has drawn attention to their particularly high concentration in *Andreas*, where compounds referring to the sea as land or as a path are especially common.[15] In these, Olsen finds a 'dichotomy of sea and land' that is 'rather easily noticed', and that a number of the terms used to describe pathways over land are ambiguous and could equally be used to refer to routes across water.[16] What is particularly important about these terms, I would add, is not simply that they make land of water, and vice versa, but that they are explicitly concerned with travel *through* land and water. Some, in addition, use terrestrial terms to refer to movement through the heavens, including the 'swegles gong/gang' (passage of the sun, ll. 208, 455, 869), which is fairly concrete, and the 'upweg' (upward road, l. 830) which angels follow on their return to heaven, which is less so. All of these terms for movement through the cosmos should, however, be understood in the context of a significant number of words for pathways that are used to express the journey of the soul. The best examples of this are the description of Andrew preaching 'lifes weg' (the way to life, l. 170) when God first speaks to him, which is as we leave him at the end of the poem, instructing the Mermedonians 'on geleafan weg' (in the way of faith, l. 1680). Similarly,

---

13 On Andrew's journey as pilgrimage, see Breen, '"What a Long, Strange Trip It's Been"'; and on the motif of journeys in the Vercelli Book in *Andreas* and elsewhere, see McBrine, 'The Journey Motif', pp. 308–13.
14 Kiser, '*Andreas* and the *lifes weg*', p. 67.
15 Olsen, 'The Dichotomy of Land and Sea', p. 385. Examples include 'bæðweg' (bath-way, ll. 223, 513), 'holmweg' (ocean road, l. 382), 'brimrad' (sea-road, ll. 1262, 1587), 'hran-rad' (orca's road or whale's road, ll. 266, 634), also 'hronrade' (l. 821), 'faroðstræt' (tidal road, ll. 311, 898), 'deop gelad' (deep-sea roads, l. 190), 'ealad' (sea-road, l. 441), 'lago-lad' (ocean path, l. 314), 'seolhpæð' (seal-path, l. 1714), 'swanrad' (swan's path, l. 196), and 'yðlad' (wave-road, l. 499).
16 Olsen, 'The Dichotomy of Land and Sea', p. 389. Terms used of roads, pathways, streets, and ways by land, some of which could also suggest ways by sea, include 'foldweg' (country road, ll. 206, 775), 'gang' (passage or course, l. 1694), 'herestræt' (army road or raiding highway, ll. 200, 831), 'mearcpæð' (path along a border, path through the march, or path between houses, ll. 788, 1061), 'stig' (pathway or trail, ll. 985, 1442), 'stræt' (street or highway, ll. 334, 774, 985, 1062, 1236, 1580), 'swaðu' (path, ll. 1422, 1441), and 'weg' (way, road, or path, ll. 198, 1234).

Andrew refers to Christ as having undertaken 'grundwæge gumena cynnes' (mankind's earthly road, l. 582) in the course of his life as a man. There are also other less distinct references to moral paths in the poem which might be considered part of this category. They are invoked in Matthew's lament after he has been blinded at the beginning of the poem, when he says that he has endeavoured to do God's will on 'wega gehwam' (every path, l. 65), and they also appear in Andrew's description of the 'soðfæstes swaðe' (path of the Righteous One, l. 673) followed by the apostles, a calling recognized by the high priest in the Temple of Jerusalem. Building on Olsen's arguments, one might suggest that the poem does not present a dichotomy between land and sea, so much as draw attention to passage through the earth and the heavens by all manner of moving creatures, including non-humans such as sea-creatures (and the sun!) and rational beings of various kinds — human, angelic, and divine. It is reference to the journeys of the soul, however, which elevates this beyond the circular limits of the mundane. The point to take away from this is that movement through all that is created — be it land, sea, or sky — is all movement across the perilous and uncertain waters of earthly life. Even dry land cannot be trusted, as the poem shows, but only 'se ðe wuldres blæd gestaðolade strangum mihtum' (He who established the bliss of glory with strong powers, ll. 535–36).

Mermedonia

There is no small irony in the fact that although the sea is explicitly described as perilous in *Andreas*, dry land is far more dangerous.[17] Mermedonia is an indistinct sort of place until we are confronted by its presence — standing before its walls, or thrown into the town lock-up with Matthew and the rest of the visitors imprisoned there. It is all the more frightening and imposing for it. Interpretations of the landscape continue to vary quite widely; some have maintained that it is located in fenland,[18] some take the descriptions of it as *ealand* and *igland* to indicate an island stronghold (perhaps like Alfred's Athelney), and others that it is a land beside waters, seemingly close to the sea over which Andrew is transported to reach Mermedonia.[19]

---

17 Thank you to Hugh Magennis for drawing my attention to this.
18 Brady, 'Echoes of Britons on a Fenland Frontier'. *Mearc*, coupled with *igland* (if it is read in a certain way) are the only evidence to suggest this in *Andreas*; until the flood takes place there is no mention of fenland or water in the descriptions of Mermedonia after Andrew has crossed the sea.
19 See discussion in Grosz, 'The Island of Exiles', p. 240; Bolintineanu, 'The Land of Mermedonia', pp. 153–54; *Andreas: An Edition*, ed. by North and Bintley, pp. 218–20. For a longer discussion of the location of island monasteries, see Della Hooke's contribution to this volume, 'The Sacred Nature of Rivers, Wells, Springs, and Other Wetlands in Anglo-Saxon England'.

Whichever of these is the case, there is precious little water to be seen in and around Mermedonia, whose inhabitants rely on 'blod' (blood, l. 23) rather than 'wæteres drync' (water to drink, l. 22), consuming blood and flesh in a blasphemous inversion of the Eucharist.[20] After his voyage by sea, Andrew is left slumbering beside a 'herestræte' (army-street, l. 831), 'burhwealle neh' (near the burh-wall, l. 833) of Mermedonia, where he is awoken by the rising sun (ll. 837–38), and looks across the 'wang […] fore burggeatum' (plain […] before the burh's-gates, ll. 839–40) towards the grey stone, towers, walls, and buildings of the city.[21] Although there is no water in front of Andrew at this point, his men are described as sleeping beside him here, 'on greote', which if *greot* is translated as 'sand' rather than 'earth', suggests that they are lying on the strand in plain sight of Mermedonia.[22] The city is dry inside save for blood and tears: the guards, 'dreore druncne' (blood-drunk, l. 1003) have 'deaðwang rudon' (reddened the plain of death, l. 1003) outside the prison, and their 'heorodreorig' (blood-soaked, l. 1083) corpses later contribute to this effect. Before Andrew's passion begins, little more is seen of Mermedonia save for the meeting place at a 'stapul ærenne' (brass pillar, l. 1062) beside a 'mearcpaðe' (path between two houses, or boundary path, l. 1061), where he awaits their assembly after freeing Matthew and the other captives.[23]

The description of Andrew's torture provides further information about the makeup of Mermedonia, but does not offer a single drop of flowing water, only more blood. The Mermedonians make good use of the landscape, dragging Andrew 'ofer landsceare' (over countryside, l. 1229), and 'æfter dunscræfum ymb stanhleoðo' (by caves in the downs, around stone cliffs, ll. 1232–33) and 'stræte stanfage' (streets paved with stone, l. 1236). The rocky landscape of Mermedonia lies somewhere between construction and deconstruction, natural and human-made, as the poet's description of it using the familiar

---

20 See discussion in Lee, *The Guest-Hall of Eden*, p. 89; Hamilton, 'The Diet and Digestion of Allegory'; Casteen, '*Andreas*'; Irving, 'A Reading of *Andreas*', p. 219; see also discussion in Magennis, *Anglo-Saxon Appetites*, pp. 143–59.
21 This suggests that he is not, as Garner writes, still in his ship; see Garner, *Structuring Spaces*, p. 106.
22 This is certainly the way in which *greot* is being used towards the beginning of the poem, when Andrew makes his way to the seashore (l. 238), discovering mariners there who have sailed from Mermedonia (l. 254), and is subsequently used of the ocean floor (l. 425). Later in the poem, it refers to the earth from which Abraham, Isaac, and Jacob arise (l. 794), the ground on which the guards at the town lock-up fall dead (l. 1084), and the earth from which God resurrects some of the dead Mermedonians (l. 1624). If they are indeed on the beach, this scene would then obey the type-scene of the 'hero on the beach' identified in Crowne, 'The Hero on the Beach'; followed by Irving, 'A Reading of *Andreas*', p. 220; and Hamilton, '*Andreas* and *Beowulf*', p. 94.
23 On Andrew's torture as an 'imitation of Christ's passion', see Biggs, 'The Passion of Andreas', p. 414; see also further discussion of this setting and its potential resonances for contemporary audiences in *Andreas: An Edition*, ed. by North and Bintley, pp. 82–83, 269–70.

phrase 'enta ærgeweorc' (ancient works of giants, l. 1235) confirms.²⁴ Its effect on Andrew's body is to bring forth waves of blood:

> Wæs þæs halgan lic
> sarbennum soden,   swate bestemed,
> banhus abrocen;   blod yðum weoll,
> haton heolfre. (ll. 1238–41)
>
>> [The saint's body was
>> sodden with sore wounds, blood bespattered,
>> bone-house broken; blood welled in waves,
>> with hot gore.]

These lines are immediately reminiscent of those in *The Dream of the Rood* which describe the blood and water that drench the cross: 'hwilum hit wæs mid wætan bestemed, beswyled mid swates gange' (sometimes it was spattered with water, drenched with flowing water, ll. 22–23).²⁵ But there is no water in Mermedonia just yet. The poet writes:

> Land wæron freorig;
> cealdum cylegicelum   clang wæteres þrym
> ofer eastreamas,   is brycgade
> blæce brimrade. (ll. 1259–62)
>
>> [Their lands were frozen;
>> with cold icicles did water's majesty shrink
>> across rivers and streams, ice bridged
>> the pale sea-roads.]

The landscape of the city, like that found in *The Ruin* and *The Wanderer*, finds itself at a still point, awaiting the sort of liberation from frost by God described in *Beowulf*.²⁶ Following further torture of this kind, and verbal sparring with

---

24   A classic study of this symbolism in *The Wanderer* is found in Frankis, 'The Thematic Significance of *Enta Geweorc*'.

25   References to *The Dream of the Rood*, ed. by Swanton. Reading also notes the parallel here between the wounds of Christ and Andrew in Reading, 'Baptism, Conversion, and Selfhood', pp. 3–5.

26   Bolintineanu, 'The Land of Mermedonia', p. 157; Frank draws attention to this 'raging' winter in 'North-Sea Soundings in *Andreas*', p. 6. The cold is read in a similarly literal fashion by Irving, 'A Reading of *Andreas*', p. 231, and Magennis, *Images of Community in Old English Poetry*, p. 177; and as literally and metaphorically applicable in Greenfield, *A Critical History of Old English Literature*, p. 106 (even if 'out of keeping with the literal situation'). Hieatt, however, considers this to be purely 'a metaphor for the apostle's sufferings in his dark prison, not a description of the climate of Mermedonia', but does not really explain why, deferring to Frey 'on the use of winter imagery in connection with the idea of exile from heaven'; Hieatt, 'The Harrowing of Mermedonia', p. 53; Frey, 'Exile and Elegy'. There is no reason why the literally frost-bound landscape of Mermedonia may not also reflect Andrew's psychological state. See also discussion in Stiles, '*Hapax Legomena* as Poetic Devices', p. 138.

the demon who taunts him in his cell, Andrew cries out to God, bemoaning the extent of his wounds, and says:

> Sint me leoðu tolocen,   lic sare gebrocen,
> banhus blodfag;   benne weallað,
> seonodolg swatige. (1404–06)
>
> [My limbs are dislocated, body sorely broken,
> bone-house blood-stained; the wounds well up,
> gory sinew-gashes.]

As Jonathan Wilcox has discussed, in both instances the poet uses the verb *weallan*, a word elsewhere used primarily of waters, to emphasize the effusiveness of Andrew's injuries.[27] God tells Andrew to look back at the trail through the city, at where his blood has nourished the surfaces of Mermedonia's earth, where 'geseh he geblowene bearwas standan blædum gehrodene, swa he ær his blod aget' (he saw groves standing in bloom and adorned with blossoms where his blood he poured before, ll. 1448–49). Following this, God heals Andrew's wounds, and the blood of his sacrifice is answered by the baptismal waters of purification.[28]

After an aside in which the poet makes plain the sad and hard task that has occupied him up to this point, Andrew addresses one of the 'stapulas' (pillars, l. 1494) in his cell, saying:

> Læt nu of þinum staþole   streamas weallan,
> ea inflede,   nu ðe ælmihtig
> hateð, heofona cyning,   þæt ðu hrædlice
> on þis fræte folc   forð onsende
> wæter windrynig   to wera cwealme,
> geofon geotende. (ll. 1503–08)
>
> [Let now from your pedestal streams well up,
> a river in flood, now that the Almighty
> King of Heaven commands you to send
> promptly forth into this apostate people
> wide waters in spate to men's destruction,
> an outpouring ocean.]

God has said nothing of the sort in the immediately preceding lines — what he does do is guarantee Andrew that he will no longer come to any physical harm. The poet also develops several details in this scene that are not found in the poem's Greek and Latin analogues. In the Greek *Praxeis* Andrew addresses a pillar topped by an alabaster statue, and in the Latin Casanatensis version,

---

27 Wilcox, 'Eating People Is Wrong', p. 213.
28 Implicit in this is the suggestion that the blood of sacrifice and the waters of purification are the same as those flowing from the pierced side of Christ (John 19. 34).

the column is made of marble, like the column in *Andreas*, and is topped by a marble statue.[29] In all of the poem's analogues, the flood erupts from the mouth of the statue, which is conspicuous by its absence in *Andreas*. The reasons for this are unclear. As Daniel Anlezark and others have noted, there is a correlation between this moment in *Andreas* and the burst of dragon's fire from the barrow in *Beowulf*, which may equate Andrew's victory with Beowulf's victory in death, whilst demonstrating the superiority of God's strength over Beowulf's.[30] The absence of the statue could reflect Andrew's rejection of idolatry, though the walking and talking statue earlier in the poem would argue against this. Lori Ann Garner has suggested that this may be down to the relatively sparse decoration of contemporary stone columns,[31] such as those from Reculver now housed in the crypt of Canterbury Cathedral. Though they may have used columns like these in their own constructions, and reused those of Roman manufacture to suit similar purposes, people in early medieval England were not ignorant of statuary, nor of ornate column pedestals.

The simplicity of the column and the removal of the statue, amongst other things, permit greater conceptual and exegetical fluidity in this moment, and allow the pillar to take on the significance of multiple symbols from Christian typology. In *Andreas*, the water emerges from the base of the pillar, from its 'staþol'. Just as Hrothgar stands 'on staþole' in *Beowulf* to address those gathered in Heorot, the message of the pillar is foundational. There are two episodes in the Old Testament in which the relationship between a pillar/rod and/or water/oil is particularly significant, upon which the *Andreas* poet may be drawing in this scene.[32] Most obviously, the staff wielded by Moses is given power to part the Red Sea (Exodus 14. 16), and brings water out of the rock at Horeb (Exodus 17. 1–7) to quench the thirst of the Israelites.[33] In a parallel episode in Numbers 20. 8–11, God instructs Moses to speak to a rock from whence water will emerge. Moses strikes the rock twice and the water pours forth. This life-giving water in Exodus and Numbers prefigures baptismal water in the Christian tradition. In Genesis, Jacob pours oil over

---

29 *Andreas: An Edition*, ed. by North and Bintley, pp. 294–95.
30 Anlezark, *Water and Fire*, pp. 350–59.
31 Garner, 'The Old English *Andreas*', p. 60. Waugh suggests that it is the voice of the city which 'speaks' here, in releasing the flood, voicing the Old Testament. If the poet did intend this, the removal of the statue, from whose mouth the water erupts in the analogues, would be doubly curious; see Waugh, 'The City as Speaker of the Old Testament'.
32 For further discussion of the significance of the pillar, particularly in relation to the tablets of the law in Exodus, and the speaking stone sculpture earlier in the poem, see Bintley, 'The Stones of the Wall Will Cry Out'; also Szittya, 'The Living Stone and the Patriarchs'. This pillar might also be compared with the 'apostle pillars' identified as sites of baptism in Lang, 'The Apostles in Anglo-Saxon Sculpture'.
33 A parallel noted by Hill, 'Figural Narrative in *Andreas*', pp. 266–67; Hieatt, 'The Harrowing of Mermedonia', pp. 49–51; Irving, 'A Reading of *Andreas*', p. 236; and Kiser, '*Andreas* and the *lifes weg*', p. 73.

the stone upon which he pillowed his head after dreaming of a ladder to heaven, calls it the 'domus dei' (Genesis 28. 10–22), and names the place Bethel.[34] Although there is no mention of oil in *Andreas*, the presence of the pillar in both episodes and the importance of chrism symbolizing the Holy Spirit in baptism invites comparison between the two. This connection is strengthened by the building of a 'domus dei', 'godes tempel' (God's temple, l. 1634), on the site where 'se flod onsprang' (the flood sprang forth, l. 1635).[35] If one is permitted to claim the pillar as part of a tree/rod/pillar/cross complex, the pillar could also be compared with not only the cross of Christ, but also the Tree of Life in Eden, the garden from whence four rivers issue (Genesis 2. 8–10). The same Tree of Life also appears in John's vision of heavenly Jerusalem, planted either side of the river flowing from the throne of God, where the leaves of the tree, memorably, are 'for the healing of the nations' (Revelation 22. 1–2).[36] Given that the jail will become God's temple, and Mermedonia an earthly city modelled upon heavenly Jerusalem, it is no great stretch to see these waters as the water of life in Revelation, which the righteous are invited to drink.[37]

The water itself in this moment is baptismal, as numerous commentators have observed, but it represents the full range of meanings found in water in Scripture. Firstly, as it is in the Noahic Deluge (Genesis 7–9), the flood is a transformative and purifying force intended to cleanse the sins of Mermedonia through destruction, as Daniel Anlezark has discussed in detail.[38] Its appearance is extremely violent:

> Næs þa wordlatu   wihte þon mare,
> þær se stan togan.   Stream ut aweoll
> fleow ofer foldan;   famige walcan
> mid ærdæge   eorðan þehton,
> myclade mereflod.   Meoduscerwen wearð
> æfter symbeldæge;   slæpe tobrugdon
> searuhæbbende.   Sund grunde onfeng,
> deope gedrefed;   duguð wearð afyrhted

---

34 Jacob did this once again shortly before his death; see Genesis 35. 14. Bethel later housed the Ark of the Covenant.
35 Irving, 'A Reading of *Andreas*', p. 236.
36 This appearance of the tree recalls the appearance of earthly fruit trees in Ezekiel 47. 12.
37 Revelation 22. 17: 'et Spiritus, et sponsa dicunt: Veni. Et qui audit, dicat: Veni. Et qui sitit, veniat: et qui vult, accipiat aquam vitae, gratis' (And the Spirit and the bride say: Come. And he who hears, let him say: Come. And he who thirsts, let him come: and he who will, let him take the water of life, freely).
38 Hill, 'Figural Narrative in *Andreas*', p. 268. Ferhatović also notes the 'constantly intertwined' processes of 'creation and destruction' associated with water in *Andreas* in Ferhatović, '*Spolia*-Inflected Poetics', p. 216. On the purifying properties of water, see, for example, numerous passages in Leviticus 1. 9, 13; 6. 28; 8. 6, 21; 11. 32, 36; 14. 5–6, 8–9, 50–55; 15. 6–27; 16. 4, 24–28; 17. 15; and Numbers 5. 17–27; 8. 7; 19.

þurh þæs flodes fær.   Fæge swulton,
geonge on geofene   guðræs fornam
þurh sealtes swelg;   þæt was sorgbyrþen,
biter beorþegu. (ll. 1522–33)

> [Not a jot slower than his speech did the stone
> obey, but yawned wide. A stream welled out,
> flooded the landscape. Foamy breakers
> covered the earth in the early part of day,
> a sea-flood swelled. It was a serving of mead
> after the feast-day, men who kept weapons
> were torn from sleep. Sea enfolded ground,
> stirred from the depths. The company took fright
> at this flood's assault. Doomed, they died,
> young men in ocean snatched by war-charge
> of salt's swallow. That was a brewing of sorrow,
> a bitter beer-tasting.]

The immediate function of the water here is to kill, and to terrify, forcing the Mermedonians to realize the injustice of their actions towards Andrew, and to seek guidance and solace in his words (ll. 1558–68).[39] As we have already seen, the water presents the opportunity for God to demonstrate his favour of Israel by opening the Red Sea to Moses and the Israelites (Exodus 14–15). As Thomas Hill notes, a version of this takes place in *Andreas*, as Andrew leaves his place of confinement:[40]

Þa se æðeling het
streamfare stillan,   stormas restan
ymbe stanhleoðu.   Stop ut hræðe
cene collenferð,   cacern ageaf,
gleawmod gode leof;   him wearð gearu sona
þurh streamræce   stræt gerymed.
Smeolt wæs se sigewang,   symble wæs dryge
folde fram flode,   swa his fot gestop. (1575–82)

> [The prince then commanded
> the torrent to be still, storms to abate
> around stone gates. He moved out quickly,
> brave audacious man, gave up the jail
> the wise man dear to God. For him a ready street
> through the driving current was at once cleared.
> Pleasant the plain of victory, ever dry was
> ground of flood wherever his foot advanced.]

---

39 See discussion of this episode in Anlezark, *Water and Fire*, pp. 210–33.
40 Hill, 'Figural Narrative in *Andreas*', p. 268.

Here Andrew has also taken on the role of Christ, as well as Moses, in calming the storm (see discussion above), but also in walking *through* the water like Jesus walking *on* it.[41] The waters are subsequently sucked down into a barrow which has not been mentioned before, taking fourteen of the Mermedonians with them: an especially apposite portal to hell, as a late-Saxon audience would have understood it.[42] It is, as John Hines writes, 'truly impossible to encounter this collocation of *beorg* and the accursed burial places of condemned criminals without evoking the parallel late Anglo-Saxon use of what they perceived as heathen barrows as execution sites'.[43]

There is only one further mention of the water after this, when Andrew resurrects those whom 'ær geofon cwealde' (the ocean had just killed, l. 1624), who 'þurh flodes fæ feorh aleton' (through flood's attack had lost their lives, l. 1629). After the flood's symbolic baptism of Mermedonia has taken place, they formally 'onfengon fulwihte ond freoðuwære' (received baptism and protective covenant, l. 1630), in what Hill sees as an allusion to the 'death of the Old Man and birth of the New Man in baptism', before the aforementioned church is built on the spot 'þær sio geogoð aras þurh fæder fulwiht ond se flod onsprang' (where the youth arose through the Father's baptism and the flood sprang forth, ll. 1634–45).[44] The transformation that Mermedonia undergoes following this is particularly striking, as Hugh Magennis and others have discussed.[45] For the first time, when people gather together in this newly minted 'winburg' (wine-town, l. 1637), it is not to consume human flesh, but to receive 'fullwihtes bæð' (the bath of baptism, l. 1640), to abandon 'diofolgild' (idolatry, l. 1641) and their 'ealde ealhstedas' (ancient sanctuary places, l. 1642), and to have 'æ godes riht aræred' (God's testament exalted as law, ll. 1644–45). Looking back to the scene of the flood, we can see how the act of baptism is anticipated by the limits imposed upon its watery embrace. The flood does not fill the whole land; it is confined within the stone walls of Mermedonia. An angel prevents the Mermedonians from fleeing the 'burh' (town, l. 1541) 'to dunscræfum' (to mountain caves, l. 1539) by encompassing it in a similarly baptismal 'blacan lige' (gleaming fire, l. 1541). They are not permitted 'of þam fæstenne fleame spowan' (to flee from the stronghold successfully, l. 1544), but sing their songs of grief 'innan burgum' (within the town, l. 1547), as 'leges blæstas weallas ymbwurpon' (blasts of flame enveloped the walls, l. 1553). When the floodwaters eventually die down, the

---

41 Matthew 14. 22–34; Mark 6. 45–53; John 5. 15–21.
42 See discussion in Semple, 'A Fear of the Past'; Semple, 'Images of Damnation in Late Anglo-Saxon Manuscripts'; Semple, *Perceptions of the Prehistoric in Anglo-Saxon England*.
43 See discussion of this episode and the landscapes of *Andreas* more broadly in Hines, *Voices in the Past*, pp. 61–62.
44 Hill, 'Figural Narrative in *Andreas*', p. 269; also Reading, 'Baptism, Conversion, and Selfhood', pp. 9, 13–14; and Bolintineanu, 'The Land of Mermedonia', p. 159.
45 Magennis, *Images of Community in Old English Poetry*, pp. 174–75; Bolintineanu, 'The Land of Mermedonia', p. 160.

caller of the storms bids them to rest 'ymbe stanhleoðu' (around stone-slopes, l. 1577), drawing attention one final time to the limits of the flood. Coupled with what we know about the baptism that subsequently takes place there, this emphasis on the form of the town, and the coupling of fire and water, may suggest that the walled city of Mermedonia becomes, in this moment of peril, an enormous baptismal font. When one considers the shape of walled Roman cities in the early medieval landscape, some of which tend towards the round or octagonal shape of fonts and octagonal baptisteries, this becomes less fanciful, and much easier to visualize.[46] This encourages us to think about Mermedonia as a place of baptism not only for its inhabitants, but also for those in the surrounding area. Imagining the baptism of large populations of heathens for a contemporary audience is likely to have called to mind places of mass baptism in sources such as Bede's *Historia ecclesiastica*.[47]

So far, we have seen that water in *Andreas* serves a variety of different purposes, some of which may seem at first to be quite disparate. We have seen how the sea is presented as a vast open territory, albeit one that can be traversed by skilful mariners. Although in this respect the sea is no different to any other earthly territory, in *Andreas* it is also used to express the uncertainty of earthly life — a trope well known in medieval literature, and one that draws on Augustine's interpretation of Christ and the Church as a form of Noah's ark intended to bear the faithful to salvation.[48] We have also seen the dearth of water in Mermedonia until the arrival of Andrew, whose gushing wounds portend the flood summoned from the pillar. This baptism of Mermedonia succeeds in collapsing the symbolism of the Noahic flood, the Mosaic parting of the waters, Christ's calming of the storm, and the healing waters of heavenly Jerusalem in a single act of tremendous power. There is nothing careless or random about this design; the shift in Mermedonia's fortunes, and the reshaping of heathendom into Christendom through this healing water, is carefully controlled. It is perhaps for this reason that the poem's central fitt, number eight of fifteen (l. 822), begins after the sea voyage on the liminal zone of the seashore, a place which is neither land nor water, facing the grey stone of the city.[49]

---

46 Good examples of places with wall circuits like this (rather than those with a more rectangular form like London or Winchester) are Canterbury, Silchester, Wroxeter, and Chichester.
47 See, for example, the baptism of Northumbrians at Yeavering in the time of Edwin in Bede, *HE*, II. 14. See also Carolyn Twomey's chapter in this volume, 'Rivers and Rituals: Baptism in the Early English Landscape'.
48 *Augustine: City of God*, ed. and trans. by Levine, XV. 26–27. For Bede's discussion of the ark found in his commentary *On Genesis*, building on Augustine, see Anlezark, *Water and Fire*, pp. 53–58.
49 For 'the converging evidence for the original text having just the fifteen fitts that the sectional divisions of the extant folios show', and a 'schema for the compositional design of *Andreas* to represent an arithmetically computed plan for the length of the principal sections (or fitts) as well as the length of the complete poem', see discussion in Stevick, 'Arithmetical Design', pp. 110, 99.

In each of these instances, I would suggest that the poet invites conscious reflection on water's role in the creation and ordering of the cosmos. Bede's commentary *On Genesis* offers some useful insight into the way in which the *Andreas* poet's thoughts about water may have been channelled, as Bede tackled the difficulties of what it meant for God to separate 'aquas ab aquis' (the waters from the waters) in Genesis 1. 6, and to create the firmament in the midst of the waters occupied by humans. Bede's commentary on this verse emphasizes the way in which God's control of this elusive element demonstrates the extent of his power:

> Si quem uero mouet quomodo aquae, quarum natura est fluitare semper atque ad ima delabi, super caelum consistere possint, cuius rotunda uidetur esse figura, meminerit scripturae dicentis de Deo, *Qui ligat aquas in nubibus suis ut non erumpent pariter deorsum*. Et intellegat quia *qui* infra caelum *ligat aquas* ad tempus cum uult ut non pariter decidant, nulla firmioris substantiae crepidine sustentatas sed uaporibus solummodo nubium retentas, ipse etiam potuit aquas super rotundam caeli spheram ne umquam delabantur, non uaporali tenuitate sed solidate suspendere glaciali.[50]
>
> [But if it puzzles anyone, how the waters, whose nature it is always to flow and sink to the lowest point, can settle above heaven, whose shape seems to be round, he should remember Holy Scripture saying about God, *He binds up the waters in his clouds, so that they break not out and fall down together*. And he should understand that God, *who* when he wishes and as occasion warrants *binds up the waters* beneath heaven, which are supported by no foundation of a firmer substance, but are held only by the vapours of the clouds so that they too may not fall, could also suspend the waters above the round sphere of the heaven, not with vaprous thinness but with ice-like solidity, so that they would never fall.][51]

Bede's comments on the authority of God here invite comparison with the author's role in shaping *Andreas*, corralling the waters and earth to shape and define the world of the poem. As Christopher Fee has noted, 'Writing [...] is a central and unifying activity throughout *Andreas*. At the conclusion of his passion, Andreas has been "re-written" into the "Word" through the application of the text of torture upon the parchment of his flesh; the source texts, too, have been rewritten in order to illustrate more fully this very same transformation.'[52] In much the same way, in the shaping of the poem's waters, the

---

50 Bede, *Libri quatuor in principium Genesis*, ed. by Jones, pp. 10–11.
51 Bede, *On Genesis*, trans. by Kendall, p. 76. See also John J. Gallagher, '"Streams of Wholesome Learning": The Waters of Genesis in Early Anglo-Saxon Exegesis', in this volume and, for further discussion of Bede's commentary on the flood, see Anlezark, *Water and Fire*, pp. 44–111.
52 Fee, 'Productive Destruction', pp. 59–60.

Andreas poet is also acting as a conduit for this continual process of creation. There are numerous passages in the Old Testament which can be interpreted in this way, but most relevant are those reflections on this idea found in the New Testament, most notably in the Gospel of John. Here, speaking with a Samaritan woman from whom he had begged a drink of water, when she questioned whether he was greater than Jacob, 'Jesus answered, and said to her: Whoever drinks of this water, shall thirst again; but he who drinks of the water that I will give him, shall not thirst forever. But the water that I will give him, shall become in him a fountain of water, springing up into life everlasting' (John 4. 13–14). Later on in John 7 (37–39):

> In novissimo autem die magno festivitatis stabat Jesus, et clamabat dicens: Si quis sitit, veniat ad me et bibat. Qui credit in me, sicut dicit Scriptura, flumina de ventre ejus fluent aquae vivae. Hoc autem dixit de Spiritu, quem accepturi erant credentes in eum: nondum enim erat Spiritus quia Iesus nondum erat glorificatus.
>
> > [And on the last, and great day of the festivity, Jesus stood and cried, saying: If any man thirst, let him come to me and drink. He that believeth in me, as the scripture says, Out of his belly shall flow rivers of living water. Now this he said of the Spirit which they should receive, who believed in him: for as yet the Spirit was not given, because Jesus was not yet glorified.]

These lines in John identify Christ as the source of living water, here explicitly identified as the Holy Spirit. Waters are not only, therefore, used by the poet to show God's shaping of the world, and the path of human life on the earthly journey over uncertain waters, but they also become a means by which the poet is able to describe his function as part of this process. As Christ here says that living waters will flow from within those who have drunk at his wellspring, the poet, like the stone in Andrew's cell, becomes a conduit for living water, that is, the Holy Spirit. This has significant implications when it comes to thinking about the function of poetry in late Saxon England, and the way in which the poet saw his role as part of the process of mediating the (apocryphal) acts of Andrew in Mermedonia to his audience. In order to do this, as Brian Shaw notes, he was required to both 'be faithful to the spirit of the original and, simultaneously, rework the texture of the story to reinforce the idea of the validity of the spoken word, as Andrew's story moves from the Latin to the vernacular'.[53] In doing so he 'was consciously deepening the liturgical dimensions of his story', as Mary Walsh observed.[54] The poet, as the conduit for the living water passed on to him as the textual tradition

---

53 Shaw, 'Translation and Transformation in *Andreas*', p. 165.
54 Walsh, 'The Baptismal Flood in the Old English *Andreas*', p. 158. Hill also draws this comparison with the liturgy ('Figural Narrative in *Andreas*', p. 271), in which he was followed by Hamilton ('The Diet and Digestion of Allegory', p. 153).

of the poem, marshals, controls, and separates *aquas ab aquis* following the example of God, Noah, Moses, Christ, and Andrew. As this water is shared with the listener, *Andreas* both describes water and *is* living water.[55]

For the *Andreas* poet and others like him in early medieval Europe and further afield who were accustomed to writing and receiving literature in this way, mindful of the relationship between the material and the spiritual, the presence of water in this and other works is likely to have been profoundly different from that of modern readers. Stripped of the distance afforded by metaphor, modern clichés of conceptual fluidity, or the watery depths of the unconscious (not to mention flesh melting, thawing, and resolving itself into a dew, or all that is solid melting into air), the ubiquity of water in *Andreas* and its absolute necessity to the life of all things reflects what its author saw as the indivisibility of creation from the Creator. Water sustains the earth, and the body, because creation is ordered in this way — and the sustenance of the Holy Spirit is no different. When water gives life, or bears the individual upon its surface, it is through God's will, and when water is terrible, it is because God is terrible. In an age when saints like Wilfrid might call upon rainclouds, or kings like Cnut set themselves against the invulnerable tide to make a point about the limited power of earthly rulers, the membrane separating the symbolic waters of Scripture from the waters of the earth grew vanishingly thin. It may pay to think of water in this world not only as a *symbol* of the living word, and the textual and scriptural tradition transmitting the word of God, but as something rather closer to the physical embodiment of this, in the same way that the fruits of the earth might become body and blood: the *thing-in-itself* surging with divine potentiality, whether in ocean, river, or inkhorn.

Works Cited

Primary Sources

The Acts of Andrew in the Country of the Cannibals: Translations from the Greek, Latin, and Old English, trans. by Robert Boenig (New York: Garland, 1991)

Ælfric's Colloquy, ed. by G. N. Garmonsway, rev. edn, Exeter Medieval Texts and Studies (Exeter: University of Exeter Press, 1991)

Andreas: An Edition, ed. by Richard North and Michael D. J. Bintley, Exeter Medieval Texts and Studies (Liverpool: Liverpool University Press, 2016)

Augustine: City of God, vol. IV: *Books 12–15*, ed. and trans. by Philip Levine, Loeb Classical Library, 414 (Cambridge, MA: Harvard University Press, 1966)

Bede, *Bede's Ecclesiastical History of the English People*, ed. and trans. by Bertram Colgrave and R. A. B. Mynors (Oxford: Clarendon Press, 1969)

---

55 This process also strongly suggests the baptism of the Holy Spirit and the beginning of apostolic evangelization in Acts 1 and 2.

―――, *Bede: On Genesis*, trans. by Calvin B. Kendall, Translated Texts for Historians, 48 (Liverpool: Liverpool University Press, 2008)

―――, *Libri quatuor in principium Genesis usque ad nativitatem Isaac et eiectionem Ismahelis adnotationum*, ed. by Charles W. Jones, vol. 1 of *Bedae Venerabilis Opera: Opera exegetica*, CCSL, 118A (Turnhout: Brepols, 1967)

The Dream of the Rood, ed. by Michael Swanton, Exeter Medieval Texts and Studies (Exeter: Exeter University Press, 1996)

The Exeter Anthology of Old English Poetry: An Edition of Exeter Dean and Chapter MS 3501, ed. by Bernard J. Muir, 2 vols (Exeter: Exeter University Press, 1994)

Holy Bible Douay-Rheims Version, with Challoner Revisions 1749–52 (Baltimore, MD: John Murphy Company, 1899)

The Junius Manuscript, ed. by George Philip Krapp, The Anglo-Saxon Poetic Records, 1 (New York: Columbia University Press, 1931)

Secondary Works

Anlezark, Daniel, *Water and Fire: The Myth of the Flood in Anglo-Saxon England* (Manchester: Manchester University Press, 2006)

Biggs, Frederick M., 'The Passion of Andreas: *Andreas* 1398–1491', *Studia Philologica*, 85 (1988), 413–27

Bintley, Michael D. J., 'The Stones of the Wall Will Cry Out: Lithic Emissaries and Marble Messengers in *Andreas*', in *Insular Iconographies: Essays in Honour of Jane Hawkes*, ed. by Meg Boulton and Michael D. J. Bintley (Woodbridge: Boydell and Brewer, 2019), pp. 61–79

Bolintineanu, Alexandra, 'The Land of Mermedonia in the Old English *Andreas*', *Neophilologus*, 93 (2009), 149–64

Brady, Lindy, 'Echoes of Britons on a Fenland Frontier in the Old English *Andreas*', *Review of English Studies*, 61 (2010), 669–89

Breen, Nathan A., '"What a Long, Strange Trip It's Been": Narration, Movement and Revelation in the Old English *Andreas*', *Essays in Medieval Studies*, 25 (2008), 71–79

Casteen, John, '*Andreas*: Mermedonian Cannibalism and Figural Narrative', *Neuphilologische Mitteilungen*, 75 (1974), 74–78

Cesario, Marilina, '*Fyrenne Dracan* in the Anglo-Saxon Chronicle', in *Textiles, Text, Intertext: Essays in Honour of Gale R. Owen-Crocker*, ed. by Maren Clegg Hyer and Jill Frederick (Woodbridge: Boydell & Brewer, 2016), pp. 153–70

Crowne, D. K., 'The Hero on the Beach: An Example of Composition by Theme in Anglo-Saxon Poetry', *Neuphilologische Mitteilungen*, 61 (1960), 362–72

Fee, Christopher, 'Productive Destruction: Torture, Text, and the Body in the Old English *Andreas*', *Essays in Medieval Studies*, 11 (1994), 51–62

Ferhatović, Denis, '*Spolia*-Inflected Poetics of the Old English *Andreas*', *Studies in Philology*, 110 (2013), 199–219

Frank, Roberta, 'North-Sea Soundings in *Andreas*', in *Early Medieval English: Texts and Interpretations. Studies Presented to Donald G. Scragg*, ed. by Elaine Treharne and Susan Rosser (Tempe: ACMRS, 2002), pp. 1–11

Frankis, P. J., 'The Thematic Significance of *Enta Geweorc* and Related Imagery in *The Wanderer*', *Anglo-Saxon England*, 2 (1973), 253–69

Frey, Leonard, 'Exile and Elegy in Anglo-Saxon Christian Epic Poetry', *Journal of English and Germanic Philology*, 62 (1963), 293–302

Garner, Lori Ann, 'The Old English *Andreas* and the Mermedonian Cityscape', *Essays in Medieval Studies*, 24 (2007), 53–63

———, *Structuring Spaces: Oral Poetics and Architecture in Early Medieval England* (Notre Dame: University of Notre Dame Press, 2011)

Greenfield, Stanley B., *A Critical History of Old English Literature* (New York: New York University Press, 1965)

Grosz, O. J. H., 'The Island of Exiles: A Note on *Andreas* 15', *English Language Notes*, 7.4 (1970), 240

Hamilton, David, '*Andreas* and *Beowulf*: Placing the Hero', in *Anglo-Saxon Poetry: Essays in Appreciation, for John C. McGalliard*, ed. by Lewis E. Nicholson and Dolores Warwick Frese (Notre Dame: University of Notre Dame Press, 1975), pp. 81–98

———, 'The Diet and Digestion of Allegory in *Andreas*', *Anglo-Saxon England*, 1 (1972), 147–58

Hieatt, Constance B., 'The Harrowing of Mermedonia: Typological Patterns in the Old English *Andreas*', *Neuphilologische Mitteilungen*, 77 (1976), 49–62

Hill, Thomas D., 'Figural Narrative in *Andreas*', *Neuphilologische Mitteilungen*, 70 (1969), 261–73

Hines, John, *Voices in the Past: English Literature and Archaeology* (Cambridge: D. S. Brewer, 2004)

Irving, Edward B., 'A Reading of *Andreas*: The Poem as Poem', *Anglo-Saxon England*, 12 (1983), 215–37

Kiser, Lisa, '*Andreas* and the *lifes weg*: Convention and Innovation in Old English Metaphor', *Neuphilologische Mitteilungen*, 85 (1984), 65–75

Lang, James, 'The Apostles in Anglo-Saxon Sculpture in the Age of Alcuin', *Early Medieval Europe*, 8.2 (1999), 271–82

Lee, Alvin A., *The Guest-Hall of Eden* (New Haven, CT: Yale University Press, 1972)

Magennis, Hugh, *Anglo-Saxon Appetites: Food and Drink and their Consumption in Old English and Related Literature* (Dublin: Four Courts Press, 1999)

———, *Images of Community in Old English Poetry* (Cambridge: Cambridge University Press, 1996)

McBrine, Patrick, 'The Journey Motif in the Poems of the Vercelli Book', in *New Readings in the Vercelli Book*, ed. by Samantha Zacher and Andy Orchard (Toronto: University of Toronto Press, 2009), pp. 298–317

Momma, Haruko, 'Ælfric's Fisherman and the *Hronrad*: A Colloquy on the Occupation', in *The Maritime World of the Anglo-Saxons*, ed. by Stacey S. Klein, William Schipper, and Shannon Lewis-Simpson, Essays in Anglo-Saxon Studies, 5 (Tempe: ACMRS, 2014), pp. 303–22

Olsen, Karin, 'The Dichotomy of Land and Sea in the Old English *Andreas*', *English Studies*, 79 (1998), 385–94

Reading, Amity, 'Baptism, Conversion, and Selfhood in the Old English *Andreas*', *Studies in Philology*, 112 (2015), 1–23

Semple, Sarah, 'A Fear of the Past: The Place of the Prehistoric Burial Mound in the Ideology of Middle and Later Anglo-Saxon England', *World Archaeology*, 30 (1998), 109–26

——, 'Images of Damnation in Late Anglo-Saxon Manuscripts', *Anglo-Saxon England*, 32 (2003), 31–45

——, *Perceptions of the Prehistoric in Anglo-Saxon England: Religion, Ritual and Rulership in the Landscape* (Oxford: Oxford University Press, 2013)

Shaw, Brian, 'Translation and Transformation in *Andreas*', in *Prosody and Poetics in the Early Middle Ages: Essays in Honour of C. B. Hieatt*, ed. by M. J. Toswell (Toronto: University of Toronto Press, 1995), pp. 164–79

Stevick, Robert D., 'Arithmetical Design of the Old English *Andreas*', in *Anglo-Saxon Poetry: Essays in Appreciation for John C. McGalliard*, ed. by Lewis E. Nicholson and Dolores Warwick Frese (Notre Dame: University of Notre Dame Press, 1975), pp. 99–115

Stiles, Laura S., '*Hapax Legomena* as Poetic Devices in the Old English *Andreas*' (unpublished doctoral thesis, University of Georgia, 2002)

Szittya, Penn R., 'The Living Stone and the Patriarchs: Typological Imagery in *Andreas*, lines 706–810', *Journal of English and Germanic Philology*, 72 (1973), 167–74

Walsh, Marie Michelle, 'The Baptismal Flood in the Old English *Andreas*: Liturgical and Typological Depths', *Traditio*, 33 (1977), 137–58

Waugh, Robin, 'The City as Speaker of the Old Testament in *Andreas*', in *Old English Literature and the Old Testament*, ed. by Michael Fox and Manish Sharma (Toronto: University of Toronto Press, 2012), pp. 253–65

Wilcox, Jonathan, 'Eating People Is Wrong: Funny Style in *Andreas* and its Analogues', in *Anglo-Saxon Styles*, ed. by Catherine Karkov and George Hardin-Brown, SUNY Series in Medieval Studies (Albany: State of New York University Press), pp. 201–22

HELEN APPLETON

Water, Wisdom, and Worldliness in the Anglo-Saxon Prose Lives of Guthlac

Introduction: The Texts and their Relations

The Mercian hermit Guthlac of Crowland (*c.* 673–714) is the subject of a Latin hagiography composed *c.* 730–40 by the monk Felix.[1] This *Vita*, dedicated to the East Anglian king Ælfwald, was translated into Old English prose prior to the mid-tenth century.[2] Water, both literal and metaphorical, plays a prominent part in Felix's *Vita sancti Guthlaci*. Guthlac's hermitage site, on the Fenland island of Crowland, was reachable only by boat, and several of Guthlac's miracles relate directly to this watery landscape. The water of the Fens also has a symbolic value: it separates Guthlac from the world, and his mastery of this hostile environment attests to his sanctity. In addition, Felix's *Vita* is infused with purely metaphorical waters. Drawing on a variety of source texts, Felix depicts ordered streams, gentle dew, and nourishing drafts that signify wisdom, while fierce whirlpools and stormy seas represent the turbulence of the world. All these waters intersect and merge to produce a

---

1 The *terminus ad quem* for the *Vita*'s composition is suggested by its dedication to Ælfwald, who died 749, and the description of Guthlac's contemporaries Cissa and Wilfrid as still living. The lack of reference to Guthlac in Bede's *Historia Ecclesiastica*, despite his interest in the region, suggests a *terminus post quem* of *c.* 730. Felix, *Life of St Guthlac*, ed. and trans. by Colgrave, pp. 18–19.
2 A number of other Guthlac materials survive from Anglo-Saxon England. The Exeter Book contains two Old English poems about Guthlac: *Guthlac A* and *Guthlac B*. Guthlac also appears in the *Old English Martyrology*, and is referenced in the pre-Conquest list of saints' resting places preserved in London, British Library, MS Stowe 944 and Cambridge, Corpus Christi College, MS 201. See *Guthlac A* and *Guthlac B*, ed. by Muir, pp. 111–59; *The Guthlac Poems*, ed. by Roberts; *The Old English Martyrology*, ed. and trans. by Rauer, p. 80; *Die Heiligen Englands*, ed. by Liebermann, p. 11.

> Helen Appleton • (helen.appleton@ell.ox.ac.uk) is a member of the Faculty of English, University of Oxford.

Meanings of Water in Early Medieval England, ed. by Carolyn Twomey and Daniel Anlezark, Studies in the Early Middle Ages, 47 (Turnhout: Brepols, 2021), pp. 211–239

text with both literal and metaphysical significance, successfully promoting Guthlac through its edifying portrait of a man of God.[3]

The Latin and Old English *vitae* of Guthlac have a complex textual tradition; they present varying treatments of these waters, related to the ways successive authors handled their sources. In its surviving form the Old English translation is generally faithful to the physical geography of Felix's text, taking pains to produce an effective description of the landscape. However, Felix's more metaphorical uses of water are not replicated with such fidelity. The intertextual and cross-referential waters that characterize Felix's text are often reduced or altogether omitted from the Old English. The result is a streamlining of the metaphorical waters, leaving only more commonplace images, well steeped in Christian tradition. This chapter will highlight hagiographic water's significance to readers in Latin and the vernacular through an examination of how water in all its forms contributes to the spiritual richness of Felix's text, as well as considering the effect that the Old English *Life of Guthlac*'s omissions and adaptations have on our reading of the saint. Differences between the two texts offer important insights into the translation's origins and development, exposing the original translator's skill, the role of copyists in further shaping the *Life*, and how the texts' respective audiences were expected to respond to water imagery.

Nothing is known of Felix beyond what is revealed to us by his work.[4] Felix, who describes himself in his Prologue as 'catholicae congregationis vernaculus' (a servant of the Catholic community, pp. 60, 61), must have been a monk, but the identity of his monastery is unknown.[5] The indebtedness of the *Vita sancti Guthlaci* to earlier works gives a good sense of the texts available to Felix. As Bertram Colgrave, who edited the *Vita*, highlights, the text is heavily dependent on other sources; even Felix's description of himself is, in fact, one of many borrowings from Aldhelm.[6] Felix also quotes Scripture extensively, and draws on Vergil, Bede's *Prose Life of Cuthbert*, Sulpicius Severus's *Vita Martini*, Evagrius's translation of Athanasius's *Vita Antonii*, Jerome's *Vita S. Pauli*, and the works of Gregory the Great.[7] Felix's use of source materials, as Britton Brooks has examined, creates both short lexical echoes and complex intertextual allusions that bolster Guthlac's standing.[8] Longer quotations locate the saint in a hagiographic tradition, while shorter

---

3 On Guthlac as 'vir Dei Guthlacus', see Kurtz, 'From St Antony to St Guthlac', pp. 126–27.
4 On attempts to identify Felix, see *Das angelsächsiche Prosa-Leben des hl. Guthlac*, ed. by Gonser, pp. 14–15.
5 Felix, *Life of St Guthlac*, ed. and trans. by Colgrave, p. 16. All quotations from the *Vita* and translations are from Colgrave. All other translations are the author's unless otherwise stated.
6 The opening exordium of Aldhelm's *Epistola ad Acircium*. Aldhelm, *Epistola ad Acircium*, ed. by Ehwald, p. 61.
7 Felix, *Life of St Guthlac*, ed. and trans. by Colgrave, pp. 16–17, 57–58. See also Weston, 'Guthlac Betwixt and Between', pp. 5–6.
8 Brooks, 'Felix's Construction of the English Fenlands'.

echoes of familiar scriptural, Aldhelmian, and Vergilian Latin serve to catalyse the reader's attention and highlight particular images. These references also afford the educated reader access to a larger interpretative framework within which Guthlac may be placed.

Felix's *Vita sancti Guthlaci* appears to have been popular in England throughout the early medieval period. Eight of the surviving thirteen medieval manuscripts are from the Anglo-Saxon period, the majority late, suggesting recopying.[9] The Old English prose *Life of Guthlac* survives in only two manuscripts: a complete text is preserved in London, British Library, MS Cotton Vespasian D.xxi, fols 18–40 (s. xi²); and two chapters, centred on Guthlac's hermitage and vision of hell, appear, somewhat reframed, as Homily 23 in the late tenth-century Vercelli Book.[10] The two versions of the Old English text appear to depend on a common source, but at some considerable remove, indicating repeated recopying.[11] Paul Gonser, in his 1909 edition of the *Life of Guthlac*, argues that the Vercelli and Vespasian texts descend independently from the original.[12] The Vespasian text is Late West Saxon but has been subject to what Donald Scragg terms 'large-scale linguistic modernization'.[13] Vercelli Homily 23 contains a larger number of Anglian words and evidences the translation's original early non–West Saxon form.[14] The retention of material related to Anglian kings and Guthlac's Mercian origins, as well as evidence that his cult was always stronger in Mercia, make a non–West Saxon origin

---

9 Felix, *Life of St Guthlac*, ed. and trans. by Colgrave, pp. 26–46. All the manuscripts are entirely Insular in origin except one: Boulogne, Bibliothèque Municipale, MS 106 (637). It is unclear which of that manuscript's contents, including the Guthlac material, were written at Bath and which at St Omer.

10 *Das angelsächsiche Prosa-Leben des hl. Guthlac*, ed. by Gonser, pp. 36–52. As Scragg observes, Vercelli 23 is 'not a homily in any conventional sense'. It faithfully excerpts the longer narrative, and 'makes no concession to an audience unfamiliar with the context'. 'Vercelli Homily 23', ed. by Scragg, p. 381, n. 3. The Vespasian text has been edited in *The Anglo-Saxon Version of the Life of St Guthlac*, ed. and trans. by Goodwin, with collations from Vercelli, and in *Das angelsächsische Prosa-Leben des hl. Guthlac*, ed. by Gonser, and Crawford [now Roberts], 'Guthlac', who both print the two texts in parallel. Vercelli Homily 23 has been edited in *Vercelli Homilies IX–XXIII*, ed. by Szarmach, Pilch, 'The Last Vercelli Homily', and 'Vercelli Homily 23', ed. by Scragg. The Vespasian text has been translated by Goodwin and in *Anglo Saxon Prose*, trans. by Swanton, pp. 39–62. Quotations from the Old English *Life* are from the Vespasian text edited by Gonser, unless otherwise specified.

11 The relationship between the two texts was first noted by Goodwin in 1848: *The Anglo-Saxon Version of the Life of St Guthlac*, ed. and trans. by Goodwin, pp. iv–v.

12 *Das angelsächsiche Prosa-Leben des hl. Guthlac*, ed. by Gonser, p. 48. See also Gonser, pp. 36–42; Crawford, 'Guthlac', pp. 124–29.

13 Scragg, 'The Corpus of Anonymous Lives', p. 210.

14 Scragg, 'The Corpus of Anonymous Lives', p. 210. On Anglian features, see *Das angelsächsiche Prosa-Leben des hl. Guthlac*, ed. by Gonser, pp. 49–51; Crawford, 'Guthlac', pp. 48, 162–66, 189–209. Further support for an Anglian origin for the translation comes from Roberts's analysis of unhistorical gender congruence in the text. Roberts, 'Traces of Unhistorical Gender Congruence', pp. 36–37. See also Crawford, 'Guthlac', pp. 218–24.

probable for cultural reasons also.[15] The spellings preserved in the Vercelli text suggest that both manuscripts' common source was relatively early, as does evidence that the original translator used the *Romanum* Psalter.[16] Jane Roberts has concluded that 'an Alfredian date cannot be proved for the making of the original translation of the *Vita sancti Guthlaci* into English, but it is the most attractive of the various possibilities open'.[17] Such an early date for an Old English prose hagiography may account for the translation's unusual presentation; it exhibits none of the deference to authority or consciousness of its own status as a translation that characterize later prose lives, such as those of Ælfric.[18] In common with one other relatively early vernacular prose hagiographic translation, *The Life of Mary of Egypt*, the *Life of Guthlac* does not indicate that is it a translated text.[19] The Old English text includes Felix's Prologue, preserves the use of first-person pronouns, and refers to Felix's contemporaries as living, so reads as if it is itself the work of Felix.

Any attempt to compare the *Life of Guthlac* to the *Vita sancti Guthlaci* with the aim of establishing the translator's approach is complicated by the fact that neither surviving version of the Old English *Life* is a good record of the original, as well as by discrepancies between Colgrave's edition of Felix's *Vita* and the translation's source text.[20] As Roberts observes, Gonser's collation of the Vespasian and Vercelli texts reveals that 'a fuller and much different original translation lay behind the Life'.[21] Roberts goes on to state: 'words, phrases and even sentences have disappeared from it [Vespasian] and from the homily, and, on the evidence of the parallel parts, at different times and to different ends'.[22] The Old English text remains an editorial challenge: both manuscript copies contain many unusual words, the Vespasian text has clearly

---

15 On evidence for Guthlac's cult, see Roberts, 'An Inventory of Early Guthlac Materials'.
16 Appleton, 'The Psalter in the Prose Lives of Guthlac'. Günter Scherer thought the *Life of Guthlac* late, but most scholars take an opposing view. Scherer, *Zur Geographie und Chronologie des angelsächsischen Wortschatzes*, p. 5.
17 Roberts, 'The Old English Prose Translation', p. 367.
18 See Whatley, 'Lost in Translation'; Stanton, *The Culture of Translation in Anglo-Saxon England*.
19 Whatley, 'Late Old English Hagiography', p. 450; Whatley, 'Lost in Translation', p. 193. Compare *The Old English Life of St Mary of Egypt*, ed. and trans. by Magennis.
20 Roberts, 'Two Readings in the Guthlac Homily', p. 201. As no manuscript of the *Vita* is especially reliable, Colgrave's is a 'reconstructive text'. The result may come close to Felix's eighth-century original but, as Whitney French Bolton highlights, differs from the source used by the Old English translation. Bolton narrows the source manuscript down to a member of Colgrave's Group 4, with C_2 (Cambridge, Corpus Christi College, MS 389, s. x^2) and N (London, British Library, MS Cotton Nero E.i, s. xi^{med}) especially like. In his critical apparatus Colgrave provides readings from these two manuscripts, enabling more accurate comparison. Felix, *Life of St Guthlac*, ed. and trans. by Colgrave, p. 52; Bolton, 'The Manuscript Source'. See also Crawford, 'Guthlac', pp. 70–78.
21 Roberts, 'Two Readings in the Guthlac Homily', p. 201. See also Crawford, 'Guthlac', p. 125.
22 Roberts, 'Two Readings in the Guthlac Homily', p. 201. For a fuller analysis of differences between the texts, see Crawford, 'Guthlac', pp. 136–50.

been extensively revised, and the Vercelli text has numerous copying errors.[23] Editors have favoured emendation in many places, obscuring readings that may be closer to the original translation.[24] As Roberts observes, 'any attempt to present a detailed comparison of the *Vita* and the *Life* to illustrate the translator's methods (as conducted by Gonser 1909, pp. 52–94) can produce little trustworthy evidence on this score, because so much of the original translation has been obscured by revision'.[25] Nevertheless, it is possible, using the editions of Gonser and Colgrave, together with the work of Roberts, to offer an analysis of how the Vespasian text differs in effect to Felix's work and, by employing the Vercelli text and critical apparatus, tentatively to suggest some changes which may be the product of the translator's labour rather than later copyists' alterations.

The Old English translation of the *Vita sancti Guthlaci* was presumably intended as a substitute for the Latin, directed to an audience unable to engage with Felix's original, yet desirous of the narrative in its entirety. As Roberts notes, it would take about an hour to read out the *Life of Guthlac*, making it too long to be used in most preaching contexts, although it could serve as the source of preaching material, as Vercelli 23 shows.[26] We might imagine the text as refectory reading, or as intended for a literate, lay audience — perhaps members of an Anglian noble family especially devoted to the saint. The nature of Felix's Latin may have encouraged the production of a translation at a relatively early point. Colgrave notes that Felix's style is difficult, and clearly caused problems even for highly educated medieval readers — Orderic Vitalis, writing in the twelfth century, described the *Vita* as 'prolixus et aliquantulum obscurus' (very long and the style somewhat obscure).[27] This complexity influences the approach shown by the Vespasian and Vercelli texts, which eliminate many of Felix's more convoluted phrases. Yet other omissions from the Vespasian text seem to be driven by more than a desire for clarity: they adjust the emphasis of the material, altering its message.[28] It is clear from discrepancies between the two texts that some of this simplification and reorientation originated not with the translator, but with later copyists. As Roberts observes 'each [manuscript] preserves structures modelled upon the Latin where the other shows simplification or an attempt at simplification'. As the Vercelli text only overlaps with a small portion of the Vespasian *Life*, scope to recover the nature of the original by comparison is severely limited.[29]

---

23 'Vercelli Homily 23', ed. by Scragg, p. 381; Crawford, 'Guthlac', p. 227.
24 On this problem, with examples, see Roberts, 'Two Readings in the Guthlac Homily'.
25 Crawford, 'Guthlac', p. 233.
26 Roberts, 'The Old English Prose Translation', p. 365.
27 Felix, *Life of St Guthlac*, ed. and trans. by Colgrave, p. 17. Latham, Howlett, and Ashdowne, *Dictionary of Medieval Latin from British Sources* evidences Felix's fondness for neologisms.
28 See for example Waugh, 'The Blindness Curse and Nonmiracles'; Whatley, 'Lost in Translation'.
29 Roberts, 'Two Readings in the Guthlac Homily', p. 202.

Despite these textual problems, the *Life of Guthlac* is generally viewed as a close translation of the Latin. Yet, for all its apparent fidelity to Felix, the *Life* has a distinct identity. The *Life* restructures the *Vita* considerably; the Prologue and fifty-three chapters of the Latin become four chapters in the Old English, with a Prologue and twenty-two sections.[30] Felix's complex descriptive style is much simplified, and, as Lisa Weston notes, the majority of the intertextual allusions which characterize Felix's prose are lost.[31] I suggest that while much of the stylistic simplification seen in the *Life of Guthlac* is due to streamlining by later copyists, the original translator recognized phrasal echoes of well-known Latin texts but, as these would be lost in the vernacular, carefully and stylishly rephrased these passages to preserve their original effect. The translator has, as Colgrave puts its, 'unusual skill', handling Felix's difficult style with assurance.[32] This excellent Latinity would have allowed the translator to perceive and respond to Felix's myriad intertextual echoes. For example, in Chapter 2, Felix quotes Vergil in his description of the origins of Penwalh, Guthlac's father: 'Huius etiam viri progenies per nobilissima inlustrium regum nomina antiqua ab origine Icles digesto ordine cucurrit' (Moreover the descent of this man was traced in set order through the most noble names of famous kings, back to Icel in whom it began in days of old, pp. 74, 75). The phrase 'antiqua ab origine' is *Aeneid* I. 642, the ancestry of Dido.[33] Section 1 of the Old English translation retains this important information but reframes it, as the complimentary *Aeneid* echo would be undetectable in Old English. Vergil is replaced with paired superlatives that assert the greatness of the descendants of Icel: 'He wæs þæs yldestan and þæs æþelstan cynnes, þe Iclingas wæron genemnede' (He was of the oldest and noblest kin, who were named the Iclings, p. 104). Similarly, Chapter 34's echo of *Aeneid* II. 303 'et arrectis auribus' (with ears alert, pp. 110, 111) becomes the paired 'and hawode and hercnode' (and looking and listening, p. 136) in *Life of Guthlac* Section 6, highlighting this moment with alliteration and homoioteleuton. Instances such as these make it probable that some of the *Life*'s alterations, including those to the water imagery, are the work of the original translator, deftly adapting Felix's echoing style in order to preserve much of its effect for a vernacular audience.

The difficulties in separating what may be the translator's response to quotation from later copyists' simplification can be exemplified by the *Life of Guthlac*'s treatment of the opening of Felix's Chapter 19, which describes the dawn of the first day of Guthlac's spiritual life. Felix's Latin reads: 'Ergo exutis umbrosae noctis caliginibus, cum sol mortalibus egris igneum demoverat

---

30 For a discussion of this restructuring, see Roberts, 'The Old English Prose Translation', pp. 364–65; Weston, 'Guthlac Betwixt and Between', p. 15.
31 Weston, 'Guthlac Betwixt and Between', p. 16. On other alterations, see Roberts, 'Guthlac of Crowland and the Seals of the Cross', pp. 118–21; Waugh, 'The Blindness Curse and Nonmiracles'; Whatley, 'Lost in Translation'.
32 Felix, *Life of St Guthlac*, ed. and trans. by Colgrave, p. 19.
33 Appleton and Robinson, 'Further Echoes of Vergil's *Aeneid*', p. 354.

ortum et matutini volucres avino forcipe pipant, tunc indutos artus agresti de spatulo surgens arrexit, et signato cordis gremio salutari sigillo' (So when the mists of the dark night had been dispersed and the sun had risen in fire over helpless mortals, while the winged tribe chirped their morning songs from the beaks that birds possess, then he dressed and raised his limbs from his rustic bed and, signing himself with the sign of salvation on his breast, pp. 82, 83). This image is rendered in Section 2 of the *Life* as: 'Mid þy þære nihte þystro gewiton, and hit dæg was, þa aras he and hine sylfne getacnode insegle Cristes rode' (When the darkness of night had departed, and it was day, then he rose and signed himself with the seal of Christ's cross, p. 110). While the use of 'mid' in the Old English neatly mimics the ablative absolute construction of Felix's Latin, almost all the associated imagery is removed. Felix's dawn is full of Vergilian phrasing — 'matutini volucres' (*Aeneid* VIII. 456) and 'mortalibus egris' (*Aeneid* II. 268, etc.).[34] The elevated language highlights this dawn as both literal and spiritual brightening. In contrast the Old English text dispenses with these Vergilian echoes, and also with the details of Guthlac's rising, making Guthlac's pious gesture appear much more immediate. The abruptness of the *Life*'s decisive gesture of piety from the nascent saint highlights the significance of this dawn, an effect Felix achieves through lyrical language. This passage exemplifies the kind of simplification seen in the Vespasian text. The absence of stylish touches, of the kind discussed above, in the Old English prose raises questions about whether this was a one-stage process. Did these changes occur simultaneously, or were Vergilian echoes rephrased by the translator, then the passage trimmed by a later copyist to streamline the narrative? Similar uncertainties arise around the omission or adaptation of water imagery in the *Life*.

Comparison of the Vespasian and Vercelli texts reveals one passage that is retained in Vercelli but omitted in Vespasian, indicating that a later copyist is responsible for the omission. In Chapter 31 Guthlac is taken by demons and given a vision of hell:

> Non solum enim fluctuantium flammarum ignivomos gurgites illic turgescere cerneres, immo etiam sulphurei glaciali grandine mixti vortices, globosis sparginibus sidera paene tangentes videbantur. (p. 104)
>
> [For not only could one see there the fiery abyss swelling with surging flames, but even the sulphurous eddies of flame mixed with icy hail seemed almost to touch the stars with drops of spray.] (p. 105)

---

34 Colgrave identifies 'mortalibus egris' as Vergilian, and 'matutini volucres' is noted by Weston. Felix, *Life of St Guthlac*, ed. and trans. by Colgrave, p. 187; Weston, 'Guthlac Betwixt and Between', p. 6. Further use of Vergilian language to describe dawn occurs in Chapter 41 where Felix writes 'excussa ergo opacae noctis caligine, cum sol aureum caelo demoverat ortum' (So when the sun had driven away the black mist of night and dispelled the golden dawn from the sky, pp. 128, 129). Felix's 'opacae noctis' is *Aeneid* X. 161–62, while 'sol aureum' recalls 'sol aureus' of *Georgics* I. 232 and IV. 51. Neither of these instances has been noted previously. This imagery is absent from Section 12 of the *Life*.

This marvellous image, which riffs off *Aeneid* III. 574 ('attollitque globos flammarum et sidera lambit arneid'), is not included in the Vespasian text.[35] But Vercelli does preserve a translation of this section: 'And nalas þæt an þæt he þær þa leglican hyðe ðæs fyres upþyddan geseah, ac eac þa fulan hrecetunge swefles þær geseah upgeotan' (And he not only saw the flaming wave of the fire swell up there, but he also saw the foul belching of sulphur welling up there).[36] As Roberts notes: 'the compiler behind the Vespasian life may have cancelled this passage either because of its complexity or because of the density in it of obsolescent words'.[37] As discussed below, in Felix's *Vita sancti Guthlaci* this image of water-like fire in the abyss connects hell to earlier depictions of the world as a watery chaos. This connection, partially preserved in Vercelli by words such as 'hyðe', is absent from Vespasian, illustrating how watery imagery inherited from Felix may have undergone multiple alterations during the creation and transmission of the *Life*.

The above passage exemplifies the way in which Felix's more metaphorical waters are handled in the *Life*: either diluted because they are a lexical echo identifiable only in the Latin, or deleted as part of a programme of simplification. Yet, although it operates on a less inter- and intra-textual level than in the *Vita*, water remains integral to the construction of Guthlac's sanctity in the *Life*, through its emphasis on the literal waters that define Guthlac's environment. While the waters of the *Life* are predominantly material, at points metaphorical imagery is actually amplified in the Old English, producing effective envelopes, absent from the Latin. These new metaphors may be attributed to the original translator's project to ensure that the spiritual force of Felix's work should remain in Old English; they suggest something about vernacular prose style, but also reveal which figurative images could be expected to resonate with a vernacular audience. The two texts reveal how their respective audiences were expected to respond to both literal and metaphorical water, highlighting the relative currency of particular images within both groups. While Felix envisages a readership able to apprehend and appreciate watery metaphors inherited from Scripture and patristic sources, allowing him to sustain a Gregorian association between water and knowledge, the Old English employs water in a more limited range of uses, with metaphors that draw on the waters of everyday life.

Prologue: The Source of the Waters

The Prologue is very important to both the *Vita sancti Guthlaci* and the *Life of Guthlac*, as it frames the reader's response to the watery material that

---

35 This parallel has not been noted before. On the image of waters reaching the sky in Anglo-Saxon texts, see Wright, *The Irish Tradition in Old English Literature*, p. 135.
36 'Vercelli Homily 23', ed. by Scragg, p. 390.
37 Roberts, 'Two Readings in the Guthlac Homily', p. 207.

follows. Felix's Prologue sets up both ordered, nourishing water as an image of learning, and chaotic water as a reflection of the world; these images aid in the interpretation of the literal landscape of Crowland and anticipate the metaphorical waters of the rest of the *Life*. The first image of water in the *Vita sancti Guthlaci* is metaphorical. Felix modestly states that his work should not be viewed as prideful, but as the product of obedience, 'dum alii plurimi Anglorum librarii coram ingeniositatis fluenta inter flores rethoricae per virecta litteraturae pure, liquide lucideque rivantur' (seeing that there are many other English scholars in our midst who make the waters of genius flow in pure and lucid streams among the flowers of rhetoric and amid the green meadows of literature, pp. 62, 63). This delightful image of literary culture as a kind of *locus amoenus* follows an explicit quotation from Gregory's *Moralia in Iob* that holy writing not be measured by the rules of Donatus. Gabriel Knappe has described this section as a 'commonplace passage on Christian eloquence', but while some of Felix's statements are conventional, the source for the image of the streams and flowers is unclear.[38] Felix's description anticipates the waters of scriptural knowledge that appear later in the text, but in a more secular context. The flowers may derive from Jerome's *Praefatio in Danielem prophetam*, where he describes enjoying the '*flores rhetoricos*' of Quintilian and Cicero.[39] There is also a reminiscence of Aldhelm's *Prosa de Virginitate* 3, describing the mental disposition of the virgins, which 'per florulenta scripturarum arva late vagans bibula curiositate decurrit' (roaming widely through the flowering fields of scripture, traverses (them) with thirsty curiosity).[40] The idea of genius as streams appears to be Felix's own, possibly influenced by Gregorian images of wisdom as water (discussed below) — a playful demonstration of rhetorical skill while performing humility.[41]

This modesty image is slightly reconfigured in the Old English translation. While the *Life* preserves Felix's initial request for pardon if readers 'her hwylc hleaterlic word onfinde' (find any ridiculous word here, p. 101), Gregory's comment on Donatus is removed, presumably because reference to the Latin grammarian would be meaningless to an audience needing the narrative in the vernacular. Felix's image of scholars directing the streams of genius among the flowers of rhetoric in the meadows of literature is rendered, with care, as: 'swa ic menige wat on Angelcynne mid þam fægerum stafum

---

38 Knappe, 'Classical Rhetoric in Anglo-Saxon England', p. 14, n. 40. See also Knappe, *Traditionen der klassischen Rhetorik im angelsächsischen England*, pp. 156–57.
39 Jerome, *Praefatio in Danielem prophetam*, ed. by Migne, col. 1291. Felix's usage is the only one in *Patrologia Latina* except Jerome's and direct quotations thereof. On this text in Anglo-Saxon England, see Lapidge, *The Anglo-Saxon Library*, p. 313.
40 Aldhelm, *Prosa de virginitate*, ed. by Ehwald, p. 232; translation from Aldhelm, *The Prose Works*, trans. by Lapidge and Herren, p. 61.
41 The image may be influenced by Gregory's streams of truth (*fluenta veritatis* in *Regula Pastoralis* I. 2, *Moralia in Iob* VII. 37, XXXIII. 10) and knowledge (*scientiae fluenta* in *Regula Pastoralis* III. 39).

gegylde, fæger and glæwlice gesette, þæt hig þas boc sylf settan mihton' (as I know many in England who, gilded with the fair letters, fairly and cleverly set, might have composed this book themselves, p. 101). This passage is not intertextual, drawing instead on the reader's experience of the material text. The professed humility seems more genuine, but the image of intellectual waters created by Felix in anticipation of later depictions of knowledge as water has been lost.

The Prologue also provides the first image of water as a symbol of the world. Felix, again reflecting on his writing process, states: 'Sed ne sensus legentium prolixae sententiae molesta defensio obnubilet, pestiferis obtrectantium incantationibus aures obturantes, velut transvadato vasti gurgitis aequore, ad vitam sancti Guthlaci stilum flectendo quasi ad portum vitae pergemus' (But for fear that my laboured defence and long drawn out periods may cast a veil over the minds of my readers, let us stop our ears against the pestiferous incantations of our detractors as though we were traversing the waters of a vast whirlpool and let us steer our pen towards the life of St Guthlac as though we were making for the haven of life, pp. 62, 63). The Old English text removes the quotation from Psalm 58. 4–5 (57. 5–6), 'et *obturantes aures* suas | quae non exaudient vocem *incantantium*' (emphasis mine), which would be inaudible in the vernacular, but retains the image of the tempest: 'Ac þylæs ic lengc þone þanc hefige þara leornendra mid gesegenum þara fremdra tælnysse, swa swa ic strange sæ and mycele oferliðe, and nu becume to þære smyltestan hyðe Guðlaces lifes' (But lest I longer weigh down the minds of those learning with speaking of the criticisms of strangers, I sail as if over a strong and mighty sea, and now come to the most tranquil of harbours, the life of Guthlac, p. 102). In the *Vita* these chaotic waters foreshadow later images of the world being like turbulent water, from which Guthlac offers sanctuary, and form an envelope with the final miracle. Although the *Life* appears to set up a similar pattern, the imagery is not sustained — probably due to the truncation of the text by copyists.

Felix's text presents itself as being a kind of ordered water; the false modesty of his discussion of rhetoric presenting a text suffused with precious waters, while the storm imagery imparts an image of the *Vita* as life-giving safe haven, just as Felix's hermitage island, complete with safe landing place, was a secure space amidst the chaos of the surrounding fens. Felix's text then becomes a sustaining draught with which readers may hydrate themselves, a model for ordered life that offers shelter from the chaos of the world. By including only the image of the rough sea and highlighting the tranquil nature of the harbour Guthlac offers, the Old English Prologue presents an idealized image of the saint, but does not establish a relationship between water and intellect, nor does it reflect as self-consciously on its own status as a sustaining text. Nevertheless, its forceful image of the *Life of Guthlac* as 'þære smyltestan hyðe' provides an emphatic direction to the reader to approach this text as a spiritually edifying work offering sanctuary from the tribulations of the world.

Water as a Desert

Water in its literal form is integral to the initial construction of Guthlac's sanctity. It is most prominent in earlier parts of the *Vita*, where Felix deals with the occupation of the island Crowland. The disordered water of the Fens provides a barrier, isolating Guthlac and testing his physical and spiritual reserves. In this way it occupies the same role as the desert in the lives of earlier hermit saints such as the desert fathers. Felix encourages the reader to perceive this connection between fenland and desert. In *Vita sancti Guthlaci* Chapter 24, Felix writes that texts about the desert fathers were a direct inspiration to Guthlac: 'Cum enim priscorum monachorum solitariam vitam legebat, tum inluminato cordis gremio avida cupidine heremum quaerere fervebat' (For when he read about the solitary life of monks in former days, then his heart was enlightened and burned with an eager desire to make his way into the desert, pp. 86, 87). These texts were also a direct inspiration to Felix; as Benjamin Kurtz highlights, Felix's *Vita* is deeply textually indebted to the Evagrian *Vita Antonii*.[42] Numerous narrative borrowings show that Felix perceived a link between Guthlac's *Life* and that of Antony of Egypt, with the fenland around Crowland, as Michael Lapidge and Rosalind Love observe, 'playing the role of Antony's Egyptian desert'.[43] Felix's other sources also shape his construction of water as desert. The *Vita sancti Guthlaci* is indebted to Jerome's *Vita S. Pauli* and Bede's *Prose Vita S. Cuthberti*, texts which share the *Vita Antonii*'s focus on the hermit saint in the landscape, and use Antonian material as a basis.[44] Cuthbert, like Guthlac, is an island-dwelling hermit, occupying a site on Farne surrounded by the turbulent waters of the sea. Felix's combination of sources shows that he perceived an Antonian tradition in England, of which Cuthbert was the exemplar, and represented Crowland within this pattern.

Cuthbert was certainly not the first to view an island surrounded by water as being a suitable substitute for a mountain in the desert. Insular Christianity, particularly in areas of Irish influence, provides many examples of island monasteries and hermitages.[45] In a landscape where an excess of water was more likely to pose difficulties than an absence thereof, high places surrounded by sea or marsh, such as Iona and Lindisfarne, were logical substitutes for high points in the desert, such as Antony's mountain. In choosing these sites, Insular Christians participated in a tradition of adapting a spiritual template

---

[42] Kurtz, 'From St Antony to St Guthlac', esp. pp. 104–27; Felix, *Life of St Guthlac*, ed. and trans. by Colgrave, pp. 16–17.
[43] Lapidge and Love, 'The Latin Hagiography of England and Wales (600–1550)', p. 212.
[44] See Felix, *Life of St Guthlac*, ed. and trans. by Colgrave, pp. 16–17; Bede, *VCP*; Jerome, *Vita Sancti Pauli primi eremitae*, ed. by Migne; *Fontes Anglo-Saxonici: World Wide Web Register*.
[45] See Pickles, 'Anglo-Saxon Monasteries as Sacred Places', pp. 40–44. See also Della Hooke, 'The Sacred Nature of Rivers, Wells, Springs, and Other Wetlands in Anglo-Saxon England', in this collection.

to the local landscape. Benedict of Nursia, whose life as related in Gregory's *Dialogi* is an influence on both Felix's text and Bede's *Prose Vita S. Cuthberti*, chose to inhabit a mountain surrounded by poorly managed pagan lands at Monte Cassino. Martin of Tours, whose life is narrated in another of Felix's sources, Sulpicius Severus's *Vita Martini*, inhabited the island of Gallinaria (Isola d'Albenga) as a hermit. Guthlac, as described by Felix, is merely the latest iteration of the tradition of the saint inhabiting a high point surrounded by waste space.

Felix is at great pains to depict Guthlac's home as a desert, repeatedly using the word *heremus* (desert) in Chapter 25 to describe the space. This image is not entirely accurate; as Kelly A. Kilpatrick highlights, the area was better connected than Felix's deliberate focus on the desolate marsh-scape conveys.[46] Yet Felix's text is in other respects a very faithful representation of the disordered watery environment of the Fens. As Helen Foxhall Forbes notes: 'the fenland landscape described by Felix is not a complete fantasy, and his references to dark and stagnant water or boggy places are matched in other contexts, such as descriptions of features and landmarks described in the boundaries of estates recorded in charters'.[47] Felix creates neologisms in order to present this watery landscape more precisely, generating words such as *riviga* (stream) used in the 'rivigarum anfractibus' (tortuous streams, pp. 86, 87) of Chapter 24.[48] The Old English *Life of Guthlac* attempts to replicate this exactitude and, as Brooks has noted, takes the *Vita*'s resemblance to charter bounds, perceived by Foxhall Forbes, much further. Felix's 'rivigarum anfractibus' is translated in Section 3 of the Life as 'fule eariþas' (p. 113); as Brooks discusses, *eariþas* is a hapax legomenon based on the vocabulary of charter bounds.[49] Both Felix's text and the Vespasian *Life* depict disordered, wet space that can be visualized in very precise terms by their readers, emphasizing Guthlac's saintly achievement in successfully inhabiting such a landscape.[50]

In Chapter 24 Felix introduces the hermitage site with an extended description of the Fenland:

> Est in meditullaneis Brittanniae partibus inmensae magnitudinis aterrima palus, quae, a Grontae fluminis ripis incipiens, haud procul a castello quem dicunt nomine Gronte, nunc stagnis, nunc flactris, interdum

---

46 Kilpatrick, 'The Place-Names in Felix's *Vita sancti Guthlaci*', esp. p. 37. Despite Felix's assertions of Crowland's remoteness, Guthlac's frequent visitors hint at the reality.
47 Foxhall Forbes, *Heaven and Earth in Anglo-Saxon England*, p. 92.
48 Felix provides the only use of this word in the *Dictionary of Medieval Latin. Anfractus* is Aldhelmian. See Latham, Howlett, and Ashdowne, *Dictionary of Medieval Latin from British Sources*; *Patrologia Latina Database*; 'Library of Latin Texts–A'; 'Library of Latin Texts–B'.
49 Brooks, *Restoring Creation*, pp. 237–38. Brooks discusses numerous examples of charter-bound vocabulary used to describe fens. Brooks, *Restoring Creation*, pp. 234–43.
50 The Vercelli text does not include a description of the fen. As it focuses primarily on Guthlac's combat with demons, water imagery is less important to its effect.

nigris fusi vaporis laticibus, necnon et crebris insularum nemorumque intervenientibus flexuosis rivigarum anfractibus, ab austro in aquilonem mare tenus longissimo tractu protenditur. (p. 86)

> [There is in the midland district of Britain a most dismal fen of immense size, which begins at the banks of the river Granta not far from the fortified settlement which is called Grantchester, and stretches from the south as far north as the sea. It is a very long tract, now overhung by fog, sometimes studded with wooded islands and traversed by the windings of tortuous streams.] (p. 87)

Felix emphasizes the scale of this watery, un-navigable waste, highlighting Guthlac's isolation. Chapter 3 of the Old English translation echoes this passage very closely, but introduces additional features, giving a more defined and evocative picture of the space:

Ys on Bretonelande sum fenn unmætre mycelnysse, þæt onginneð fram Grante ea, naht feor fram þære cestre, ðy ylcan nama ys nemned Granteceaster. Þær synd unmætre moras, hwilon sweart wætersteal, and hwilon fule eariþas yrnende, and swylce eac manige ealand, and hreod, and beorhgas, and treowgewrido, and hit mid menigfealdan bignyssum widgille andlang þurhwunað on norðsæ. (p. 113)

> [There is in Britain a fen of immeasurable greatness that begins from the river Granta, not far from the city, which by the same name is named Grantchester. There are immeasurable moors, sometimes dark standing water, and sometimes foul running streams, and also many islands, and reeds, and mounds, and thickets of trees, and it, the vast expanse, with manifold largeness, extends continuously to the North Sea.]

In the *Life of Guthlac* this description opens Section 3, rather than coming part way through a chapter as it does in Felix, emphasizing the significance of this space to the construction of Guthlac's sanctity.

The water of the fen is not wholesome or ordered, as the Old English text makes especially clear. It is chaotic water, which renders the environment hostile. Fens are repeatedly presented as negative spaces in the literary corpus: they are the abode of Grendel in *Beowulf*, and *Maxims II* notes: 'Þyrs sceal on fenne gewunian | ana innan lande' (A monster shall live in the fen, alone within the land, ll. 42b–43a).[51] The *Rune Poem*, punning on *secg* (m. 'sedge'; 'man'; f. 'sword'), presents even the plants of a fen as hostile:

ᛉ eolhx secg eard hæfþ   oftust on fenne,
wexeð on wature,   wundaþ grimme,

---

51 *The Anglo-Saxon Minor Poems*, ed. by Dobbie, p. 56.

> blode breneð   beorna gehwylcne
> ðe him ænigne   onfeng gedeð. (ll. 41–44)[52]

> [Elk's sedge is mostly to be found in a fen; it grows in the water, wounds fiercely, blood burns every warrior, anyone who seizes it.]

The fen's association with monsters and chaos makes it spiritually equivalent to spaces occupied by the desert fathers, so Guthlac's ability to live contentedly in this environment demonstrates his sanctity.[53]

Felix presents Guthlac not only as successfully inhabiting the remote Fenland site, but also as having some control over its waters, navigating channels and offering protection from threats. For example, Chapter 37 relates Guthlac's recovery of a parchment stolen by birds.[54] Guthlac successfully directs the document's owner to where it balances on a reed in the middle of a pool. The principal miracle is, as Felix tells us, the preservation of the parchment from water damage: 'Mirabile dictu! tangi, non tactae, contiguis videbantur ab undis' (marvellous to relate, they were apparently being touched by the waves around them and yet were intact, pp. 118, 119). The use of 'mirabile dictu' highlights Guthlac's power over water as the principal miracle, but as Robin Waugh has shown in detail, the Vespasian *Life* omits this aspect, and Guthlac's miracle becomes the decidedly less wonderous finding of the document.[55]

In the concluding section of Chapter 38 Felix emphasizes Guthlac's harmonious relation with the natural world in a passage that quotes Bede's *Prose Vita S. Cuthberti*, Isaiah, and Matthew, but draws in more of the environment by adding air and water to the list of things that obey Guthlac:

> Non solum vero terræ aerisque animalia illius iussionibus obtemperabant, immo etiam aqua aerque ipsi veri Dei vero famulo oboediebant. Nam qui auctori omnium creaturarum fideliter et integro spiritu famulatur, non est mirandum si eius imperiis ac votis omnis creatura deserviat. At plerumque idcirco subiectae nobis creaturae dominium perdimus, quia Domino universorum creatori servire negligimus, secundum illud: 'Si oboedieritis et audieritis me, bona terrae comedetis', et reliqua: item: 'Si abundaverit fides vestra ut granum sinapis' et reliqua. (p. 120)

---

52 *The Anglo-Saxon Minor Poems*, ed. by Dobbie, p. 29.
53 For further exploration of the mutable relationship between water and land in Anglo-Saxon England, see Wickham-Crowley, 'Living on the *Ecg*'.
54 The Latin term used for the birds is *corvi*, usually translated 'raven' (*corvus corax*). Colgrave translates as 'jackdaw' (*corvus monedula*), because he sees the mischievous behaviour described in the *Vita* as more typical of that species. Felix, *Life of St Guthlac*, ed. and trans. by Colgrave, p. 187. However, Felix specifies that there is only a pair of naughty birds, 'duo alites corvi' (Chapter 38), whereas jackdaws are highly sociable and usually live in larger family groups. The fact that there are only two birds points either to ravens or to carrion crows (*corvus corone*), which are intelligent, bold, and opportunistic.
55 Waugh, 'The Blindness Curse and Nonmiracles', pp. 407–10.

[Not only indeed did the creatures of the earth and sky obey his commands, but also even the very water and the air obeyed the true servant of the true God. For if a man faithfully and wholeheartedly serves the Maker of all created things, it is no wonder if all creation should minister to his commands and wishes. But for the most part we lose dominion over the creation which was made subject to us, because we ourselves neglect to serve the Lord and Creator of all things, as it is said: 'If ye be willing and obedient ye shall eat the good of the land', and so on; and, 'If ye have faith as a grain of mustard seed', and so on.] (p. 121)

This passage looks back to the preceding bird miracles, flatteringly models Guthlac after Cuthbert, and provides a scriptural framework incorporating the idea that keeping covenant with God can improve man's place in creation. The waters of the fen obey Guthlac because, through his virtue, he establishes a relationship with God so close that it undoes aspects of Adam's curse and restores creation around the saint to a more prelapsarian state.

The absence of this passage from the Vespasian text and its treatment of the parchment miracle mean that Guthlac is not shown to have dominion over the waters; this compromises the *Life*'s construction of Guthlac's sanctity by weakening the connection between literal waters and piety. It is impossible to determine whether the original translator or a later copyist omitted this passage, but as a result the act of inhabiting the watery landscape, while still a defining feature of Guthlac's sanctity, is not so clearly and explicitly connected to his piety in the *Life* as it is in the *Vita*.[56] For Felix, Guthlac's seclusion, although very literal, is nonetheless a textual construction that works to place the saint in a hagiographic tradition; this aligns the literal and metaphorical waters of the text, with piety enabling Guthlac to resist the chaotic waters of the word and benefit from the vital liquids of faith and doctrine.

The Waters of Wisdom

Felix employs metaphorical water imagery that works in conjunction with the literal waters of his text to construct Guthlac's sanctity. Felix associates ordered, gentle water with knowledge, and represents the troubled world as a turbulent sea. Like the physical water of the Fens, these metaphorical waters are more prominent in the earlier sections of the text. These traditional metaphors interact with the more concrete geographical setting of Guthlac's hermitage to ensure that Crowland is viewed by the reader as a spiritually significant space: Guthlac's successful ordering and transformation of the fenland reveals his wisdom and unworldliness.

---

56 The Vercelli text does not include this section, so no comparison can be made.

The image of wisdom as water that must be ordered is famously explored in the *Metrical Epilogue* to the Old English translation of Gregory's *Pastoral Care*:

Ðis is nu se wæterscipe ðe us wereda god
to frofre gehet foldbuendum.
He cwæð ðæt he wolde ðæt on worulde forð
of ðæm innoðum a libbendu
wætru fleowen, ðe wel on hine
gelifden under lyfte. Is hit lytel tweo
ðæt ðæs wæterscipes welsprynge is
on hefonrice, ðæt is halig gæst.
Ðonan hine hlodan halge and gecorene,
siððan hine gierdon ða ðe gode herdon
ðurh halga bec hider on eorðan
geond manna mod missenlice.
Sume hine weriað on gewitlocan,
wisdomes stream, welerum gehæftað,
ðæt he on unnyt ut ne tofloweð.
Ac se wæl wunað on weres breostum
ðurh dryhtnes giefe diop and stille.
Sume hine lætað ofer landscare
riðum torinnan; nis ðæt rædlic ðing,
gif swa hlutor wæter, hlud and undiop,
tofloweð æfter feldum oð hit to fenne werð. (ll. 1–21)[57]

> [This is now the portion of water that the God of hosts promised us to comfort earth dwellers. He said that he wished that in the world ever living waters would flow forth from the innards of those under the sky who thoroughly believed in him. It is little doubt that the portion of water's wellspring is in the kingdom of heaven, that is the Holy Ghost. From thence the saints and the chosen drew it, afterwards they who obeyed God directed it through holy books hither on earth variously through the minds of men. Some guard it in their mind, wisdom's stream, detain it with lips, so that it does not flow out useless. But the deep pool dwells in man's breast through God's grace, deep and still. Some let it run over land-shares in streams; that is not an advisable thing, if such clear water, loud and shallow, flows over fields until it becomes a fen.][58]

---

57 *The Anglo-Saxon Minor Poems*, ed. by Dobbie, pp. 111–12. See Daniel Anlezark's essay in this volume, 'Drawing Alfredian Waters: The Old English *Metrical Epilogue* to the *Pastoral Care*, Boethian *Metre* 20, and *Solomon and Saturn II*'.

58 *Wæterscipe* is usually translated as 'body of water', but as Atherton highlights, it is used in the Old English Benedictine Rule to refer to channelled water associated with water mills, and the image of controlled water would be a more apt rendering of Gregory's metaphors. See Atherton, *The Making of England*, pp. 82–84.

The opening section of this poem neatly summarizes the kind of water and wisdom images that Felix presents. The poet, as Malcolm Godden notes, 'evidently develops passages in Chapters 38 and 39 in the [Old English] *Pastoral Care*, which are reworkings of Gregory's metaphors'.[59] It is possible that Felix, encountering the same metaphors in his reading of Gregory, was likewise encouraged to present images of knowledge as spiritual waters profitably absorbed by the virtuous, and also to connect Guthlac's ordering of the waters of the fen to his mental discipline.

Regula Pastoralis III. 14 is one of the sections lying behind the Old English verse:

> Humana etenim mens, aquae more circumclusa ad superiora colligitur, quia illud repetit unde descendit; et relaxata deperit, quia se per infima inutiliter spargit. Quot enim supervacuis verbis a silentii sui censura dissipatur, quasi tot rivis extra se ducitur.[60]
>
>> [For the human mind behaves like water; when closed up it collects into higher levels, in that it seeks again that height from which it descended. And when let loose, it loses itself, in that it disperses itself uselessly through the lowest places. For by as many superfluous words as it is dissipated from the censorship of its silence, by so many channels is it led away out of itself.]

Regula Pastoralis has not been identified as one of Felix's sources, but the same image of water and wisdom occurs with almost identical phrasing in *Moralia in Iob* VII. 37.[61] This text is used in the *Vita*, so Felix knew the image and may have been influenced by its depiction of wisdom as ordered water, related to ordered words.

Drawing on a tradition of scriptural exegesis, Gregory, like Felix, presents water as multivalent: it can represent both positive and negative forces. As Gregory states in *Moralia in Iob* XIX. 6:

> Aquae in scriptura sacra aliquando sanctum spiritum, aliquando scientiam sacram, aliquando scientiam prauam, aliquando tribulationem, aliquando defluentes populos, aliquando mentes fidem sequentium designare solent.[62]
>
>> [Waters in Holy Scripture are wont sometimes to denote the Holy Spirit, sometimes sacred knowledge, sometimes wrong knowledge, sometimes calamity, sometimes drifting peoples, sometimes the minds of those following the faith.]

---

59 Godden, 'Prologues and Epilogues', p. 467.
60 Gregory the Great, *Regula Pastoralis*, ed. by Migne, LXXVII, col. 73A.
61 Gregory the Great, *Moralia in Iob*, ed. by Adriaen.
62 Gregory the Great, *Moralia in Iob*, ed. by Adriaen.

The variety of water imagery in the Book of Job provides an opportunity for Gregory to explore all these metaphorical readings of water, articulating ideas that can be seen to flow into Felix's work. Although exact verbal correspondences are hard to find, and many of the images have their origins in Scripture, Gregory's fondness for aqueous imagery is arguably a significant influence on Felix's own varied watery metaphors.

Felix presents gentle or nourishing waters as a positive force representing knowledge or grace, Gregory's *scientia sacra*. In Chapter 22 he describes Guthlac's progression to the state of Psalteratus: 'Cum enim litteris edoctus psalmorum canticum discere maluisset, tunc frugifera supra memorati viri praecordia roscidis roris caelestis imbribus divina gratia ubertim inrigabat' (When indeed, after having been taught his letters, he set his mind to learning the chanting of the psalms, then the divine grace sprinkled this same man's fertile heart copiously with the moist showers of heavenly dew, pp. 84, 85). The collocation 'roris caelestis' draws on the scriptural 'rore caeli', used twice in Genesis, and more pertinently three times in Daniel, where being wetted with the 'dew of heaven' leads to Nebuchadnezzar recognizing the power of God (Daniel 5. 21).[63] The water imagery is lost at this point in Section 2 of the Old English. Felix's echo of scriptural Latin, 'rore caeli', is not retained, but the fertile heart remains: 'Mid þy he þa wæs in stafas and on leornunge getogen, þa girnde he his sealmas to leornianne; þa wæron þa wæstmberendan breost þæs eadigan weres mid godes gife gefyllede' (When he was educated in letters and learning, then he yearned to learn his psalms; then the fruitful breast of this blessed man was filled with God's grace, p. 112). This omission removes the idea of grace as water but was probably triggered by the translator's recognition of a scriptural echo that would not work in Old English, whereas the metaphor of a fertile breast is easily comprehended.

Elsewhere the *Life* is at pains to conserve water imagery. In Chapter 43 Felix contrasts spiritual liquids with more worldly ones. Recalling Gregorian images, he presents Scripture as a liquid that may be drunk. Felix describes how when Guthlac and a visiting abbot were in conference, 'divinarum Scripturarum haustibus inebriarent' (they were drinking deep draughts from the holy scriptures, pp. 132, 133).[64] Guthlac, employing visionary powers, interrupts their discussion to reveal the true whereabouts of the abbot's two absent servants: 'Dicebat enim illos ad cuiusdam viduae casam devertisse et, dum non adhuc tertia hora esset, in delicatis viduae fulcris inebriari coepisse' (For he said that they had turned into the house of a

---

63 Although Weston suggests that the source is Aldhelm's *De metris*: 'Roscidis sacrorum dogmatum umbribus ubertim perfudit'. Weston, 'Guthlac Betwixt and Between', pp. 7, 25. This image is given considerable importance in the Old English poem *Daniel*. See Portnoy, '*Daniel* and the Dew-Laden Wind'.
64 Cf. 'haustibus evangelici nectaris' (draughts of Gospel nectar, pp. 144, 145) in Chapter 46. Retained in the Old English *Life*, Section 17, p. 156.

certain widow, and, though it was not yet the third hour, had begun to drink deep draughts at the widow's luxurious table, pp. 134, 135). The repetition of *inebrio* connects and contrasts these two incidents. Section 14 of the Old English text helps the reader to understand the initial drinking metaphor by adding a literal interpretation prior to Guthlac's interruption: 'mid þan hi þa sylfe betweonum drencton of þam willan haligra gewrita, þa betwyx þa halgan gewritu þe hi spræcon [...]' (while they gave themselves mutually to drink from the well of holy scriptures, then betwixt their discussion of holy scriptures [...], p. 149).[65] The translation retains the complex image of the waters of Scripture as they contrast so beautifully with the liquid imbibed by the servants who 'wæron ondrencte mid oferdrynce' (were drunk with overdrink, p. 150). In both versions of the life the virtuous metaphorical waters drunk in an orderly way by the hermit and the abbot are juxtaposed with the problematic worldly liquid immoderately consumed by the two servants, and fruitful spiritual irrigation with unprofitable fen-like drunkenness.

Metaphorical water is not always a symbol of good in the *Vita*: reflecting Gregory's waters of *scientia prava*, wickedness can also be drunk. In Chapter 35 Guthlac's servant Beccel attempts to slay him. Guthlac realizes that Beccel is afflicted by an evil spirit and says 'Quare amari veneni mortiferas limphas non vomis?' (Why do you not spew out the deadly draught of bitter poison?, pp. 112, 113). Felix elsewhere uses *lympha* for holy water (Chapters 11 and 53), but here the liquid is metaphorical and represents the evil Beccel has absorbed. This imagery of evil as a liquid is much expanded in the Old English, creating an envelope that gives greater clarity and context to Guthlac's urging of Beccel to reject the poison. Felix describes Beccel's initial corruption by an evil spirit which entered him and 'pestiferis vanae gloriae fastibus illum inflare coepit' (began to puff him up with pestiferous arrogance and vainglory, pp. 112, 113). In the Old English Section 7 Beccel's heart and thought are 'mid his searwes attre geond sprengde and mengde' (sprinkled and mixed with his poison of treachery, p. 137), anticipating Guthlac's question 'For hwon nelt þu þæs biteran attres þa deaþberendan wæter of þe aspiwan?' (Why do you not spew the death-bearing water of that bitter poison from you?, p. 138). This amplification of the imagery of evil as infectious poisonous liquid, reminiscent of the image of disease as a poison in some Old English medical charms, skilfully makes the image clearer for a vernacular audience, giving greater impression of Guthlac's prophetic insight, the focus of this episode.[66]

---

65 The manuscript reading is *dremdon*, but both Goodwin and Gonser emend to *drencton*. *Drencton* preserves the verbal parallel with the servants' drunkenness and is to be preferred. *Dremdon* might easily have been substituted by a scribe unfamiliar with the image employed.
66 A similar addition of an envelope in the Old English occurs with the spears of the evil spirits in Chapter 6.

Water and Worldliness

For Felix, the most problematic waters by far are those that represent life in the world. These metaphorical waters flow into the literal ones surrounding Guthlac's hermitage, which, as discussed above, are a desert-like space representing the world at its most disordered. In Chapter 27, which describes Guthlac's arrival on Crowland and his taking up of spiritual arms, Felix comments: 'sic et sanctae memoriae virum Guthlac de tumido aestuantis saeculi gurgite, de obliquis mortalis aevi anfractibus, de atris vergentis mundi faucibus ad perpetuae beatitudinis militiam, ad directi itineris callem, ad veri luminis prospectum perduxit' (so also He led Guthlac a man of saintly memory from the eddying whirlpool of these turbid times, from the tortuous paths of this mortal age, from the black jaws of this declining world to the struggle for eternal bliss, to the straight path and to the vision of the true light, pp. 92, 93).[67] These metaphorical obstacles through which Guthlac is guided by God recall the literal landscape obstructions navigated in Chapters 24 and 25 to reach the hermitage site, such as the 'rivigarum anfractibus' (tortuous streams) of Chapter 24. The Old English *Life* does not make the same connection; Section 3 simply describes Guthlac as being led 'of þære gedrefednysse þissere worulde' (from the confusion of this world, p. 117). By using this image of life as chaotic water, through which Guthlac finds a path, Felix connects Guthlac's successful crossing of the fen with his transition from a secular life spent in the pursuit of worldly glory to a higher spiritual existence as a hermit. The physical and spiritual movements become reflections of one another.

Felix's use of the noun *gurges* (whirlpool) in the above passage connects back to the storm imagery of the Prologue, where Felix resolves to ignore potential detractors, 'velut transvadato vasti gurgitis aequore' (as though we were traversing the waters of a vast whirlpool, pp. 62, 63), and to make for the life of Guthlac 'quasi ad portum vitae pergemus' (as though we were making for the haven of life, pp. 62, 63). The noun *gurges* is repeatedly employed by Felix in the *Vita* to describe the world. The collocation 'saeculi gurgites' occurs twice: once in the above passage, and once in Chapter 18 where Felix describes Guthlac as being 'inter saeculi gurgites iactaretur' (tossed amid the whirling waves of the world, pp. 80, 81), a troubled state leading to his calling to the religious life. Section 2 of the Old English echoes this image with 'betweox þises andweardan middaneardes wealcan' (betwixt the tumult of this present world, p. 109), capturing the watery associations with *wealcan*.[68] For both texts, the tumult of waters provides an effective image of the chaotic nature of worldly life, connecting its troubles to the fen through which Felix navigates

---

67 'atris [...] faucibus' is from *Aeneid* VI. 240–41.
68 *Wealcan* is commonly used of water. See Bosworth and Toller, *An Anglo-Saxon Dictionary*, p. 1171.

to his spiritual and physical anchorage at Crowland, but only in the Latin are these links sharpened by intra-textual allusions.

As well as using water to link literal and metaphorical landscapes, Felix also employs repeated vocabulary to connect the worldly life to the horrors of hell. Felix's repeated use of *gurges* for the worldly life associates the metaphorical tumult of the world with the fiery chaos of hell. While the Old English connects water with worldliness, it does not connect worldliness so explicitly with hell, presenting the terrors of that space as much more separate from Guthlac's life. As discussed above, in Chapter 31 Guthlac is abducted by demons and shown hell, which Felix describes using *gurges*: 'fluctuantium flammarum ignivomos gurgites […] turgescere' (the fiery abyss swelling with surging flames, pp. 104, 105). The translation is only preserved in Vercelli 'þa leglican hyðe ðæs fyres upþyddan' (the flaming wave of the fire swelling up).[69] Although the Old English images used to translate *gurges* have some connotations of water, they lack the precise lexical connection made in the Latin, which ties the cares of the world to the place where they might well lead. As a result, hell is a much more separate and unknowable space in the Old English — effectively terrifying, but harder to interpret.

Extending this particular metaphor, the monastic may be seen as Guthlac's port offering shelter from the storm, just as his *Vita* is a haven for Felix in the Prologue. This image of the monastic life as a harbour reflects Aldhelm's *Prosa de Virginitate* 10, which may be one of Felix's inspirations:

> dum illi periculoso saeculi naufragio et grassante dirae tempestatis turbine uelut inter Scillam Siciliae et barathrum uoraginis nauigantes ad portum coenubialis uitae festinantes, licet aliquantulum quassatis cymbae compagibus, Christo gubernante feliciter peruenerunt.[70]
>
> > [while the others, sailing (as it were) near the perilous shipwreck of this world with the whirlwind of a dreadful tempest raging, as though between the Sicilian Scylla and the gulf of the whirlpool [i.e. Charybdis], hasten towards the harbour of the monastic life, and with Christ as their pilot arrive safely, even though the timbers of their ship are somewhat shaken.][71]

As Andy Orchard highlights, this image recurs towards the end of the text, where Aldhelm reflects on his own literary endeavours, perhaps influencing Felix's Prologue.[72] This idea of the world as a stormy sea on which man is tossed occurs repeatedly in Felix's work. As Orchard notes, this image is commonplace

---

69 'Vercelli Homily 23', ed. by Scragg, p. 390. On the difficulties of editing this passage, and alternate possibilities which may bring the Old English closer to the Latin, see Roberts, 'Two Readings in the Guthlac Homily', pp. 206–10.
70 Aldhelm, *Prosa de virginitate*, ed. by Ehwald, p. 238.
71 Aldhelm, *The Prose Works*, trans. by Lapidge and Herren, p. 67.
72 Orchard, *Pride and Prodigies*, p. 96; image discussed pp. 96–98.

in medieval Christian texts, and there are many possible authorities from whom Felix might have absorbed it; for example, Gregory the Great employs it in the epistle to *Moralia in Iob* and Homily 29.[73] But whatever this image's conventional origins, Felix handles it with nuance, using repeated vocabulary to ensure that metaphorical waters mirror the hazardous literal waters surrounding Guthlac, and the dangers of succumbing to the worldly life are made clear.

This image of life as a whirlpool from which we must seek safe harbour is repeated in Chapter 41, in which Guthlac heals the possessed youth Hwætred by holy water and exsufflation.[74] Literal water and air are followed by a metaphorical echo, as Felix states: 'Ipse autem, velut qui de aestuantis gurgitis fluctibus ad portum deducitur, longa suspiria imo de pectore trahens, ad pristinae salutis valitudinem redditum se esse intellexit' (And the youth, like one who is brought into port out of the billows and the boiling waves, heaved some deep sighs from the depth of his bosom and realized that he had been restored to his former health, pp. 130, 131). The use of *gurges* and *fluct-* echoes the earlier description of hell, emphasizing the danger from which Hwætred has been delivered. Recalling the Prologue, Guthlac is again associated with providing a harbour — controlled, sheltered, and managed water, where one is safe from the disordered violence of the world. The Old English translation presents a very different image for Hwætred's recovery, leaving the Prologue un-echoed. The focus in Section 12 is instead on awakening, as Felix's image of Pega's recovery from paralysing grief in Chapter 50 of the *Vita* is transferred to Hwætred, who revives 'swa he of hefegum slæpe raxende awoce' (as if he awoke from a heavy sleep, p. 148).[75] Rather than presenting possession as a storm, the Old English imagines unconsciousness. The image is an effective way of conveying Hwætred's loss of self but does not participate in a broader metaphorical framework. In the *Vita* Hwætred finds safe harbour in Guthlac, as Felix aims to do at the start of the writing process. Literary *vita* and historical events become productively blurred, allowing the reader's consumption of the life of Guthlac to offer access to the haven of the saint himself, an association strengthened by Felix's final miracle account.

Conclusions

The final miracle of the *Vita sancti Guthlaci* is much reduced in the surviving text of the Old English translation; in the *Vita* this miracle contains water

---

73 Orchard, *Pride and Prodigies*, p. 97; Gregory the Great, *Moralia in Iob*, ed. by Adriaen; Gregory the Great, 'Homily 29', ed. by Migne. See also Schmidtke, 'Geistliche Schiffahrt'; Schmidtke, 'Geistliche Schiffahrt II'; Smithers, 'The Meaning of *The Seafarer* and *The Wanderer*'; Smithers, 'The Meaning of *The Seafarer* and *The Wanderer* (Continued)'.
74 On exsufflation, see Hill, 'When God Blew Satan out of Heaven'.
75 The corresponding description of Pega in Section 20 of the Old English removes the image of awakening from Felix's Chapter 50.

imagery that connects back the Prologue, but this is omitted in the Vespasian text. Felix's Chapter 53 concludes the *Vita* with the healing of the blind Wissa, whose sight is restored by Pega's application of a mixture of holy water and salt that her brother Guthlac had blessed. Wissa is then able to guide those who guided him to Guthlac on the journey home. In the Vespasian text, a brief Section 22 describes the same blindness cure, effected only by salt, being given to an unnamed boatman of Æthelbald; this boatman then returns home, but no mention is made of guiding others.[76] As Waugh has highlighted, the blindness envelope with the Prologue is maintained, but the water imagery that accompanies it is lost.[77] This abrupt truncation of the final miracle may well be the work of a later copyist rather than the original translator. The transformation of Wissa, whose trade is never specified in the *Vita*, into a *scipesman* (boatman) is intriguing; perhaps the conclusion of the Old English was originally fuller, and the restoration of a boatman's sight, enabling him to navigate through the Fens to safe anchorage and to convey others with him, effectively recalled the Prologue's image of sailing through tribulation to safe harbour in Guthlac. But this is merely speculative — in its surviving form the Old English *Life of Guthlac* presents a rather abrupt ending, removing the potential for textual reflection available in the Latin, before offering a final exhortation to pray to Guthlac.

In Felix's *Vita* the healing of Wissa is an apposite and carefully handled conclusion to Guthlac's hagiography. This post-mortem miracle not only demonstrates Guthlac's continued power, but also connects the reader back to the water and blindness images of Felix's Prologue, encouraging a response to the text itself, as well as the material it contains: the reader and Wissa share in mutual enlightenment granted by the edifying waters associated with Guthlac. The instant of Wissa's cure is described by Felix as follows:

> Illa quoque partem glutinati salis a sancto Guthlaco ante consecratam arripiens, in aquam offertoriam levi rasura mittebat; ipsam denique aquam, cum intra palpebras caeci guttatim stillaret, mirabile dictu! ad primum tactum primae guttae, detrusis caecitatis nubibus, oculis infusum lumen redditum est; priusquam enim alterius oculi palpebris salutaris limpha infunderetur, quicquid domi esset in ordine narrabat, visumque sibi in eodem momento donatum fatebatur. Deinde, postquam diu clausas gratia per gratiam frontis reclusit fenestras, cognovit inventum olim quod perdidit lumen, dux se ducentibus factus est revertens rursus. (pp. 168–70)
>
>> [She [St Pega] also took a piece of glutinous salt which had previously been consecrated by St Guthlac and, grating it lightly, let the scrapings fall into consecrated water. She made this water drip, drop by drop,

---

76 Æthelbald gets an arguably more flattering treatment in the Old English text. See Whatley, 'Lost in Translation'.
77 Waugh, 'The Blindness Curse and Nonmiracles'.

under the blind man's eyelids and, marvellous to relate, at the first touch of the first drop, the clouds of blindness were scattered and the light returned, pouring into his eyes. Now, before the heading water had been poured into the lid of his other eye, he described in detail all that was in the house, and said that sight had at that moment been given to him. Then after grace had, by grace, opened the windows of his head which had been so long closed, he realized that the light which he had once lost was found, and returning home he became a guide to those who were his guides.] (pp. 169–71)

The *guttae* that Felix depicts here recall the dew of grace and streams of knowledge presented earlier in the *Vita*. When Wissa's blindness has been cured by these consecrated drops of saline, he is able to navigate the fenlands and testify to the power of Guthlac. As the chaotic water of the landscape of the Fens related to the tumult of the world earlier in the text, the power of Guthlac, associated here with holy, ordered water, enables the enlightened Wissa to traverse the world successfully in both a physical and metaphorical sense, and to transmit both literal and spiritual guidance to others.

Wissa's absorption of these healing waters at Guthlac's shrine, leading to illumination, is a suitable conclusion for the *Vita* as it reflects the reader's own absorption of Felix's text. By extending the images applied to edifying texts in the *Vita*, Felix's own work can be seen as a restorative draught to be imbibed in the same manner, then shared with others. Just as the ordered streams of genius watered the meadows of literature in the Prologue, the streams of wisdom contained in Felix's text, which convey the power of Guthlac, irrigate the reader and restore them. With this final image of the once blind man successfully guiding others because of the saint's aid, reflecting how faith in Guthlac offers the reader a guide through the tribulations of the world, Felix concludes his text in a manner reminiscent of his Prologue. Wissa, who is led by his friends 'ad portum insulae Crugland' (to the landing-place of that island of Crowland, pp. 168, 169) to be healed, reminds the reader that they were included in the statement: 'ad vitam sancti Guthlaci stilum flectendo quasi ad portum vitae pergemus' (let us steer our pen towards the life of St Guthlac as though we were making for the haven of life, pp. 62, 63). Felix, like Wissa's friends, has guided the reader to Guthlac, and if faithful like Wissa, they will emerge restored and better able to navigate the trials of the world. Felix's final statement on Wissa: 'grates Deo persolvens dignas, quas nullus reddere nescit' (and he returned fitting thanks to God, which none could fail to give, pp. 170, 171), models the reader's own response to Guthlac and the *Vita*. Felix concludes with literal waters that recall and mingle with the metaphorical waters of the Prologue: the completion of the *Vita sancti Guthlaci*, for both author and reader, represents arriving, refreshed with restorative streams, at the safe haven of the waters of life, to which the reader has been safely navigated.

Although water is deeply significant in both treatments of Guthlac's life, it is clear that the two texts' projected audiences are expected to think about it in different ways. Felix, with his fondness for Gregory and Aldhelm, views Guthlac as an orderer of water, a harbour to which his readership may be guided. The Old English *Life*, as preserved in the Vespasian manuscript, makes no such demands: its readers are not envisaged as possessing the necessary intellectual framework to navigate complex intertextual water metaphors. Given images such as that found in the *Metrical Epilogue* to the *Pastoral Care*, it would be an error to presume that figurative waters were unattractive or meaningless to those who consumed literature in the vernacular. It appears from the *Life*'s retention of the harbour metaphor in the Prologue, and the handling of episodes such as Beccel's affliction and the abbot's drunken servants, that the original translator did make efforts to preserve and explain some of Felix's metaphorical waters, particularly where the images drew on familiar literal waters. This exploitation of the cultural associations of watery spaces, such as fens, within a spiritual context suggests a reasonably sophisticated primary audience. However, the actions of subsequent copyists, presumably driven by a context that prioritized simplicity, have reduced the importance of figurative waters in the *Life* to a few key incidents. The result is that water in the *Life of Guthlac* is more literal than metaphorical and primarily serves as one of several markers of Guthlac's power; it does not work to place him in a broader tradition of spiritual waters. In contrast, Felix's waters explicitly connect Guthlac to saintly forebears and place his hagiography as part of a tradition of spiritually edifying texts. Felix is evidently writing for a readership able to navigate Gregorian and Aldhelmian echoes in order to place both saint and text within an established tradition. The popularity of Gregory's *Moralia in Iob* in eighth-century England suggests that Felix's audience would have been conversant with his metaphors, allowing the *Vita sancti Guthlaci* to merge the material waters of the Fens with their metaphorical counterparts in broader Christian tradition to drive forward its spiritual purpose — the praise of Guthlac.[78]

Works Cited

Primary Sources

Aldhelm, *Epistola ad Acircium*, in *Aldhelmi Opera*, ed. by Rudolf Ehwald, MGH, Auctores Antiquissimi, 15 (Berlin: Weidmann, 1919), pp. 33–204
———, *Prosa de virginitate*, in *Aldhelmi Opera*, ed. by Rudolf Ehwald, MGH, Auctores Antiquissimi, 15 (Berlin: Weidmann, 1919), pp. 226–323
———, *Prosa de virginitate*, in *Aldhelm: The Prose Works*, trans. by Michael Lapidge and Michael Herren (Cambridge: Brewer, 1979), pp. 59–135

---

[78] On Anglo-Saxon manuscripts of and quotations from Gregory's *Moralia in Iob*, see Lapidge, *The Anglo-Saxon Library*, pp. 305–06.

Das angelsächsiche Prosa-Leben des hl. Guthlac, ed. by Paul Gonser, Anglistische Forschungen, 27 (Heidelberg: Carl Winters Universitätsbuchhandlung, 1909)

The Anglo-Saxon Minor Poems, ed. by Elliot Van Kirk Dobbie, Anglo-Saxon Poetic Records, 6 (London: Routledge, 1942)

The Anglo-Saxon Version of the Life of St Guthlac, Hermit of Crowland, ed. and trans. by Charles Wycliffe Goodwin (London: John Russell Smith, 1848)

Bede, *Vita S. Cuthberti prosaica*, in *Two Lives of Saint Cuthbert: A Life by an Anonymous Monk of Lindisfarne and Bede's Prose Life*, ed. and trans. by Bertram Colgrave (Cambridge: Cambridge University Press, 1940), pp. 141–307

Felix, *Felix's Life of St Guthlac*, ed. and trans. by Bertram Colgrave (Cambridge: Cambridge University Press, 1956)

Gregory the Great, 'Homily 29', in *Sancti Gregorii Papae I. Cognomento Magni, Opera Omnia*, in *Patrologia cursus completes, series Latina*, ed. by Jacques-Paul Migne, 221 vols (Paris: Migne, 1841–65), LXXVI (1849), cols 1213B–1219D

———, *S. Gregorii Magni Moralia in Iob*, ed. by M. Adriaen, CCSL, 143, 143A, 143B, 3 vols (Turnhout: Brepols 1979–85)

———, *Regula Pastoralis*, in *Sancti Gregorii Papae I. Cognomento Magni, Opera Omnia*, in *Patrologia cursus completus, series latina*, ed. by Jacques-Paul Migne, 221 vols (Paris: Migne, 1849), LXXVII (1849), cols 9–148

Guthlac A and *Guthlac B*, in *The Exeter Anthology of Old English Poetry: An Edition of Exeter Dean and Chapter MS 3501*, ed. by Bernard J. Muir, Exeter Medieval Texts and Studies, 2nd edn, 2 vols (Exeter: University of Exeter Press, 2000), I, pp. 111–59

The Guthlac Poems of the Exeter Book, ed. by Jane Roberts (Oxford: Clarendon Press, 1979)

Die Heiligen Englands: angelsächsisch und lateinisch, ed. by F. Liebermann (Hannover: Hahnsche Buchhandlung, 1889)

Jerome, *Praefatio in Danielem prophetam*, in *Sancti Eusebii Hieronymi Stridonensis Presbyteri Opera Omnia*, in *Patrologia cursus completus, series Latina*, ed. by Jacques-Paul Migne, 221 vols (Paris: Migne, 1841–65), XXVIII (1845), cols 1291B–1294B

———, *Hieronymus: Vita Sancti Pauli primi eremitae*, in *Patrologia cursus completes, series Latina*, ed. by Jacques-Paul Migne, 221 vols (Paris: Migne, 1841–65), LXXIII (1849), cols 105–16

'Library of Latin Texts–A', *Library of Latin Texts – online* (Brepols), <http://clt.brepolis.net/llta> [accessed 3 February 2020]

'Library of Latin Texts–B', *Library of Latin Texts – online* (Brepols), <http://clt.brepolis.net/lltb/> [accessed 3 February 2020]

The Old English Life of St Mary of Egypt: An Edition of the Old English Text with Modern English Parallel-Text Translation, ed. and trans. by Hugh Magennis (Exeter: University of Exeter Press, 2002)

The Old English Martyrology: Edition, Translation and Commentary, ed. and trans. by Christine Rauer, Anglo-Saxon Texts, 10 (Cambridge: D. S. Brewer, 2013)

Patrologia Latina Database, <http://pld.chadwyck.co.uk/> [accessed 3 February 2020]

Pilch, Herbert, 'The Last Vercelli Homily: A Sentence Analytical Edition', in *Historical Linguistics and Philology*, ed. by Jacek Fisiak (Berlin: Mouton de Gruyter, 1990), pp. 297–336

The Prose Life of Guthlac, in *Anglo-Saxon Prose*, trans. by Michael J. Swanton, 2nd edn (London: Everyman, 1993), pp. 88–113

Vercelli Homilies IX–XXIII, ed. by Paul E. Szarmach (Toronto: University of Toronto Press, 1981)

'Vercelli Homily 23', in *The Vercelli Homilies and Related Texts*, ed. by Donald Scragg, EETS, o.s. 300 (Oxford: Oxford University Press, 1992), pp. 381–94

Secondary Works

Appleton, Helen, 'The Psalter in the Prose Lives of Guthlac', in *Germano-Celtica*, ed. by Anders Ahlqvist and Pamela O'Neill, Sydney Series in Celtic Studies, 16 (Sydney: Celtic Studies Foundation, University of Sydney, 2017), pp. 61–86

Appleton, Helen, and Matthew Robinson, 'Further Echoes of Vergil's *Aeneid* in Felix's *Vita sancti Guthlaci*', *Notes and Queries*, 64 (2017), 353–55

Atherton, Mark, *The Making of England: A New History of the Anglo-Saxon World* (London: I. B. Tauris, 2017)

Bolton, W. F., 'The Manuscript Source of the Old English Prose Life of St Guthlac', *Archiv für das Studium der neueren Sprachen und Literaturen*, 197 (1961), 301–03

Bosworth, Joseph, and T. Northcote Toller, *An Anglo-Saxon Dictionary* (Oxford: Clarendon Press, 1898)

Brooks, Britton, 'Felix's Construction of the English Fenlands: Landscape, Authorizing Allusion, and Lexical Echo', in *Guthlac: Crowland's Saint*, ed. by Jane Roberts and Alan Thacker (Donington: Shaun Tyas, 2020), pp. 55–71

—— , *Restoring Creation: The Natural World in the Anglo-Saxon Saints' Lives of Cuthbert and Guthlac* (Cambridge: D. S. Brewer, 2019)

Crawford [now Roberts], Jane, 'Guthlac: An Edition of the Old English Prose Life Together with the Poems in the Exeter Book' (unpublished doctoral thesis, University of Oxford, 1967)

Fontes Anglo-Saxonici: World Wide Web Register, <https://www.st-andrews.ac.uk/~cr30/Mercian/Fontes> [accessed 18 March 2017]

Foxhall Forbes, Helen, *Heaven and Earth in Anglo-Saxon England: Theology and Society in an Age of Faith* (Farnham: Ashgate, 2013)

Godden, Malcolm, 'Prologues and Epilogues in the Old English *Pastoral Care*, and their Carolingian Models', *Journal of English and Germanic Philology*, 110 (2011), 441–73

Hill, Thomas D., 'When God Blew Satan out of Heaven: The Motif of Exsufflation in Vercelli Homily XIX and Later English Literature', *Leeds Studies in English*, 16 (1985), 132–41

Kilpatrick, Kelly A., 'The Place-Names in Felix's *Vita sancti Guthlaci*', *Nottingham Medieval Studies*, 58 (2014), 1–56

Knappe, Gabriele, 'Classical Rhetoric in Anglo-Saxon England', *Anglo-Saxon England*, 27 (1998), 5–29

———, *Traditionen der klassischen Rhetorik im angelsächsischen England*, Anglistische Forschungen, 236 (Heidelberg: C. Winter, 1996)

Kurtz, Benjamin P., 'From St Antony to St Guthlac: A Study in Biography', *University of California Publications in Modern Philology*, 12 (1926), 103–46

Lapidge, Michael, *The Anglo-Saxon Library* (Oxford: Oxford University Press, 2006)

Lapidge, Michael, and R. Love, 'The Latin Hagiography of England and Wales (600–1550)', in *Hagiographies: International History of the Latin and Vernacular Literature in the West from its Origins to 1550*, ed. by Guy Philippart, vol. III (Turnhout: Brepols, 2001), pp. 203–325

Latham, R. E., D. R. Howlett, and R. K. Ashdowne, eds, *Dictionary of Medieval Latin from British Sources* (Oxford: British Academy, 1975–2013), <http://www.dmlbs.ox.ac.uk/> [accessed 3 February 2020]

Orchard, Andy, *Pride and Prodigies: Studies in the Monsters of the 'Beowulf'-Manuscript* (Cambridge: D. S. Brewer, 1995)

Pickles, Thomas, 'Anglo-Saxon Monasteries as Sacred Places: Topography, Exegesis and Vocation', in *Sacred Text, Sacred Space: Architectural, Spiritual and Literary Convergences in England and Wales*, ed. by Joseph Sterrett and Peter Thomas (Leiden: Brill, 2011), pp. 35–56

Portnoy, Phyllis, '*Daniel* and the Dew-Laden Wind: Sources and Structures', in *Old English Literature and the Old Testament*, ed. by Michael Fox and Manish Sharma (Toronto: University of Toronto Press, 2012), pp. 195–228

Roberts, Jane, 'Guthlac of Crowland and the Seals of the Cross', in *The Place of the Cross in Anglo-Saxon England*, ed. by Catherine E. Karkov, Sarah Larratt Keefer, and Karen Louise Jolly (Woodbridge: Boydell, 2006), pp. 113–28

———, 'An Inventory of Early Guthlac Materials', *Mediaeval Studies*, 32 (1970), 193–233

———, 'The Old English Prose Translation of Felix's *Vita Sancti Guthlaci*', in *Studies in Earlier Old English Prose*, ed. by Paul E. Szarmach (Albany: State University of New York, 1986), pp. 363–80

———, 'Traces of Unhistorical Gender Congruence in a Late Old English Manuscript', *English Studies*, 51 (1970), 30–37

———, 'Two Readings in the Guthlac Homily', in *Early Medieval Texts and Interpretations: Studies Presented to Donald G. Scragg*, ed. by Elaine Treharne and Susan Rosser (Tempe: Arizona University Press, 2002), pp. 201–10

Scherer, Günter, *Zur Geographie und Chronologie des angelsächsischen Wortschatzes, im Anschluss an Bischof Waerferth's Übersetzung der 'Dialoge' Gregors* (Leipzig: Mayer und Müller, 1928)

Schmidtke, Dietrich, 'Geistliche Schiffahrt – zum Thema des Schiffes der Buße im Spätmittelalter', *Beiträge zur Geschichte der deutschen Sprache und Literatur*, 91 (1969), 357–85

———, 'Geistliche Schiffahrt – zum Thema des Schiffes der Buße im Spätmittelalter II', *Beiträge zur Geschichte der deutschen Sprache und Literatur*, 92 (1970), 115–77

Scragg, Donald, 'The Corpus of Anonymous Lives', in *Holy Men and Holy Women: Old English Prose Saints' Lives and their Contexts*, ed. by Paul E. Szarmach (Albany: State University of New York Press, 1996), pp. 209–30

Smithers, G. V., 'The Meaning of *The Seafarer* and *The Wanderer*', *Medium Ævum*, 26 (1957), 137–53

——, 'The Meaning of *The Seafarer* and *The Wanderer* (Continued)', *Medium Ævum*, 28 (1959), 1–22

Stanton, Robert, *The Culture of Translation in Anglo-Saxon England* (Woodbridge: D. S. Brewer, 2002)

Waugh, Robin, 'The Blindness Curse and Nonmiracles in the *Old English Prose Life of Saint Guthlac*', *Modern Philology*, 106 (2009), 399–426

Weston, Lisa M. C., 'Guthlac Betwixt and Between: Literacy, Cross-Temporal Affiliation, and an Anglo-Saxon Anchorite', *Journal of Medieval Religious Cultures*, 42 (2016), 1–27

Whatley, Gordon E., 'Late Old English Hagiography, *ca.* 950–1150', in *Hagiographies: International History of the Latin and Vernacular Literature in the West from its Origins to 1550*, ed. by Guy Philippart, vol. II (Turnhout: Brepols, 1996), pp. 429–99

——, 'Lost in Translation: Omission of Episodes in Some Old English Prose Saints' Legends', *Anglo-Saxon England*, 26 (1997), 187–208

Wickham-Crowley, Kelley M., 'Living on the *Ecg*: The Mutable Boundaries of Land and Water in Anglo-Saxon Contexts', in *A Place to Believe In: Locating Medieval Landscapes*, ed. by Clare A. Lees and Gillian R. Overing (University Park: Pennsylvania State University Press, 2006), pp. 85–110

Wright, Charles D., *The Irish Tradition in Old English Literature*, Cambridge Studies in Anglo-Saxon England, 6 (Cambridge: Cambridge University Press, 1993)

DANIEL ANLEZARK

Drawing Alfredian Waters: The Old English *Metrical Epilogue* to the *Pastoral Care*, Boethian *Metre* 20, and *Solomon and Saturn II*

In this essay, I will examine the related treatments of the concept of 'living waters' in three Old English poems, all of which date from the end of the ninth century or the opening decades of the tenth. The first of these is the poetic *Epilogue* to the Alfredian *Pastoral Care*, the creation of which is indisputably datable to the 890s, which was in all likelihood composed in Wessex, and was preserved in Worcester until the sixteenth century, on the last page of what is now Oxford, Bodleian Library, MS Hatton 20. The date of the *Epilogue*'s composition can be determined with an unusual degree of precision for an Old English poem, as it is closely associated with the royal court of Alfred the Great (d. 899). The poem could well be the creative work of one of the translators referred to in the Preface to the Old English *Pastoral Care*, also preserved in MS Hatton 20. Among these, only two were native speakers of Old English — Alfred himself and the Mercian Plegmund, archbishop of Canterbury, either of whom might have been the poet. It is also possible, however, that the *Epilogue* is a scribal product, and the work of no named historical figures. The poem's dialect presents features of both early West Saxon and Mercian (*gegiered* and *giefe*, beside *diop/undiop* and *iowrum*), symptomatic of the vernacular literary dialect of Alfred's court circle.[1] Given the possible mediation of at least one scribe between the poet and our copy, it would be unsafe to speculate on the original Old English dialect of the poet himself.

---

1 *The Old English Boethius*, ed. and trans. by Irvine and Godden, pp. 410 and 412. All quotations are from this edition.

> **Daniel Anlezark** • (daniel.anlezark@sydney.edu.au) is the McCaughey Professor of Early English Literature and Language at the University of Sydney, where he is currently working on a project integrating the study of early medieval science and literature.

Meanings of Water in Early Medieval England, ed. by Carolyn Twomey and Daniel Anlezark, Studies in the Early Middle Ages, 47 (Turnhout: Brepols, 2021), pp. 241–266

The second poem with a close interest in 'spiritual' waters to be studied here is *Solomon and Saturn II*, a poem of less certain date than the *Epilogue*, but its early West Saxon language (also incorporating some Mercian features) most likely dates it to somewhere in the period from 890 to about 940 — early or later dates seem improbable.[2] This conclusion harmonizes with the date of the poem's unique manuscript, Cambridge, Corpus Christi College, MS 422, which dates from the second quarter of the tenth century, but probably not earlier than 930.[3] The poem's provenance is unknown, though, as I will argue, the learning on show suggests an educated circle in which Alfredian works were known and studied. The clearest evidence of this is the poet's borrowing from the discussion of the elements found in *Metre* 20 of the Old English prosimetrical version of Boethius's *De consolatione philosophiae*. I will argue that the anonymous poet of *Solomon and Saturn II* was not only a close reader of this *Metre*, but also of the *Metrical Epilogue* to the Alfredian *Pastoral Care*. His use of waters as wisdom from both Alfredian poems evokes a scholarly context in which these works were read and studied, and their thought developed.

The *Metrical Epilogue* to the *Pastoral Care*

Metaphorical understandings of water pervade the *Pastoral Care*'s *Metrical Epilogue*. Studies by James E. Cross and William T. Whobrey have identified important biblical precedents for the poem's water imagery, though, as will be seen, these sources do not account for one particular aspect of the metaphor developed across the *Epilogue*.[4] This short poem may be read in three movements.[5] The first section of the poem presents a metaphor for Gregory's *Pastoral Care* as a conduit (or aqueduct), leading the living waters of the Holy Spirit from heaven to the book's readers (*Epilogue* ll. 1–8):

> Ðis is nu se wæterscipe   ðe us wereda god
> to frofre gehet   foldbuendum.
> He cwæð ðæt he wolde   ðæt on worulde forð
> of ðæm innoðum   a libbendu
> wætru fleowen,   ðe wel on hine
> gelifden under lyfte.   Is hit lytel tweo
> ðæt ðæs wæterscipes   welsprynge is
> on hefonrice,   ðæt is halig gæst.

---

2   See *The Old English Dialogues of Solomon and Saturn*, ed. and trans. by Anlezark, pp. 49–57.
3   See *The Old English Dialogues of Solomon and Saturn*, ed. and trans. by Anlezark, p. 3; Dumville, 'English Square Minuscule Script', pp. 143–44 and 158–59.
4   Cross, 'The *Metrical Epilogue*'; Whobrey, 'King Alfred's *Metrical Epilogue* to the *Pastoral Care*'.
5   See Whobrey, 'King Alfred's *Metrical Epilogue* to the *Pastoral Care*', p. 176.

> [This is now the aqueduct which the God of hosts promised us for the comfort of earth-dwellers. He said that he desired that ever-living waters should flow continually into the world from the inner parts of those under the sky who well believed in him. There is little doubt that the well-spring of the conduit is in the kingdom of heaven, that is, the Holy Spirit.][6]

The second movement in the poem explores first the process of inspired authorship, by which writers such as Gregory transmit God's spirit of wisdom into and through words into the minds of their readers, and what these readers might do with it for good or ill (*Epilogue* ll. 9–21):

> Ðonan hine hlodan   halge and gecorene,
> siððan hine gierdon   ða ðe gode herdon
> ðurh halga bec   hider on eorðan
> geond manna mod   missenlice.
> Sume hine weriað   on gewitlocan,
> wisdomes stream,   welerum gehæftað,
> ðæt he on unnyt   ut ne tofloweð,
> ac se wæl wunað   on weres breostum
> ðurh dryhtnes giefe   diop and stille.
> Sume hine lætað   ofer landscare
> riðum torinnan;   nis ðæt rædlic ðing,
> gif swa hlutor wæter,   hlud and undiop,
> tofloweð æfter feldum   oð hit to fenne werð.

> [From thence [the *welsprynge*] the saints and the chosen ones drew it, then those who obeyed God directed it through holy books [the *wæterscipe*] here on earth in many ways throughout the minds of men. Some dam up wisdom's stream for themselves within their minds, keep it captive with their lips, so that it does not flow away useless, but the pool remains deep and still in the man's breast through the Lord's grace. Some let it run away over terrain in small streams; that is not an advisable course of action, if clear water should pour out loud and un-deep across fields until it becomes marsh.]

In the third and final movement of the *Metrical Epilogue*, the poet addresses his readers, advising them to use vessels to draw the waters that Gregory has brought directly to their homes with the aqueduct of his *Pastoral Care* (ll. 22–30):

> Ac hladað iow nu drincan,   nu iow dryhten geaf
> ðæt iow Gregorius   gegiered hafað
> to durum iowrum   dryhtnes welle.

---

6   Translations are my own unless otherwise stated. See also Helen Appleton's chapter in this volume, 'Water, Wisdom, and Worldliness in the Anglo-Saxon Prose Lives of Guthlac'.

> Fylle nu his fætels,   se ðe fæstne hider
> kylle brohte,   cume eft hræðe.
> Gif her ðegna hwelc   ðyrelne kylle
> brohte to ðys burnan,   bete hine georne,
> ðy læs he forsceade   scirost wætra,
> oððe him lifes drync   forloren weorðe.

> [But draw yourselves water to drink, now that the Lord has granted you that Gregory has directed the Lord's spring to your doors. He who has brought here a watertight flask may now fill his vessel, and may come back quickly. If any man has brought here a leaky flask to his stream, let him repair it speedily, so that he may avoid spilling the clearest of waters or losing the drink of life.]

The metaphor of water across the poem presents an image of the process of writing and reading, focusing on how divine wisdom is mediated by inspired authors into the minds of readers, and the ways in which this wisdom dwells in the mind.

As Cross points out, the point of departure for the poet's water metaphor is John 7. 37–38:

> In novissimo autem die magno festivitatis stabat Jesus et clamabat dicens, Si quis sitit veniat ad me et bibat. Qui credit in me sicut dicit Scriptura, flumina de ventre eius fluent aquae vivae. Hoc autem dixit de Spiritu quem accepturi erant credentes in eum; nondum enim erat Spiritus datus, quia Iesus nondum erat glorificatus.[7]

> [On the last, great day of the festival, Jesus stood and called out, saying: If anyone thirst, let them come to me and drink. The one who believes in me, as scripture says, Out of his belly shall flow rivers of living water. Now he said this of the Spirit which those who believed in him should receive, because the Spirit was not yet given, because Jesus was not yet glorified.]

Whobrey notes that patristic and medieval commentators understood these 'living waters' as *aqua doctrinae* (waters of doctrine), emphasizing the gift of the Holy Spirit in teachers, and especially in the teaching authority of the apostles and later of bishops.[8] Both Whobrey and Malcolm Godden have pointed out that the imagery in Proverbs is taken up extensively by Gregory across the Latin *Pastoral Care*, though its metaphors are modified systematically according to the Old English author's understanding and interests.[9] Whobrey notes that the *Epilogue*'s imagery has been influenced

---

[7] Biblical citations are from *Biblia sacra iuxta vulgatam versionem*, ed. by Weber, with punctuation added.
[8] Whobrey, 'King Alfred's *Metrical Epilogue* to the *Pastoral Care*', pp. 177–78.
[9] Whobrey, 'King Alfred's *Metrical Epilogue* to the *Pastoral Care*', pp. 179–80.

by Proverbs 5. 15:[10] 'Bibe aquam de cisterna tua, et fluenta putei tui; deriventur fontes tui foras, et in plateis aquas tuas divide' (Drink water out of thy own cistern, and the streams of thy own well: Let thy fountains be conveyed abroad, and in the streets divide thy waters). This passage is cited by Gregory in the Latin *Pastoral Care*, who interprets the verse (III. 24):[11] 'Aquam quippe praedictor de cisterna sua bibot, cum ad cor suum rediens, prius audit ipse quod dicit. Bibit sui fluenta putei, si sui irrigatione infunditur uerbi' (For indeed the preacher drinks out of his own cistern, when, returning to his own heart, he first listens himself to what he has to say. He drinks the running waters of his own well, if he is watered by his own word). The poetic *Epilogue*, however, says something slightly different: 'the well-spring' of wisdom is not so much the preacher's (or bishop's) own, as the Holy Spirit. Malcolm Godden has noted the application of a very similar image to Gregory's writing in the Prologues to each of the books of the Old English translation of his *Dialogi*. The first of these is indicative:

> Her ongynneð se æresða stream þære clænan and hlutran burnan þurh þone halegan breosð ures fæder and larheowes þæs apostolican papan sanctus Gregorius up aspringan ond forþ yrnan to lare and to bisene eallum þam, þe lysteð feran on liues weg.[12]

> [Here begins the first stream of the clean and pure brook [or spring] to well up through the holy breast of our father and teacher Saint Gregory the apostolic pope, and to flow for the instruction and example of all those who intend to travel on the path of life.]

A similar image of patristic authors and texts as living streams, closer to that found in the *Metrical Epilogue*, referring to the works of commentary of Basil (Eustathius), Ambrose, and Augustine, is employed by Bede in the Preface to his own commentary *On Genesis*:

> prolixa legentibus doctrinae salutaris fluenta manarunt, completo in eis promisso ueritatis quo dicebat, Qui credit in me, sicut dicit scriptura, flumina de uentre eius fluent aquae uiuae.

---

10 Other water imagery found in Proverbs probably also informs the poem (and also the expression of John's Gospel): 'Lex sapientis fons vitae, ut declinet a ruina mortis' (13. 14, The law of the wise is a fountain of life, so that he may turn away from death's ruin); 'Fons vitae eruditio possidentis, doctrina stultorum fatuitas' (16. 22, Knowledge is a fountain of life to the one who possesses it, the instruction of fools is foolishness); 'Aqua profunda verba ex ore viri, et torrens redundans fons sapientiae' (18. 4, Words from a man's mouth are like deep water, and the fountain of wisdom like an overflowing stream). See Whobrey, 'King Alfred's *Metrical Epilogue* to the *Pastoral Care*', p. 177.

11 *Grégoire le Grand: Règle pastorale*, ed. by Rommel, II, 424, 426, ll. 94–97.

12 Godden, 'Prologues and Epilogues', p. 467; *Bischof Wærferths von Worcester Übersetzung der Dialoge Gregors des Grossen*, ed. by Hecht, pp. 2, 94, 179, 260.

[[they] poured forth abundant streams of the doctrine of salvation to readers. In them was fulfilled the promise of truth, in regard to which Christ said, He who believes in me, as the Scripture says, Out of his belly shall flow rivers of living water.][13]

Against this background, it is perhaps surprising that the imagery of living waters as inspired patristic teaching that we find in the *Epilogue* is not, instead, found in a poetic Prologue to the *Pastoral Care*.

Earlier in Book III (Ch. 14) of his *Pastoral Care*, Gregory develops further his image of the mind as a place for storing the water of wisdom. The effect of trapping water is positive, so that it collects at 'humana etenim mens aquae more et circumclusa ad superiora colligitur, quia illud repetit unde descendit' (higher levels, in that it seeks again the height from which it descended).[14] However, this water 'et relaxata deperit, quia se per infima inutiliter spargit' (when let loose, falls away so that it disperses itself unprofitably through the lowest places).[15] This signifies the dissipation of wisdom in idle chatter, depriving the mind of its inner source. The Old English prose author extends Gregory's metaphor into the image of a dam:

> Ac ðæt mennisce mod hæfð wætres ðeaw. Ðæt wæter, ðonne hit bið gepynd, hit miclað and uppað and fundað wið ðæs ðe hit ær from com, ðonne hit flowan ne mot ðider hit wolde. Ac gif sio pynding wierð onpennad, oððe sio wering wirð tobrocen, ðonne toflewð hit eall, and ne wierð to nanre nytte, buton to fenne.[16]

> [But the human mind has the character of water. Water, when it is dammed, increases and rises and probes towards where it came from before, when it is not able to flow to where it wants. But if the pond is opened, or the dam is broken, then it flows away, and becomes of no use at all, but becomes a marsh.]

The elaboration of the overflowing waters of the dam as a negative image, signifying the dispersal of wisdom through empty chatter, is developed further in Chapter 38 of the Old English *Pastoral Care*:

> Be ðæm wæs suiðe wel gecweden ðurh ðone wisan Salomon, ðætte se se ðæt wæter ut forlete wære fruma ðære towesnesse. Se forlæt ut ðæt wæter, se ðe his tungan stemne on unnyttum wordum lætt toflowan. Ac se wisa Salomon sæde ðætte suiðe deop pol wære gewered on ðæs wisan monnes

---

13 Bede, *Libri quatuor in principium Genesis*, ed. by Jones, p. 1, ll. 11–14; Bede, *On Genesis*, trans. by Kendall, p. 65.
14 *Grégoire le Grand: Règle pastorale*, ed. by Rommel, II, 344, ll. 66–68.
15 *Grégoire le Grand: Règle pastorale*, ed. by Rommel, II, 344, ll. 68–69.
16 *King Alfred's West-Saxon Version of Gregory's Pastoral Care*, ed. by Sweet, c. 38, p. 277; Godden, 'Prologues and Epilogues', pp. 467–68. On the term *wering*, see Anlezark, 'Old English *Exodus* 487 *werbeamas*', pp. 502–03.

mode, and suiðe lytel unnyttes utfleowe. Ac se se ðe ðone wer bricð, and
ðæt wæter ut forlæt, se bið fruma ðæs geflites.[17]

> [Concerning this it was very well spoken through the wise Solomon,
> that he who let the water out was the instigator of quarrelling. He lets
> out the water who lets the voice of his tongue flow away in useless
> words. But the wise Solomon said that a very deep pool was dammed
> up in the wise man's mind, and very little that was useless flowed out.
> But he who breaks the dam and lets the water out is the instigator of
> conflict.]

The Old English author's metaphor is reiterated in Chapter 39, and again
elaborates on the Latin source:

> ne hie nellað hie gehæftan and gepyndan hiora mod, swelce mon deopne
> pool gewerige, ac he læt his mod toflowan on ðæt ofdele giemelieste and
> ungesceadwisnesse æfter eallum his willum.[18]

> [they are not willing to restrain and dam their minds, like one dams a
> deep pool, but he lets his mind flow away in the descent of unheedfulness
> and irrationality after all his intentions.]

The close verbal debt of the water metaphor of the central section of the *Metrical
Epilogue* reveals the poem's close relationship to the prose text's expression
(*weriað, gehæftað, unnyt, tofloweð, diop, lætað, to fenne werð*). However, the
poet also develops both the metaphor and its underlying concept concerning
wisdom and the role of the inspired writer as teacher. By the end of the poem
the sustained imagery of living waters depends on an understanding of how
an elaborate system of water distribution might work in a city, bringing water
from a remote spring, via a conduit (*wæterscipe*), to the doors of people's
houses, where, as it flows by they are able to scoop water from it with pitchers.
Yet, there is no evidence that the hydraulic engineering underpinning this
metaphor was employed in early medieval England.

The term *wæterscipe* is used more often in prose than in poetry; the only
two surviving examples from verse are the *Epilogue*'s and in *Daniel*, in the
context of the poem's rendering the liturgical *Canticum trium puerorum*
(an abbreviating adaptation of Daniel 3. 56–88) recited weekly at Lauds on
Sunday. While the term can simply mean 'a body of water', as it generally
does in prose, in *Daniel* it translates *fontes* (springs), the terms shared by the
biblical canticle and its liturgical version. A similar dynamic sense is also
apparent in the *Epilogue*, where the poet draws a further distinction between
the spring (*welsprynge*, the Holy Spirit) and Gregory's work, the *Pastoral*

---

17 *King Alfred's West-Saxon Version of Gregory's Pastoral Care*, ed. by Sweet, p. 279; Godden,
'Prologues and Epilogues', p. 470.

18 *King Alfred's West-Saxon Version of Gregory's Pastoral Care*, ed. by Sweet, p. 283; Godden,
'Prologues and Epilogues', p. 470.

Care, which is the aqueduct (*wæterscipe*) which brings the waters from this spring from heaven to earth.[19] The imagery harmonizes fully with — and may even intentionally express — the medieval belief that Gregory's works were written under the direct inspiration of the Holy Spirit, giving them a quasi-scriptural status.[20]

It is also apparent that the poet is familiar with some of both the simple and more complex principles that underpin hydraulic engineering. The poet knows that waters flow downwards — in his example from heaven to earth. But he also understands how water is moved from a natural spring via a conduit or aqueduct to a distant place, where it might be distributed to many users at their doors ('to durum iowrum'). This understanding of engineering may draw on the same understanding of the properties of water included in the prose text, which as we have seen, ascribes an intentionality to water ('ðonne hit flowan ne mot ðider hit wolde'). This property of water to climb back towards its source when dammed constitutes a crucial aspect of the image: as trapped water reaches towards its source, so the mind which dams the waters of wisdom reaches upwards to heaven. The waters which are let loose, however, descend, moving away towards the earth. Behind the metaphor lies the scientific belief that 'living' water is restless by nature, with an inherent tendency to movement, either by ascent or descent. While this theory was available to the translator and the poet, the water technology implied in the *Epilogue*'s metaphor was well beyond the Anglo-Saxons of Alfred's time, though Roman aqueducts were then still in use in many places in Italy and Gaul, and were undoubtedly seen by Alfred and probably also by at least some of his scholarly helpers.

It is possible that the aptness of this imagery focused on the technology of directing and distributing water, when applied particularly to the inspired work of Gregory the Great, is both acute and closely informed. The Aqua Claudia was one of the four great aqueducts of ancient Rome, bringing water a distance of seventy kilometres from two springs to the east of the city, with a massive flow of 190,000 cubic metres per day.[21] The Aqua Claudia was extended by Nero as far as the Caelian Hill, and later by Domitian to the Palatine, from where it could provide all fourteen districts of the imperial city with water. Before he became pope, Gregory converted his family's palace on the Caelian Hill into a monastery, and from this community Augustine came to bring the gospel (and the *Pastoral Care*) to the English in 597. By Gregory's time the aqueducts had become unreliable, 'patchily repaired after being frequently cut, and improperly

---

19 See 'wæterscipe', in Clark Hall, *A Concise Anglo-Saxon Dictionary*, p. 395: 'sheet of water, waters: conduit'. See *The Antwerp-London Glossaries*, ed. by Porter, p. 124, l. 2759: *Colimbus .i. aquaductus . wæterscipes hus*.
20 See Smalley, *The Study of the Bible in the Middle Ages*, p. 12.
21 Blackman, 'The Volume of Water Delivered by the Four Great Aqueducts of Rome'.

maintained, [they] leaked and formed marshes under their junctions'.[22] As pope, Gregory wrote to the prefect of Italy demanding their repair. There can be no doubt that the young Alfred visited the Caelian Hill, and the church of Saint Andrew and the monastery on it, during both his visits to Rome as a child in the 850s, when the Aqua Claudia still struggled to supply the city with water.

The possibility that the imagery of the water conduit in the *Epilogue* is informed directly by an appreciation of the urban landscape of Rome, and the place of Gregory's ancestral home within it, is invited by its development of the imagery and emphases of the *Metrical Preface* to the *Pastoral Care*, also included in MS Hatton 20, but undoubtedly by a different poet:

Þis ærendgewrit   Agustinus
ofer sealtne sæ   suðan brohte
iegbuendum,   swa hit ær fore
adihtode   dryhtnes cempa,
Rome papa.   Ryhtspell monig
Gregorius   gleawmod gindwod
ðurh sefan snyttro,   searoðonca hord.
Forðæm he monncynnes   mæst gestriende
rodra wearde,   Romwara betest,
monna modwelegost,   mærðum gefrægost.
Siððan min on englisc   ælfred kyning
awende worda gehwelc,   and me his writerum
sende suð and norð,   heht him swelcra ma
brengan bi ðære bisene,   ðæt he his biscepum
sendan meahte,   forðæm hi his sume ðorfton,
ða ðe lædenspræce   læste cuðon.[23]

> [Augustine brought this message from the south across the salty sea to the island-dwellers, as the Lord's champion, the pope of Rome, ordered beforehand. Sagacious Gregory thoroughly understood many true discourses through a wise mind, a treasury of sagacity. Therefore he, the best of the citizens of Rome, gained most of mankind for the Guardian of the heavens, the most brilliant of men, the most famous. Afterwards King Alfred translated each word of me into English, and sent me to all his scribes, north and south, commanded each of them to make more from the exemplar, so that could send them to his bishops, because some of them needed it, those who knew Latin the least.]

---

22   See Llewellyn, *Rome in the Dark Ages*, pp. 95–97.
23   See *The Old English Boethius*, ed. and trans. by Irvine and Godden, p. 408. See also Godden, 'Prologues and Epilogues', pp. 461–64.

The two poems are essentially about the same thing: how the divine wisdom of the *Pastoral Care* had been brought to English readers. Here the historical process begins in Rome and in Latin, and ends in England in English. The poetic *Preface* also leaves no room for doubt that the translation was originally made specifically for episcopal readers, and it was the need for their growth in wisdom that is embodied in the Old English text. The dynamic of textual dissemination unites the *Metrical Preface* and the *Epilogue*, but is recapitulated in the latter by a poet in a move back to the source, transforming the salty waters across which the *Pastoral Care* text is brought from Rome, into the living waters which flow across the Caelian Hill into the city. The *Epilogue* emerges as the more sophisticated literary product reflecting on the textual process: in the imagery of movement and water, as the *Epilogue* develops a complex metaphorical movement of waters from the *Preface*'s literal travel across them; the concept of wisdom, which is simply stated in the *Preface*, but explored in psychological detail in the later poem; the focus on Gregory as the author of the text, simply put in the *Preface*, where the second poem dwells on the author's intimacy with the Holy Spirit; and finally on the city of Rome as the source of the text, where the *Preface*'s general comprehension of geography yields to the later poet's intimate understanding and metaphorical deployment of Roman topography and Gregory's place in it.

Solomon and Saturn II

The dialogues of Solomon and Saturn are among the most mysterious and obscure texts in Old English. Three of them appear together in an anthology in Cambridge, Corpus Christi College, MS 422, though they were probably composed some decades earlier. The anthology contains two poems, *Solomon and Saturn I* and *II*, separated by a prose dialogue. The first poetic dialogue and the prose dialogue concern the powers and properties of the personified Lord's Prayer (the Pater Noster). The second poetic dialogue (which was probably composed before the Pater Noster poem, despite the current manuscript arrangement) is of a very different character, and introduces strange creatures such as the *Vasa mortis*, besides mythic heroes, though the poem is also concerned with philosophical questions, including the relationship between the physical world and fate. All three texts lean heavily on Irish learning and lore of the early medieval period, though the discussion of the properties of water in *Solomon and Saturn II* reveals close affinities with the understanding of its behaviour in both the *Metrical Epilogue* to the *Pastoral Care* and the Old English *Metre of Boethius* 20.

The dialogues are set in the aquatic world of the rivers and sea of the eastern Mediterranean and Mesopotamia. The two poetic dialogues are contextualized within Saturn's prolonged search for wisdom, which has seen him travelling across rivers and seas (*Solomon and Saturn II* ll. 7–32), ultimately reaching Jerusalem.[24] In *Solomon and Saturn I*, Saturn promises, should he lose their debate (ll. 19–20): 'wende mec on willan on wæteres hrigc | ofer Coferflod Caldeas secan' (I will depart safe and sound, turn myself willingly onto the water's height over the river Chobar to seek out the Chaldeans). A similar localization is offered by Solomon, who reflects where Saturn will boast should he win their debate in *Solomon and Saturn II*: 'on Wendelsæ ofer coforflod cyðõe seccan' (across the Mediterranean beyond the river Chobar). In *Solomon and Saturn I* aquatic creatures present dangers to the traveller:[25]

> Hwilum flotan gripað
> [...]
> Hwilum he on wætere   wicg gehnægeð,
> hornum geheaweð,   oððæt him heortan blod
> famig flodes bæð   foldan geseceð. (ll. 151, 155–57)
>
>> [Sometimes they grab the sailor [...] Sometimes in the water he makes the horse stumble, cutting it with horns, until its heart's blood seeks out the current's foamy bath, seeks the riverbed.]

Solomon and Saturn II has its own terrifying creature associated with water, in the form of the mysterious *Vasa mortis*, a bird-like creature that the wise king has brought across the sea to Palestine in chains:

> Nyste hine on ðære foldan   fira ænig
> eorðan cynnes   ærðon ic hine ana onfand,
> ond hine ða gebindan het   ofer brad wæter,
> ðæt hine se modega heht   Melotes bearn
> Filistina fruma,   fæste gebindan,
> lonnum belucan   wið leodgryre.
> Ðone fugel hatað   feorbuende
> Filistina fruma,   uasa mortis. (ll. 96–103)
>
>> [No man in the world, of the earthly race, knew about it before I alone found it, and across the broad sea ordered it bound, so that the brave son of Melot, the leader of the Philistines, commanded it to be firmly bound, locked in chains against the people's terror. The distantly dwelling leaders of the Philistines call the bird *Vasa mortis*.]

The bird's name — 'instruments of death' — presents an opaque recollection of Psalm 7. 14: 'et in eo paravit *vasa mortis* sagittas suas ardentibus effecit'

---

24 See *The Old English Dialogues of Solomon and Saturn*, ed. and trans. by Anlezark, pp. 38–40. All quotations are from this edition.
25 The passage recalls elements of the dangerous rivers found in the Old English *Alexander's Letter to Aristotle*; see *The Old English Dialogues of Solomon and Saturn*, ed. and trans. by Anlezark, p. 110.

(And in it [sc. God's bow] he has prepared the instruments of death, he has made his arrows ready for those who burn).[26] The naming of the bird in the poem implicitly invites a metaphoric reading, and while this is difficult for the modern reader to decipher, the poetic passage and Psalm 7 are both focused on the judgement. This slippage between the literal and the metaphoric is characteristic of the poet's technique and constitutes an important aspect of both the playful evasiveness of the dialogue between Solomon and Saturn, and the reader's attempts to understand it.

The cosmic waters above the middle region of the world appear in all three Solomon and Saturn dialogues. In a cryptic reference, probably to the unleashing of cosmic waters at the Last Judgement, *Solomon and Saturn I* describes the fate of the foolish sinner on the final day (ll. 28–29): 'ealle beoð aweaxen | of edwittes iða heafdum' (he will be completely washed away by the waves of disgrace from the heights). This interpretation of a difficult reading is supported by the reference to the heavenly waters and the Doomsday flood in the two other dialogues, whose origins are closely related to *Solomon and Saturn I*. In the *Pater Noster Prose Dialogue*, Solomon observes:

> Pater Noster hafað gylden heafod ond sylfren feax, ond ðeah ðe ealle eorðan wæter sien gemenged wið ðam heofonlicum wætrum uppe on ane ædran, ond hit samlice rinan onginne eall middangerd, mid eallum his gesceaftum, he mæg under ðæs Pater Nosters feaxe anum locce drige gestandan.
>
>> [The Pater Noster has a golden head and silver hair, and even if all earth's waters be mixed with the heavenly waters above in one channel and it begin to rain together, middle-earth — with all its creatures — it could stand dry under one lock of the Pater Noster's hair.][27]

In *Solomon and Saturn II*, Saturn describes the sea's power and terror at the judgement (ll. 145–48). In *Solomon and Saturn I*, the Pater Noster's power over these and all waters and the creatures in them forms an important, if cryptic, element of his characterization (ll. 77–83).

Physical water in the dialogues is given an important metaphoric dimension by the poet of *Solomon and Saturn II*, who not only develops an extensive discourse on the elemental properties of water, but also invites the reader to associate the behaviour of water with the mind and, ultimately in the context of a poem debating wisdom, with wisdom itself. This invitation into the metaphoric waters is offered in the context of the first riddling exchange between the interlocutors, as Solomon asks (*Solomon and Saturn II* ll. 32–33): 'Sæge me from ðam lande | ðær nænig fyra ne mæg fotum gestæppan' (Speak to me concerning the land where no man can step with his feet). Solomon's

---

26 See *The Old English Dialogues of Solomon and Saturn*, ed. and trans. by Anlezark, p. 125.
27 *The Old English Dialogues of Solomon and Saturn*, ed. and trans. by Anlezark, p. 74, ll. 52–56.

question seems straightforward and literal, and the literal-minded Saturn answers it in this way, telling the story of the mysterious dragon-slayer 'the surging Wolf', defined by his status as a 'great sea-traveller' (*mereliðende*, l. 34).[28] Wolf's sea journeys finally led to a toxic place populated by dragons, where he died — this is the place where no man can walk with his feet:

> forðan ða foldan   ne mæg fira ænig,
> ðone mercstede,   mon gesecan,
> fugol gefleogan   ne ðon ma foldan nita. (ll. 39–41)
>
> [therefore no man can seek that land, no one that border-land, nor bird fly there, more than any of the beasts of the earth.]

Saturn's answer is over the top — not only can't a person or beast walk there, birds can't even fly there. Solomon's reply-in-reply suggests a more conventional and literal answer — the sea-bed — upon which no one can walk:

> SALOMON cwað:
> Dol bið se ðe gæð   on deop wæter,
> se ðe sund nafað,   ne gesegled scip,
> ne fugles flyht,   ne he mid fotum ne mæg
> grund geræcan;   huru se Godes cunnað
> full dyslice,   Dryhtnes meahta. (ll. 47–51)
>
> [Solomon said: 'Foolish is he who goes into deep water, he who can't swim, nor has a sail-rigged ship, nor the flight of a bird, who cannot reach the bed with his feet. Indeed, he very foolishly tests God, the Lord's might.']

Saturn's suggestion might be rooted in literal-mindedness, but he includes the fact that Wolf, like himself, was a great sea traveller. The joke may go further, as it was commonly believed in the early Middle Ages that Saturn (like Wolf's friend Nimrod) was a giant, and perhaps Wolf himself was too — the giants of Antiquity, it was also believed, could walk across the sea, with their feet touching its bed.[29]

Saturn's playfully allusive answer threatens to get the better of Solomon, by simultaneously answering his question pointing to an obscure geographic location where no foot may tread, but also anticipating the conventional answer of 'the sea-bed', even as he mockingly refutes this conventional solution to the riddle with his evocation of the prowess of giants, like himself. Solomon will have none of this, and refuses to engage with Saturn's games. His answer in reply is superficially banal in the wake of Saturn's layered wit, as he asserts that the correct answer is indeed 'the sea-bed'. However, at the same moment

---

28 See *The Old English Dialogues of Solomon and Saturn*, ed. and trans. by Anlezark, pp. 118–21; and Anlezark, 'All at Sea', pp. 237–39.
29 See *The Old English Dialogues of Solomon and Saturn*, ed. and trans. by Anlezark, p. 120.

Solomon makes a leap beyond the literal, which Saturn, even with all his allusiveness, does not. The wise king's shift from the literal incorporates the opposition of wisdom and folly into the discourse. Solomon engages Saturn at a deeper level — the giant Nimrod tested the Lord's power at Babel, and was a fool whose actions resulted in God's division of the world into the languages and nations that Saturn himself has recently traversed. Saturn himself had best be careful about how language works — deep waters are also the waters of wisdom, and the literal-minded fool best beware of their depths. Solomon, as the author of Proverbs (a work whose content and format were undoubtedly familiar to the poet of *Solomon and Saturn II*),[30] had plenty to say about the symbolism of waters and wisdom, a latent association developed across the poem.[31] In a parallel to the *Metrical Epilogue*'s association of water with books and wisdom, Solomon goes on to assert that books provide protection for the righteous and the wise (ll. 67–68): 'Sige hie onsendað soðfæstra gehwam, | hælo hyðe, ðam ðe hie lufað' (They present victory to each of the righteous, a harbour of safety for those who love them).

In a far closer parallel to the *Epilogue*, *Solomon and Saturn II* discusses the elemental properties and powers of water in the context of an ongoing, and often submerged, discourse on wisdom and knowledge. Water is the focus of two extended discussions in the poem, though both now are incomplete:

SATVRNVS CVÆÐ:
Ac forhwon fealleð se snaw,   foldan behydeð,
bewrihð wyrta cið,   wæstmas getigeð,
geðyð hie ond geðreatað,   ðæt hie ðrage beoð
cealde geclungne?   Full oft he gecostað eac
wildeora worn,   wætum he oferbricgeð,
gebryceð burga geat,   baldlice fereð,
reafað   ***
swiðor micle   ðonne se swipra nið
se hine gelædeð   on ða laðan wic
mid ða frǽcnan   feonde to willan. (ll. 124–33)

> [Saturn said: But why does snow fall — it covers the earth, encloses the shoots of plants, binds things that grow, crushes and inhibits them, so that for a long while they are withered with cold? Very often it distresses many wild animals too, makes a bridge over water, breaches the gate of the citadel, boldly proceeds, robs ... <Solomon said:> ... much stronger than the deceitful malice which will lead him into the hateful abode with the terrible ones to the Enemy's delight.]

---

30 See *The Old English Dialogues of Solomon and Saturn*, ed. and trans. by Anlezark, pp. 12, 18, 20.
31 See above, note 10.

We cannot now know for certain if the text that resumes after the textual loss also belongs with the poem's discussion of the ice question, though if it does, because of the alternating logic of the debate we have the tail-end of Solomon's reply, and not more of Saturn's question. If this is the answer to Saturn's water-ice question, then two passages at the poem's end would be related. The first describes a cold, wet version of hell (ll. 290–91): 'wælcealde wic, wintre beðeahte, | wæter in sende ond wyrmgeardas' (a place of deadly cold covered in winter, sent water in there, and snake-pits). The second is *Solomon and Saturn II*'s damaged ending '… swice, ær he soð wite, | ðæt ða sienfullan saula sticien | mid hettendum helle tomiddes' ([to?] deceit, before he knows the truth, that the sinful souls should be stuck in hell, with tormentors in the middle of hell).[32] The paradoxical — or contrary — standpoint that Saturn takes in his question, that water brings death to plants and destruction to towns, probably led into an answer which concludes with the sinner locked in a wet hell. Solomon undoubtedly refers to the infernal region in his reply, so that if this is understood as his answer to Saturn's question about icy water, then the wise king has once again taken Saturn's literal-minded question and provided an answer which incorporates the elemental into the tropological. The scientific underpinning of Saturn's question is important for our understanding of the poem: both snow and ice were understood to be water, congealed and compressed. Bede notes that in the case of snow this compression happens at altitude in the sky, with the coalescing of raindrops.[33] This point is clearly understood and exploited by the poet, just as the verbal imagery that frames ice in the poem reflects its own compressed nature: enclosing, binding, and crushing. The comment that ice breaks gates suggests natural observation — trapped water when frozen expands, and indeed can break gates, or even stone.

Soon after, the dialogue Solomon resumes and extends the discourse on water in its liquid and solid states, as the wise king muses about the relationship between cold (a property of water) and another element, fire. One of the two, fire or cold, must be more powerful, because the two cannot coexist for long:

SALOMON CVÆÐ:
Ne mæg fyres feng   ne forstes cile,
snaw ne sunne   somod eardian,

---

[32] It is difficult to read the damaged passage. OE *swice* could be read as a subjunctive of *swician*, 'to weaken'. However, the immediate reference in the passage to the sinner's failure to realize the truth makes it more likely that *swice* is a 'deceit' that has just been practiced on him by a devil, tricking him into hell.

[33] In his *De natura rerum*, ed. by Jones, pp. 173–234, Bede passes on the theory found in the Hiberno-Latin *Liber de ordine Creaturarum* about the formation of snow (Ch. 35): 'Niues aquarum uapore, necdum densao in guttas, sed gelu praeripiente formantur' (Snow is formed from the vapour of water that, being forestalled by cold, is not yet condensed into drops). See Bede, *On the Nature of Things*, trans. by Kendall and Wallis, p. 93. See *Liber de ordine Creaturarum*, ed. by Díaz y Díaz, p. 130.

> aldor geæfnan,   ac hira sceal anra gehwylc
> onlutan ond onliðigan   ðe hafað læsse mægnn. (ll. 177–80)

> [Solomon said: 'Fire's grasp and frost's chill, snow and sun, can neither dwell nor endure life together, but either one of them must submit and yield, that which has less power.']

At this moment in the dialogue, the observation seems to go nowhere. Solomon's statement has come as a reply to an apparently unrelated question from Saturn (ll. 175–76): 'Forhwon ne moton we ðonne ealle mid onmedlan | gegnum gangan in Godes rice?' (Then why can't we all go forwards with pomp into God's kingdom?). This query itself emerges from noting the paradox that weeping and laughter (*wop ond hleahtor*, l. 171) are often companions. Saturn replies to Solomon's comment on the contest between fire and cold with another question (l. 181): 'Ac forhwon ðonne leofað se wyrsa leng?' (But why then does the worse person live longer?). The poet develops his discussion of elements and their properties in a context that ties them to questions of morality and wisdom in a way that at times becomes dumbfounding. It is difficult to determine, however, how much the modern reader's confusion is a legitimate reflection of the author's own. For example, how significant are the precise terms of Saturn's blasphemous question about all people equally entering 'God's kingdom' (*in Godes rice*)? In medieval scientific theory, derived ultimately from the thought of Aristotle, any understanding of the elements and their properties in 'heaven' would not be the same as of their behaviour in the world below.

The importance of this careful distinction emerges soon after, when the discussion of two elements (fire and water) and their properties (heat and cold) resumes:

> Saturnus cwæð:
> Ac forhwam winneð ðis wæter   *geond woroldrice*,
> dreogeð deop gesceaft?   Ne mot on dæg restan,
> neahtes neðyð,   cræfte tyð,
> cristnað ond clænsað   cwicra manigo,
> wuldre gewlitigað.   Ic wihte ne cann
> forhwan se stream   ne mot stillan neahtes *** (ll. 215–20)

> [Saturn said: But why does this water struggle *throughout the kingdom of this world*, undergo a profound destiny? It cannot rest by day, goes boldly by night, it drags with force; it christens and cleanses many a living person, wondrously beautifies. I do not understand at all why the current cannot be still at night ***.] (emphasis added)

Both the distinctions being made and the subject matter under discussion evoke, not coincidentally, the discussion of the same elements and their properties in the Old English Boethius. Old English *Metre* 20 extends and elaborates the Old English prose Chapter 33, a version of the Latin *Metre* 9 of Book III of Boethius's Latin *De consolatione philosophiae*.³⁴ The ideas and phrasing of the Old English Boethius represent a significant expansion of the Latin text's passing reference to the elements (water, fire, earth, and air), and *Metre* 20 offers a fuller version than the prose, placing a greater emphasis on Christian doctrine in the role of the creator:

> Hwæt, þu, wuldres god,
> þone anne naman   eft todældes,
> fæder, on feower;   wæs þara folde an
> and wæter oðer   *worulde dæles,*
> and fyr is þridde   and feowerðe lyft;
> *þæt is eall weoruld   eft togædere.*
> Habbað þeah þa feower   frumstol hiora,
> æghwilc hiora   agenne stede,
> þeah anra hwilc   wið oðer sie
> miclum gemenged,   and mid mægne eac
> fæder ælmihtiges   fæste gebunden
> gesiblice   softe togædre
> mid bebode þine,   bilewit fæder,
> þætte heora ænig   oðres ne dorste
> mearce ofergangan   for metodes ege;
> ac geþweorod sint   ðegnas togædre,
> cyninges cempan,   cele wið hæto,
> wæt wið drygum,   winnað hwæðre.
> [...]
> Fela monna wat
> þætte yfemest is   eallra gesceafta
> fyr ofer eorðan,   folde neoðemest.
> Is þæt wundorlic,   weroda drihten,
> þæt ðu mid geþeahte   þinum wyrcest
> þæt ðu þæm gesceaftum   swa gesceadlice
> mearce gesettest,   and hi gemengdest eac. (*Metre* 20, ll. 57–74, 83–89)

> [Oh, God of glory, father, you later divided that one name into four: one of those was earth and the second water, *parts of the world*, and fire is the third and the fourth air; *together again that is the whole world*. Yet each of the four have their own birthplace, their own station, though each of them may be greatly mingled with the other by the might of the father almighty also bound fast, peaceably, smoothly together by your command, merciful father, so that none of them dared to cross the other's boundary out of fear of the creator, but the thegns, the champions of the king, are kept

---

34   *The Old English Boethius*, ed. by Godden and Irvine, ch. 33, I, 310–18, and *Metre* 20, I, 463–70; these are translated at II, 48–53 and II, 149–52, respectively.

in agreement together, cold with heat, wet with dry; they compete, however. [...] Many men know that highest above the earth of all creations is fire, and land the lowest. It is wonderful, Lord of hosts, that by your design you ensure that you thus appropriately set a boundary for those creatures and also mingle them.] (emphasis added)

The tension between the perfect separateness of the elements on their 'original thrones' (*frumstol*) on the one hand,[35] and their dynamic interrelatedness in the world of creatures on the other, lies at the heart of the Old English Boethius author's understanding. The Latin text of Book III, Metre 9 was the most heavily glossed section of the whole work in medieval manuscripts, and Malcolm Godden and Susan Irvine in their edition of the Old English *De consolatione* note that identifying either the prose author's or the poet's direct sources would be difficult; rather, both can be seen sharing in and developing the scientific ideas current in the early Middle Ages.[36]

What is most striking is not only that the poet of *Solomon and Saturn II* also shares in these ideas, but that he does so in a way which indicates close familiarity with their elaboration and expression in the Old English Boethius. Coupled with his sense of the fundamental separateness of the elements and their properties *in Godes rice*, is the poet's interest in the way these elements behave in the created world. With the Old English Boethius, the *Solomon and Saturn II* poet shares a focus on the restless behaviour of water:

> Hwæt, ðu þæm wættere   wætum and cealdum
> foldan to flore   fæste gesettest,
> *forðæm hit unstille   æghwider wolde*
> *wide toscriðan   wac and hnesce.*
> Ne meahte hit on him selfum,   soð ic geare wat,
> æfre gestandan,   ac hit sio eorðe hylt
> and swelgeð eac   be sumum dæle,
> þæt hio siðþan mæg   for ðæm sype weorðan
> geleht lyftum.
> [...]
> Ne meahte on ðære eorðan   awuht libban,
> ne wuhte þon ma   wætres brucan,
> on eardian   ænige cræfte
> for cele anum,   gif þu, cyning engla,
> wið fyre hwæthwugu   foldan and lagustream
> ne mengdest togædre,   and gemetgodest
> cele and hæto   cræfte þine. (*Metre* 20, ll. 90–98, 107–13)

---

35 On the term *frumstol*, see Huppé, *The Web of Words*, p. 46.
36 See *The Old English Boethius*, ed. by Godden and Irvine, II, 378, see also II, 510–12.

[Oh, you firmly established lands as a floor for the wet and cold water because it, *being restless, would disperse widely everywhere,* fluid and yielding. It could never stand by itself — I know for a certain truth — but the earth contains it and absorbs it to some extent, so that because of the absorption it can again be water from the air [...]. No creature could live on the earth nor moreover could anything enjoy water, live there in any way because of the cold alone, if you, king of the angels, had not mingled together land and sea with fire, and tempered cold and heat by your skill.] (emphasis added)

In the thought of the Old English Boethius, the unstillness of water as an element is tamed only through its combination with other elements. If the undamaged text of *Solomon and Saturn II* ever contained a full discussion of this phenomenon, it is now largely — but not entirely — lost.

In *Solomon and Saturn II*, Saturn's question about restless water and its properties, which encompass the physical as well as spiritual, is broken off by damage to the manuscript. The poem returns with Solomon's damaged answer, in which he is discussing fire. This passage concludes with observations of the behaviour of fire interpreted through a scientific lens:

> Leoht hafað heow ond had   Haliges Gastes,
> Cristes gecyndo;   hit ðæt gecyðeð full oft.
> *Gif hit unwitan   ænige hwile*
> *healdað butan hæftum,*   hit ðurh hrof wædeð,
> bryceð ond bærneð   boldgetimbru,
> seomað steap ond geap,   *stigeð on lenge,*
> *clymmeð on gecyndo,   cunnað hwænne mote*
> *fyr on his frumsceaft   on Fæder geardas,*
> <u>*eft to his eðle*</u>,   ðanon hit æror cuom.
> Hit bið eallenga   eorl to gesihðe,
> ðam ðe gedælan can   Dryhtnes ðecelan,
> forðon nis nænegu gecynd   cuiclifigende,
> ne fugel ne fisc   ne foldan stan,
> ne wæteres wylm   ne wudutelga,
> ne munt ne mor   ne ðes middangeard,
> ðæt he forð ne sie   fyrenes cynnes. (ll. 231–46)

> [Light has the hue and form of the Holy Spirit, Christ's nature — it makes that known very often. *If an unwise person holds it for any length of time without encasing it,* it proceeds through the roof, breaks and burns the house timbers, swings steep and high, *ascends in height, climbs according to nature, it probes the moment when fire might come* <u>back to its homeland</u>, *to its point of origin in the courts of the Father*. It is entirely visible to a man, to him who is able to share in the Lord's lantern, because *there is not any living kind — neither*

bird nor fish nor stone of the earth, nor surge of water nor tree-branch, nor mountain nor moor nor this middle-earth — that is not of the fiery race.] (emphasis added)

When read beside the discussion of the elements in the Old English *Metre* 20, and in the context of the poem's earlier discussion of the elements and the properties, the movement from the discussion of restless water in Saturn's question to Solomon's reflections on fire is not only unsurprising, but fully expected. In fact, there is clear evidence that the poetic dialogue is directly dependent on the discussion of fire which follows the account of water in *Metre* 20:

> Þæt is agen cræft   eagorstreames,
> wætres and eorþan,   and on wolcnum eac,
> and efne swa same   uppe ofer rodere.
> Þonne is þæs fyres   frumstol on riht,
> eard ofer eallum   oðrum gesceaftum
> gesewenlicum   geond þisne sidan grund;
> þeah hit wið ealla sie   eft gemenged
> weoruldgesceafta,   þeah waldan ne mot
> þæt hit ænige   eallunga fordo
> butan þæs leafe   þe us þis lif tiode,
> þæt is se eca   and se ælmihtga.
> [...]
> þeah hi unsweotole   somod eardien,
> swa nu eorðe and wæter   earfoðtæcne
> unwisra gehwæm   wuniað on fyre,
> þeah hi sindan   sweotole þæm wisum.
> *Is þæt fyr swa same   fæst on þæm wætre*
> *and on stanum eac   stille geheded*
> *earfoðhawe,   is hwæðre þær.*
> Hafað fæder engla   fyr gebunden
> efne to þon fæste   þæt hit fiolan ne mæg
> *eft æt his eðle,   þær þæt oðer fyr*
> up ofer eall þis   eardfæst wunað.
> Sona hit forlæteð   þas lænan gesceaft,
> mid cele ofercumen,   gif hit on cyððe gewit,
> and þeah wuhta gehwilc   wilnað þiderweard
> þær his mægðe bið   mæst ætgædre. (ll. 122–32, 146–60)

> [It is the special homeland of the sea, of water, [to be] on earth and also in the clouds and likewise in the heavens. The original place, the homeland, of fire is properly above all other creatures visible across this specious earth. Though it is mixed again with all worldly creatures, yet it may not manage to destroy anything entirely except by the leave of the one who granted us this life: that is the eternal

and almighty one [...]. Though they may dwell indistinguishably together, just as now earth and water dwell in fire (difficult to demonstrate to each unwise man), yet they are clear to the wise. *Fire is likewise firmly fixed in the water and also hidden immobile in stones, hard to see*; yet it is there. The father of the angels has bound fire so firmly that it cannot take itself *back to its homeland* where that other fire dwells, settled up above all this. It will immediately leave this transitory creation overcome by cold if it departs to its homeland, and yet each thing desires to go there where most of its kindred is together.]

The dialogue poet has made the material his own, but the parallels are striking and the debt undoubtable. *Solomon and Saturn II* shares with *Metre* 20 not only the specific assertion that fire dwells in both water and stone, but also the explanation that if freed, it would seek its homeland above in the heavens, which the two poems express in a shared formula: *eft to his eðle; eft æt his eðle*. The common elements and expression in what is a uniquely shared discussion in Anglo-Saxon literature proves the dependence of *Solomon and Saturn II* on *Metre* 20 beyond doubt. This debt is to the *Metre* and not its Old English prose source; the prose, while it shares the claim that fire dwells in both water and stone, does not use the poetic term *eðel*, and describes fire's home as *his agenne eard*.[37]

Significantly, there is also evidence that the *Solomon and Saturn II* poet was familiar with another Alfredian work — the *Metrical Epilogue* to the Old English *Pastoral Care*. In a striking modification to the thought of *Metre* 20, the dialogue poet identifies the element of fire with the Holy Spirit: 'Light has the hue and form of the Holy Spirit, Christ's nature — it makes that known very often'. This parallels the poetic equation of water with the Holy Spirit in the *Metrical Epilogue*: 'There is little doubt that the well-spring of the conduit is in the kingdom of heaven, that is, the Holy Spirit'. This curious parallel, however, becomes all the more noteworthy when in these two different contexts the imagery associating the elements of fire and water with the Holy Spirit are also metaphors for the mind, which must contain and avoid spilling them. With an image of failed containment, in the *Epilogue*'s metaphor the foolish reader's lack of wisdom is epitomized in the way he allows his water to spill.[38] The wise person locks up the waters of wisdom with his lips (*Epilogue* l. 14, 'wisdomes stream, welerum *gehæftað*'). The same metaphoric failure (and one element of its expression) lies at the heart of the escape of fire in *Solomon and Saturn II*: 'Gif hit unwitan ænige hwile | healdað butan hæftum' (If an unwise person holds it for any length of time without encasing it). In both poems the respective element escapes by behaving according to its natural properties,

---

37   See *The Old English Boethius*, ed. by Godden and Irvine, I, 316, l. 200.
38   Gregory and his Old English translator also agree that trapped water also moves upwards, seeking its natural place in the heavens; see above, note 16.

as these were understood by medieval science. It seems likely that the 'Lord's lantern' ('Dryhtnes ðecelan', l. 241) needed by the wise person to comprehend and contain the mysteries of the Holy Spirit and fire in *Solomon and Saturn II* is a playful but thoughtful inversion of the *kylle* (leather bottle, ll. 26, 27) used by the wise person to contain the waters of the Holy Spirit drawn from the 'Lord's well' ('dryhtnes welle', l. 23) in the *Pastoral Care Metrical Epilogue*.

Conclusion

The clear debt on the part of the *Solomon and Saturn II* poet to *Metre* 20 in his discussion of water and fire, for both his ideas and their expression, provides the poem with a place in the relative chronology of Old English literature, but also provides new evidence for the date of the composition of the Boethian *Metres*. The authorship of the Alfredian prose Boethius and the *Metres* (two related but distinct questions) continues to be a matter of debate. Godden and Irvine hypothesize

> that the OE [prose] Boethius was the work of an unknown writer of substantial learning, not necessarily connected with King Alfred or his court, but working some time in the period 890 to about 930, probably in southern England.[39]

They date the *Metres* within a wider timeframe, 'probably from around 890 to around 950 (the date of the earliest extant manuscript)'.[40] As has been noted above, *Solomon and Saturn II* is unlikely to have been composed as late as 950, and the sole surviving manuscript (CCCC, MS 422) could have been copied as early as 930. Given the probable distance in time, and perhaps also place, between the making of the Old English prose Boethius and the *Metres*,[41] the dates of the composition of the *Metres* and *Solomon and Saturn II* would appear to be only a few decades apart, at most; they could also be much closer.

The evidence provided by the Solomon and Saturn dialogue for the careful study of the Old English Boethius radically alters our understanding of the place of the Alfredian Boethius as a curriculum text in the Anglo-Saxon classroom. Borrowings from the Old English prose Boethius in the 990s by Ælfric parallel those by the anonymous author of the Old English *Distichs of Cato*, written any time between the mid-tenth century and the mid-eleventh.[42] The *Distichs* are undoubtedly associated closely with the Anglo-Saxon classroom, but any impression that these are isolated instances of knowledge of the Old English Boethius emerging from personal interest are undermined by the confident

---

39 *The Old English Boethius*, ed. by Godden and Irvine, I, 146.
40 *The Old English Boethius*, ed. by Godden and Irvine, I, 150.
41 *The Old English Boethius*, ed. by Godden and Irvine, I, 146–51.
42 See *The Old English Boethius*, ed. by Godden and Irvine, I, 207–12.

and playful engagement with the Alfredian text in *Solomon and Saturn II*, a text that also most certainly emerged in a pedagogic setting. Between them, these three authors and texts suggest that the Old English Boethius, either in its prose or prosimetric version, was carefully studied in the Anglo-Saxon schoolroom from the early tenth century and well into the eleventh. Nicholas Trevet's extensive and fully acknowledged borrowings from the prosimetric version of the Old English text translated into Latin for his commentary on the *De consolatione philosophiae*, besides the unacknowledged borrowings in the three earlier works, suggest that this tradition was an enduring one. The original thought found in the Old English version was as much valued around the year 1300 as it had been by the poet of *Solomon and Saturn II* four centuries earlier.

The intellectual milieu which produced not only the Solomonic dialogues, but also Alfred's *Pastoral Care*, is not fully appreciated, and still labours under assumptions about the personal interests of the philosopher king. Malcolm Godden has pointed out that the *Metrical Preface* to the *Pastoral Care* owes an indisputable debt to the Latin poem written by Alcuin to Preface to his own *De dialectica*, a work which draws on the Pseudo-Augustinian *Categoriae decem*, a popular text on dialectic studied and glossed across the Carolingian ninth century.[43] There is no reason to believe that Alfred — not the author of the Old English *Metrical Preface*, which contradicts statements about the authorship of the *Pastoral Care* found in the prose *Preface* — knew such a work, but someone in his circle did.[44] Against this background, the discussion about the existence of darkness in *Solomon and Saturn II* (ll. 160–65), a problem which troubled the early ninth-century Carolingian thinker Fredegisus, emerges as one more likely to be fully informed than not.[45]

The simple fact that Alfredian works were copied tells us that they were read across the tenth and eleventh centuries, as do more precise references to translations made by Alfred in the works of Ælfric in the 990s (who besides a general reference, mentions the Old English Bede and Gregory's *Dialogues* by name, though neither is by Alfred) and the praise of Alfred's Boethius in the Latin *Chronicon* of Æthelweard (of the early 970s), who refers to the work's emotional impact. *Solomon and Saturn II* establishes that the study of Alfredian texts began early, and with an interest that fused science and religion; in the context of the dialogue poem, there is also a clear interest in questions of wisdom, reading, authorship, and authority, which evoke the precise interests

---

43 Godden, 'Prologues and Epilogues', pp. 463–64.
44 The prose *Preface* presents a nuanced image of the *Pastoral Care*'s production, involving Alfred in the complex process of translation with his 'helpers'; the *Metrical Preface* is more flattering still: 'Siððan min on englisc ælfred kyning | awende worda gehwelc' (ll. 11–12, Afterwards King Alfred translated each word of me into English). The reliability of the details of both attributions is called into question by such a divergence so close to the moment of production of the Old English *Pastoral Care*.
45 See O'Brien O'Keeffe, 'Source, Method, Theory, Practice', pp. 66–68.

of the *Metrical Epilogue* to the *Pastoral Care*. The anonymous dialogue poet was not the only early reader to engage creatively with the *Metrical Epilogue*. In MS Hatton 20, the poem is immediate followed by a colophon, the joint work of 'Koenwald (Cenwald) the Monk and Ælfric the Cleric'.[46] Cenwald was bishop of Worcester from 928 to 958, and the colophon represents his own learned response to the poetic *Epilogue*. The colophon, among other texts, cites John 4. 13–14 — in which Jesus promises the Samaritan woman that whoever drinks 'sed aqua quam ego dabo ei fiet in eo fons aquae salientis in vitam aeternam' (the water that I will give him, [it] shall become in him a fountain of water, springing up into everlasting life).[47] Cenwald's colophon — the bishop's own augmentation of a text written for teachers like himself — also includes an experiment in Latin, Greek, and Hebrew, thoughtfully evoking the interest in the three sacred languages articulated in the prose Preface to the *Pastoral Care*.[48] It is not known when the colophon was included, but during the 950s seems likely.[49] The poet of *Solomon and Saturn II* was not the only early reader to study the Alfredian work and its verse *Epilogue*. The evidence suggests that Alfredian waters, and their powerful association with wisdom, reached many doors in Anglo-Saxon England.

Works Cited

Manuscripts

Cambridge, Corpus Christi College, MS 422
Oxford, Bodleian Library, MS Hatton 20

Primary Sources

The Antwerp-London Glossaries: The Latin and Latin-Old English Vocabularies from Antwerp, Museum Plantin-Moretus 16.2 — London, British Library Add. 32246, vol. I, ed. by David W. Porter (Toronto: Pontifical Institute of Mediaeval Studies, 2011)

Bede, *De natura rerum*, ed. by Charles W. Jones, CCSL, 123A (Turnhout: Brepols, 1975)

——, *Libri quatuor in principium Genesis usque ad nativitatem Isaac et eiectionem Ismahelis adnotationum*, vol. I of *Bedae Venerabilis Opera: Opera exegetica*, ed. by Charles W. Jones, CCSL, 118A (Turnhout: Brepols, 1967)

---

46 See Anlezark, 'The *tres linguae sacrae* in Bodleian Library MS Hatton 20'.
47 The version cited in the colophon is that used as an antiphon during the Mass for the third Sunday of Lent; see Anlezark, 'The *tres linguae sacrae* in Bodleian Library MS Hatton 20', pp. 64–66.
48 See Anlezark, 'The *tres linguae sacrae* in Bodleian Library MS Hatton 20'.
49 See Anlezark, 'The *tres linguae sacrae* in Bodleian Library MS Hatton 20', pp. 76–78.

———, *Bede: On Genesis*, trans. by Calvin B. Kendall, Translated Texts for Historians, 48 (Liverpool: Liverpool University Press, 2008)

———, *Bede: On the Nature of Things and On Times*, trans. by Calvin B. Kendall and Faith Wallis, Translated Texts for Historians, 56 (Liverpool: Liverpool University Press, 2010)

Biblia sacra iuxta vulgatam versionem, ed. by Robert Weber, 4th edn (Stuttgart: Deutsche Bibelgesellschaft, 1994)

Bischof Wærferths von Worcester Übersetzung der Dialoge Gregors des Grossen, ed. by Hans Hecht, Bibliothek der angelsächsischen Prosa, 5 (Leipzig: Georg H. Wigland, 1900–07; repr. 1965)

Grégoire le Grand: Règle pastorale, ed. by Floribert Rommel, trans. by Charles Morel, Sources chrétiennes, 381–82, 2 vols (Paris: Cerf, 1992)

King Alfred's West-Saxon Version of Gregory's Pastoral Care, ed. by Henry Sweet, EETS 45, 50, 2 vols (London, 1871; repr. 1958)

Liber de ordine Creaturarum: un anónimo irlandés del siglo VII, ed. by Manuel C. Díaz y Díaz (Santiago de Compostella: Universidad, 1972)

The Old English Boethius: An Edition of the Old English Versions of Boethius's De Consolatione Philosophiae, ed. by Malcolm Godden and Susan Irvine, with Mark Griffith and Rohini Jayatilaka, 2 vols (Oxford: Oxford University Press, 2009)

The Old English Boethius, with Verse Prologues and Epilogues Associated with King Alfred, ed. and trans. by Susan Irvine and Malcolm R. Godden, Dumbarton Oaks Medieval Library (Cambridge, MA: Harvard University Press, 2012)

The Old English Dialogues of Solomon and Saturn, ed. and trans. by Daniel Anlezark, Anglo-Saxon Texts, 7 (Cambridge: D. S. Brewer, 2009)

Secondary Works

Anlezark, Daniel, 'All at Sea: Beowulf's Marvellous Swimming', in *Myths, Legends and Heroes: Essays on Old Norse and Old English Literature in Honour of John McKinnell*, ed. by Daniel Anlezark (Toronto: University of Toronto Press, 2011), pp. 225–41

———, 'Old English *Exodus* 487 werbeamas', *Notes and Queries*, 62 (2015), 497–508

———, 'The *tres linguae sacrae* in Bodleian Library MS Hatton 20', in *The Embroidered Bible: Studies in Biblical Apocrypha and Pseudepigrapha in Honour of Michael E. Stone*, ed. by Matthias Henze and William Adler (Leiden: Brill, 2017), pp. 64–78

Blackman, Deane R., 'The Volume of Water Delivered by the Four Great Aqueducts of Rome', *Papers of the British School at Rome*, 46 (1978), 52–72

Clark Hall, John R., ed., *A Concise Anglo-Saxon Dictionary*, 4th edn with Herbert D. Meritt (Cambridge: Cambridge University Press, 1960)

Cross, J. E., 'The *Metrical Epilogue* to the Old English Version of Gregory's *Cura Pastoralis*', *Neuphilologische Mitteilungen*, 70 (1969), 381–86

Dumville, D. N. 'English Square Minuscule Script: The Mid-Century Phases', *Anglo-Saxon England*, 23 (1994), 133–64

Godden, Malcolm, 'Prologues and Epilogues in the Old English *Pastoral Care*, and their Carolingian Models', *Journal of English and Germanic Philology*, 110 (2011), 441–73

Huppé, Bernard F., *The Web of Words: Structural Analyses of the Old English Poems Vainglory, the Wonder of Creation, the Dream of the Rood, and Judith* (Albany: State University of New York Press, 1970)

Llewellyn, Peter, *Rome in the Dark Ages* (London: Constable, 1993)

O'Brien O'Keeffe, Katherine, 'Source, Method, Theory, Practice: On Reading Two Old English Verse Texts', *Bulletin of the John Rylands University Library of Manchester*, 76 (1994), 51–73

Smalley, Beryl, *The Study of the Bible in the Middle Ages* (Oxford: Basil Blackwell, 1952)

Whobrey, William T., 'King Alfred's *Metrical Epilogue* to the *Pastoral Care*', *Journal of English and Germanic Philology*, 90 (1991), 175–86

JILL FREDERICK

Modor is monigra mærra wihta: Watering the World in Exeter Book Riddle 84

Many scholars have noted the ways in which the collection of Old English riddles in the tenth-century Exeter Book provides information about the more ordinary aspects of Anglo-Saxon life, quotidian areas seemingly irrelevant to the more learned and aristocratic concerns of the other poetic texts remaining to us.[1] The riddles present such homely items as bread and onions, locks and keys, pens and ink; agricultural implements like sickles and ploughs; familiar creatures both domestic and wild, like oxen, chickens, badgers, and swans; natural forces like storms and water. And yet, these enigmatic entities are never what they seem, of course, by virtue of the very genre which creates them, its need for a disguise that challenges the audience to look beyond the artful play of language to determine their true physical natures. More profound than a simple guessing game, however, the riddles have transfigurative power: as John Niles has observed, 'The riddles remind us that no object is merely an object. When singled out by a poet's or artist's attention, an ordinary thing becomes luminous with spirit'.[2]

Even otherwise dangerous entities acquire a transcendence that elevates them beyond the physical realm. That the Anglo-Saxons did not perceive nature as welcoming has become a commonplace: people's survival depended on careful attention to the details of elemental forces coupled with a profound respect for the potential disaster that a lack of attention could bring.[3] Such attention and respect often ameliorates the riddles' descriptions of air and water in their various manifestations — oceans, rivers, tempestuous air — that otherwise might be presented as purely destructive. While this synthesis of

---

1 Exeter, Exeter Cath. Lib., MS 3501.
2 Niles, *Old English Enigmatic Poems*, p. 53.
3 See Neville, *Representations of the Natural World*, for a thoughtful study of how the earliest medieval English viewed the forces of nature and presented them in poetic texts.

> **Jill Frederick** • (frederck@mnstate.edu) is Professor of English Emerita at Minnesota State University Moorhead, continuing her research interests in Old English riddles and saints' lives.

Meanings of Water in Early Medieval England, ed. by Carolyn Twomey and Daniel Anlezark, Studies in the Early Middle Ages, 47 (Turnhout: Brepols, 2021), pp. 267–281

dread and reverence infuses many of the riddles, it manifests itself exceptionally strongly in Exeter Book Riddle 84, solved by consensus as 'water'.[4] Despite this consensus, however, Riddle 84 has received little in the way of scholarly attention, perhaps because its folios are so damaged. Two significant studies, one by Dieter Bitterli and another by Patrick J. Murphy,[5] have omitted any reference to the riddle despite their comprehensive attention to the rest of the Exeter Book collection. Only Corinne Dale's recent volume analyses it in any detail, placing Riddle 84 within a larger argument that uses ecocriticism and ecotheology to reject conventional anthropomorphic readings of the Exeter Book riddles and 'to focus on what the riddles have to say about the created world as an entity in itself'.[6] In service of her argument, Dale deemphasizes the nuances of the riddles' diction and structure, and consequently, in the case of Riddle 84, underplays its controlling metaphor, water as a maternal force. It is the language and design of Riddle 84, despite the gaps in its text, that provide strong evidence for a perspective that in fact acknowledges the creature's own agency. The riddle, with its solution of 'water', seems — paradoxically — to set the destructive potential of an elemental force within the creative genesis of a feminine power, leading ultimately to a joyful and eternal home with God, a process and an end well outside human control.

In a previous study, I have discussed water in Old English poetry as a dangerous power that can simultaneously function as a physical conduit of emotional and spiritual states of being, and as a channel to eternal life.[7] It begins with the axiomatic premise that nature was no friend to the Anglo-Saxons: exile into the wilderness often equated with both physical and spiritual death, and the power of water was equally frightening. We have only to look to the quintessential locus for the image of the Anglo-Saxons' distrust and fear of the sea, lines 850–66 of *Christ II*, which presents life as a journey over difficult seas. In *Beowulf*, the mere of Grendel's mother gapes as a kind of hell-mouth, while even the fragmentary description of the mysterious baths in *The Ruin* uses language reminiscent of the destructive flame of Beowulf's dragon. On its face, Old English poetry provides no evidence of practical uses for water except as a transportation method, but it speaks deeply 'on a metaphorical and symbolic level, representing waters both as figurative boundaries between the natural world and the otherworldly, and channels to the other world'.[8]

For instance, consider the lines of *Christ II*, alluded to above: they begin by presenting the 'cald wæter' (cold water) over which ships must travel,

---

4 *The Exeter Book*, ed. by Krapp and Dobbie, pp. 236–38. For references to riddle solutions through 1980, see Fry, 'Exeter Book Riddle Solutions'; this list has been extended in *The Exeter Anthology of Old English Poetry*, ed. by Muir. In the article, 'A Modest Proposal', pp. 122–23, n. 21, Neville promises an updated list in her forthcoming book, *Truth Is Trickiest*.
5 Bitterli, *Say What I Am Called*; Murphy, *Unriddling the Exeter Riddles*.
6 Dale, *The Natural World in the Exeter Book Riddles*, p. 22.
7 Frederick, 'From Whale's Road to Water under the Earth'.
8 Frederick, 'From Whale's Road to Water under the Earth', p. 17.

the 'frecne stream' (dangerous current), and the 'yða ofermæta' (unending waves).⁹ They do not turn away from 'se drohtað strong' (the difficult journey) (ll. 850–56). In fact, the sea's bleak conditions are the very ones that allow for a safe homecoming into the heavenly harbour; they seem to be a kind of grace in that

> godes gæstsunu,   […] us giefe sealde
> þæt we oncnawan magun   ofer ceoles bord
> hwær we sælan sceolon   sundhengestas,
> ealde yðmæras,   ancrum fæste.
> Utan us to þære hyðe   hyht staþelian. (ll. 860–64)
>
> [God's ghostly son […] gave us the gift so that we might understand over the ship's beams where we should secure our sea-stallions, our old foam-mares, fast with anchors. Let us to that harbour moor our hope.]¹⁰

The intermingling of fear and joy found in these lines from *Christ II* also permeates Riddle 84 and, like this passage in *Christ II*, demonstrates that water offers an important passageway to God and an eternal home, both literally — by drowning — and symbolically through baptism. Once again, we see the paradox of destruction overcome by redemption.¹¹ While Riddle 84 certainly presents water as a hazardous element, within its lines water again acquires an otherworldly life of its own and on its own terms. The riddle's remarkably bright — even lapidary — language suggests a beauty and potential for nurture at odds with the dangers commonly associated with this primal element, and certainly in contrast with the language attached to water found in other Old English poetry. Even in its fragmentary state, the riddle situates water within an image of feminine power, commingling maternal nurturance with creative force that — at the same time — minimizes the destructive quality of the element.

The riddle's lines begin conventionally enough: 'An wiht is on eorþan wundrum acenned' (A certain creature on earth is born from wonders) (l. 1), since a number of the riddles begin with some variation of this statement, 'I am a wondrous creature' or 'I saw a wondrous creature', using variations of *wundor* (wonder) or *wrætlic* (ornamented) as the descriptor.¹² These terms,

---

9 *The Exeter Book*, ed. by Krapp and Dobbie, p. 27.
10 All translations from the Old English are my own unless otherwise noted.
11 Frederick, 'From Whale's Road to Water under the Earth', p. 32.
12 For instance, 'Ic eom wunderlicu wiht' (Riddles 18, 20, 24, 25,), 'Ic wiht geseah wundorlice | hornum bitweonum huþe lædan' (Riddle 29), 'Wiht cwom æfter wege wrætlicu liþan' (Riddle 33), 'Ic seah wyhte wrætlice twa | undearnunga' (Riddle 42), 'Ic seah wrætlice wuhte feower' (Riddle 51), 'Ic seah in healle […] wrætlic wudutreow' (Riddle 55), 'Ic on þinge grefrægn þeodcyninges | wrætlice wiht' (Riddle 67), 'Ic þa wiht geseah on weg feran; | heo wæs wrætlice wundrum gegierwed' (Riddle 68), 'Wiht is wrætlic' (Riddle 70), 'Ic seah wundorlice wiht' (Riddle 87).

unsurprisingly, suggest that its audience might be unfamiliar with the riddle creature, perhaps because it is highly unusual. In this instance the creature's singularity, however, stems from its physical force:

> hreoh ond reþe,   hafað ryne strongne,
> grimme grymetað   ond be grunde fareð. (ll. 2–3)
>
> [fierce and violent, it has a strong current, roars grimly, and travels from [or along] the ground.]

These lines establish the background against which the rest of the creature's description is set: it is not just wondrous to look at but to hear, as well. It rages and roars; it moves quickly and with purpose as it travels alongside or from the ground. The creature's sense of power is immediate, even without specific words for that power (which occur soon enough). In addition, the word *ryne* in line 2 adds another level of meaning in that it seems to be a play on the word 'rune', suggesting the idea of both 'mystery' and the written letter or word.[13] The *DOE* corpus provides 113 instances of the word *geryne* (mystery),[14] almost always in the context of the divine, that is, the mysteries of the Trinity, the mysteries of Christ.[15] This word, in fact, situates the core significance of the whole riddle as it will unfold to create a kind of transcendent family.

The composition of the family begins in the next line, 'Modor is monigra mærra wihta' (It is the mother of many greater creatures) (l. 4), which firmly establishes the riddle creature as female and calibrates her force by denoting her as a mother. In his commentary on this riddle, Craig Williamson notes that a variant of the same phrase, 'moddor monigra cynna' (mother of many kinds) occurs in line 2 of Riddle 41, also solved as 'water', and he reminds us that this maternal and generative motif finds its source in the Anglo-Latin riddles of Aldhelm.[16] In Aldhelm's Riddle 29, *Aqua* (Water), the creature states, 'Nam volucres caeli nantesque per aequora pisces | Olim sumpserunt ex me primordia vitae' (The birds of the sky and the fish swimming in the sea once drew from me the beginnings of their life).[17] In the same way, Aldhelm's Riddle 73, *Fons* (Fountain), expresses the same birth imagery as Riddle 29, asking 'Quis numerus capiat vel quis laterculus aequet, | Vita viventum generem quot milia partu' (What number could embrace or what calculation encompass the many thousands of living creatures which I engender through

---

13 See 'rūn', in Bosworth and Toller, *An Anglo-Saxon Dictionary*, II. a mystery, cf. gerýne; IV. of that which is written, with the idea of mystery or magic; V. a rune, a letter.
14 Cameron, Amos, and Healey, *Dictionary of Old English*.
15 In the notes to his translation of the riddle, Williamson also identifies this pattern: 'water is the great mother of all creation, allied (however primitive her nature) with a brooding Trinity' (*A Feast of Creatures*, p. 212).
16 *The Old English Riddles*, ed. by Williamson, pp. 369, 276.
17 Aldhelm, *Riddles*, ed. by Pitman, p. 16, ll. 4–5; Aldhelm, *The Poetic Works*, trans. by Lapidge and Rosier, pp. 75–76.

birth?).[18] Nevertheless, the direction in which the Old English riddle takes this fundamental image demonstrates a profoundly different perspective on the nature of this nurture. It moves far away from the connotations of maternal affection and protection to the idea of ferocity, even violence, in service of her children.

In line 5 ('Fæger ferende fundað æfre', the beautiful one setting out hastens always), the verbs attached to the maternal image, *ferende* (setting out) and *fundað* (hastens), reiterate the energetic movement of the first three lines, and the rest of the riddle text reinforces that sense of energy. Comparing the kinetic quality of Old English Riddle 84, however, with its Latin analogues (if not sources), the distinction between them is apparent, even allowing for the greater development of the subject in the Old English riddle. Aldhelm begins his description of *Aqua* by asking 'Quis non obstupeat nostri spectacula fati' (Who would not be astonished at the spectacle of my nature?), mirroring the sense of *wundor* of the Old English, but the second line, 'Dum virtute fero silvarum robora mille' (With my strength I support [the weight of] a thousand trees),[19] provides merely a sense of inert reinforcement. In the Latin, just the water-mother's offspring move: 'Nam volucres caeli nantesque per aequora pisces' (Only the birds of the sky and the fish swimming in the sea). The creature of Aldhelm's riddle, *Fons*, initially claims, 'Per cava telluris clam serpo celerrimus antra | Flexos venarum girans anfractibus orbes' (I creep stealthily and speedily through empty hollows of the earth, winding my twisted route along the curves of its arteries),[20] a bit contradictorily, since the verb *serpere* means to move slowly and yet the fountain apparently also moves *celerrimus* (very swiftly). However, the following line describes a static creature: 'caream vita sensu quoque funditus expers' (I am devoid of life and utterly lacking in sensation).[21] The waters of Aldhelm's *enigmata* are more tranquil than not, while the Old English riddle presents a protean force, filled with constant motion and passion.

In fact, it would seem that the very protean quality of the Old English creature makes her difficult even to describe, never mind understand:

> Nænig oþrum mæg
> wlite ond wisan   wordum gecyþan,
> hu mislic biþ   mægen þara cynna,
> fyrn forðgesceaft. (ll. 6–9a)

---

18 Aldhelm, *Riddles*, ed. by Pitman, p. 42, ll. 4–5; Aldhelm, *The Poetic Works*, trans. by Lapidge and Rosier, pp. 85–86.
19 Aldhelm, *Riddles*, ed. by Pitman, p. 16, ll. 1–2; Aldhelm, *The Poetic Works*, trans. by Lapidge and Rosier, p. 75.
20 Aldhelm, *Riddles*, ed. by Pitman, p. 42, ll. 1–2; Aldhelm, *The Poetic Works*, trans. by Lapidge and Rosier, p. 85.
21 Aldhelm, *Riddles*, ed. by Pitman, p. 42, l. 3; Aldhelm, *The Poetic Works*, trans. by Lapidge and Rosier, p. 86.

[No one can explain to another with words her appearance and nature, how various is the power of that kind, that creation of old.]

Despite their claim that no one can explain the feminine creature's fluidity, these lines nevertheless call attention to the characteristics emphasized by the remainder of the riddle: luminosity, wisdom, and a venerable lineage.[22] In particular, she draws power from her ancient ancestry which she in turn passes on to her progeny, as line 4 has already established. While the manuscript is damaged at this point, the lines still have an underlying coherence and sense:

> fæder ealle bewat
> or ond ende,   swylce an sunu,
> mære meotudes bearn,   þurh [… … … … .]ed,
> ond þæt hyhste mæge[… ..]es gæ [… .
> … … … … … … … … … … ..]   dyre cræft. (ll. 9b–13)

> [the father watches over all, beginning and end, likewise a son, great offspring of the creator, through … … … … …., and the highest [power] … … … … … … … … … … … … … … a precious skill.]

Evocatively, these lines link the creature's motherhood with both a father and a son, who is characterized as the great issue of a creator, 'mære meotudes bearn'.[23] Even more suggestively, they begin to underscore the sense in which the riddle creature is part of a kind of cosmic family in which the father watches and guides all the offspring produced by the mother.[24]

While the word *fæder* obviously has several usages, figurative and literal, the term *meotud* is semantically uniform. In the over two hundred instances of the word *meotud* (taking into consideration its variant spellings) collected in the *DOE* corpus, all seem to function, as the Bosworth and Toller *Dictionary* defines it, as 'an epithet of the Deity'.[25] The phrase, *meotudes bearn*, occurs in just one other instance, line 126a of *Christ I*.[26] However, a number of instances of

---

22 The adjective *neol* [*neowol*] is difficult to pin down; it suggests depth and profundity, and is sometimes attached to the idea of hell's abyss, but doesn't seem always to carry a sense of evil. Given that the phrase *neol is nearograp* follows on the idea of water's rushing current, it might suggest the variability of the water's strength (the word *mislic* in line 8 and line 56, front and back of the riddle).

23 It is worth noting that in line 52b, amidst the destruction of the text, the phrase *hrif wundigen* appears. The word *hrif*, while not exclusively linked to the womb of Christ's mother, often appears in that context, according to the *DOE*. The presence of this word does seem to reinforce the implications of divinity that infuse the riddle.

24 In their note concerning line 12 of this riddle, Krapp and Dobbie observe, 'The evidence of the MS. favors the restoration *mæge[n halg]es gæ[stes]*' (*The Exeter Book*, ed. by Krapp and Dobbie, p. 374). I have tried to not depend on emendations in my argument, but this phrase — 'the power of the holy spirit' — is congruent with the implication of the Trinity in the phrase 'mære meotudes bearn'.

25 See 'metod', in Bosworth and Toller, *An Anglo-Saxon Dictionary*.

26 *The Exeter Book*, ed. by Krapp and Dobbie, p. 6.

meotudes sunu also occur, always in connection with Christ,[27] either as a phrase or the words in close proximity to one another. The allusion in these lines is unmistakable, especially in conjunction with line 10a, 'or ond ende' (beginning and end), suggesting the sweep of biblical and human history.[28] The lines in which the phrases occur are so obliterated that the actions involved with the father and son cannot be determined. Nevertheless, the closing descriptors, 'þæt hyhste mæge' (highest power) (l. 12a) and 'dyre cræft' (precious skill) (l. 13b) reinforce the sense that they represent the idea of God, the highest power, with the measured craft to create and monitor elemental forces (for example, to separate water from land, as in Genesis 1. 9–10).

The damaged nature of the folios makes it impossible to find direct links among the focal points of the riddle — the mother, the father, and the son — set up in the first complete section of the riddle. However, they are clearly interconnected by virtue of the diction (*moddor, fæder, bearn*) as well as the generative power emphasized in the text. Yet the riddle focuses on the creature's feminine power, not the role of the paternal: the bulk of the riddle describes her beauty, her utility, the affection of those who look on her. Following line 19b the riddle again gathers together a critical mass of information in which here the word *wlitig* appears: 'þon ær wæs | wlitig ond wynsum' (that was earlier beautiful and joyful) (ll. 19b–20a). In fact, the riddle uses some variant of the word *wlitig* five times in the extant fifty-six lines, its first use asserting the inadequacy of words for the appearance and nature ('wlite ond wisan', l. 4) of the creature. In this phrase, *wlite* does denote simply appearance or countenance,[29] but its usages clearly indicate that the word emphasizes the idea of beauty and form, aligning it with the adjectival *wlitig*, meaning beautiful, glorious.[30] The word's collocation with *wynsum* seems to be formulaic,[31] but that designation does not minimize the exactness of the image. In their juxtaposition the two words suggest the same quality of swift intensity implied in the riddle's opening lines, especially as the pairing leads into the reference to the *moddor* as 'mægene eacen' (l. 21). This time she is 'endowed with strength' of her own; she is not simply the begetter of powerful offspring.

The word *mægen* (power) occurs five times throughout the riddle,[32] the same number of times as the word *wlitig*. This equivalence helps to embody

---

27 See 'meotud + sunu', in Cameron, Amos, and Healey, *Dictionary of Old English*.
28 Revelation 1. 8, 'I am Alpha and Omega, the beginning and the end, saith the Lord God, who is, and who was, and who is to come, the Almighty'. All references to Scripture in the body of this chapter are taken from the Douay-Rheims translation of the Latin Vulgate (1914).
29 See 'wlite', in Bosworth and Toller, *An Anglo-Saxon Dictionary*.
30 See 'wlitig', in Bosworth and Toller, *An Anglo-Saxon Dictionary*.
31 The phrase occurs in the *DOE* corpus twenty-two times; see Cameron, Amos, and Healey, *Dictionary of Old English*.
32 Other variants of the idea of power — the words *mære* (in ll. 4 and 11) and *meaht* (in l. 24) — also appear.

a creature infused with the beauties of elemental force, reinforced by the multiplicity of adjectives and other words that build the poem's central image. In the next lines alone, she is

> wundrum bewreþed,   wistum gehladen,
> hordum gehroden,   hæleþum dyre. (ll. 22–23)
>
>> [wrapped with wonders, laden with sustenance, adorned with treasure, dear to men.]

This description certainly contradicts the opening lines' claim that no one can explain the creature's appearance or nature: she seems to glow with light, as the words bestow a sense of riches and abundance. The next lines again reiterate her splendour and force, noting that her power increases when it is put into service:

> Mægen bið gemiclad,   meaht gesweotlad,
> wlite biþ geweorþad   wuldornyttingum,
> wynsum wuldorgimm   wloncum getenge,
> clængeorn bið ond cystig,   cræfte eacen;
> hio biþ eadgum leof,   earmum getæse,
> freolic, sellic. (ll. 24–29)
>
>> [Her strength will increase, her power revealed, her countenance will be honoured by its glorious use, a joyful glory-jewel close to splendid things, she is bountiful and ardent for purity, endowed with skill; she is beloved by the fortunate, useful to the wretched, wondrous, admirable.]

Mægen in this context may suggest 'virtue' rather than physical force, as the word has been glossed as such.[33] This link between the quality of righteousness and revealed strength continues to impart a deific quality to the creature, especially in conjunction with the language that follows this statement. It acknowledges again the water-mother's beautiful visage, this time honoured by (or with?) glorious service. This passage further describes her as a 'joyful glory-jewel' situated close to splendid things, bountiful and yearning for purity. Line 27b of the passage, 'cræfte eacen', contains a parallel with line 21b, 'mægene eacen', that emphasizes the water-mother's volition and agency in that *cræft*, while also denoting strength or virtue, has secondary meanings of knowledge and skill.

The ideas that shape the water-mother of Riddle 84 are not unique to it. Williamson has observed that the water is a 'shape-shifter', appearing as 'ice, snow, rain, hail, stream, lake, sea';[34] thus in Riddles 33 and 68/69 (each solved

---

33   See 'mægen', in Bosworth and Toller, *An Anglo-Saxon Dictionary*.
34   Williamson, *A Feast of Creatures*, p. 212.

by appreciable consensus as 'ice' or 'iceberg'[35]), the image of the water as both mother and offspring also appears. Both creatures are also *wrætlic* (wondrous, 33:1; 68:2a), and in the case of Riddle 33, with a destructive bite:

> Wæs hio hetegrim,   hilde to sæne,
> biter beadoweorca;   bordweallas grof,
> heardhiþende. (ll. 5–7b)
>
>> [It was fierce, slow in sea combat, bitter in battle work; it carved the shield-walls, hard-ravaging.]

And as Riddle 84's creature is both mother and child, so too is the creature of Riddle 33:

> Is min modor   mægða cynnes
> þæs deorestan,   þæt is dohtor min
> eacen up liden,   swa þæt is ældum cuþ,
> firum on folce,   þæt seo on foldan sceal
> on ealra londa gehwam   lissum stondan. (ll. 9–13)
>
>> [My mother is of woman-kind the dearest, she is my daughter grown up vast, as it is known to people, men among their folk, that she must in each of all the lands on earth stand in grace.]

While Riddles 68 and 69 contain the same quality of elegance and beauty as occurs in Riddle 84, Riddle 84 moves beyond the creature's appearance. It emphasizes and pairs the creature's beauty with her utility, marrying her bountiful nature with her strength and potential for destruction; the base form of *gifre* can mean both 'useful' and 'greedy'. *From* can mean both 'bold' and 'abundant'; *graedgost* is 'greediest'; *swithost* is 'most powerful'. This language clearly demonstrates that she can be destructive as well as giving, as she treads upon or tramples the ground:

> fromast ond swiþost,
> gifrost ond grædgost   grundbedd trideþ,
> þæs þe under lyfte   aloden wurde
> ond ælda bearn   eagum sawe. (ll. 29b–32)
>
>> [Most abundant and most powerful, most useful and most greedy, she treads the ground which under the heaven has grown up and that the children of men saw with their eyes.]

The association between water and maternal violence found here is not unique in the canon of Old English poetry. It most obviously occurs in *Beowulf*, the

---

35 A difference of opinion exists as to whether Riddles 68 and 69 are one or two texts; the consensus solution for both options, however, is 'ice'.

description of Grendel's mother clearly presenting her as a water-woman, who 'wæteregesan wunion scolde | cealde streamas' (must live in terrible water, cold currents) (*Bwf* ll. 1260–61a).[36] Her motivation, revenge, manifests itself in the diction associated with her physical actions: Grendel's mother is 'gifre and galgmod' (greedy and gloomy) (*Bwf* l. 1277a), and 'floda begong | heorogifre beheold' (she beheld the waters' expanse, eager for destruction) (*Bwf* 1497b–1498), images that echo line 30 of the riddle which describes the water-mother as 'gifrost ond grædgost' (greediest and most ravenous). In addition, Grendel's mother has the same intensity of movement as the water-mother has demonstrated, moving *on ofste* (in haste) (*Bwf* l. 1292a) and *hraðe* (quickly) (*Bwf* l. 1294a).

The difference between these two female entities, of course, is that while Grendel's mother begets a son, her offspring is purely destructive; she is herself destructive and she begets destruction. In Riddle 84, however, the destructive feminine force of the water is balanced, indeed outweighed, by her creative power, creation flowing from seeming destruction, with the sense of creation acquiring a particularly feminine quality in the next lines:

swa þæt wuldor wifeð,   worldbearna mægen,
þeah þe ferþum gleaw   *   *   *
mon mode snottor   mengo wundra.[37] (ll. 33–35)

> [In this way she weaves that glory, the strength of the world's sons, although the prudent in spirit [lacuna?] a man wise in mind with a multitude of wonders.]

In her recent volume, *Weaving Words and Binding Bodies*, Megan Cavell devotes a long footnote to these verses of the riddle, asserting that they present an 'arguably bleached example of weaving' attached to the picture of water 'as a powerful woman who *þæt wuldor wifeð, worldbearna mægen* (33) (weaves glory, the might of world-children)'.[38] Although the cases of *þæt wuldor* and *mægen* are arguable — as neuters, these words could be either nominative or accusative — Cavell takes them as accusative, as does Williamson's translation of the riddle (in which he transforms *wifeð* into a noun: She is the weaver | Of world-children's might). Nevertheless, she concludes that water 'is depicted as a creator who is responsible for creating the glory and might of men'.[39]

Cavell's reading agrees with my own understanding of the riddle as depicting not just a creative force but a distinctly feminine creative force. I have noted elsewhere that the creature of Riddle 84 is miraculous because she exists as both mother and child, born from wonders (l. 1) and mother of

---

36 *Klaeber's Beowulf*, ed. by Fulk, Bjork, and Niles, p. 44.
37 Krapp and Dobbie note that the manuscript is not damaged at line 34b, so it is likely a half-line was lost from the text (*The Exeter Book*, ed. by Krapp and Dobbie, p. 375).
38 Cavell, *Weaving Words and Binding Bodies*, p. 275 n. 102.
39 Cavell, *Weaving Words and Binding Bodies*, p. 275 n. 102.

greater wonders (l. 4).[40] Such a paradox is also reflected in the way that almost everything in the riddle connected with destruction is counterbalanced with, indeed outweighed by, creation and beauty. The beauty and the destructive power of the water-mother merge to create her worldly and other-worldly functions. In particular, lines 33–35 contain overtones of baptismal imagery: the glory she weaves — the strength of the world's children — is the salvation the water offers by means of the 'mære meotudes bearn' to whom the riddle's initial verses refer. This image of baptism and salvation exists as well, and developed further, in lines 36–43:

> Hrusan bið heardra,   hæleþum frodra,
> geofum bið gearora,   gimmum deorra;
> worulde wlitigað,   wæstmum tydreð,
> firene dwæsceð,
> oft utan beweorpeð   anre þecene,
> wundrum gewlitegad,   geond werþeode,
> þæt wafiað   weras ofer eorþan,
> þæt magon micle   [… … … … … … … … … … …]sceafte.[41]

> > [She is harder than earth, wiser than men, readier with gifts, more precious than gems; she beautifies, nourishes the world with her fruits, destroys sin, [lacuna?] often casts over the surface a certain covering, decorated with wonders, throughout the nation, that are amazed, men over the earth, that they are greatly able… … … … … … … … … … creation.][42]

Again we recognize that the water gives by taking away, offering the world the benefits of salvation by expunging sin. In this context the verb *dwæscan* is especially appropriate to the water's action, as Bosworth and Toller define it as 'to extinguish, to put out'.[43] The *DOE* shows that the verb is more often than not used in two ways, to express the destruction of sin as well as the idea of putting out fires. The image created, in spite of the lacuna in line 39, almost suggests smothering, the idea that water covers the earth in order to effect the redemptive destruction.

While nowhere does the riddle contain language suggesting that the water described in its text equates with the *event* of a flood, the central image of water over the earth easily evinces Noah's Flood, about which Bede writes at length in Book II of his commentary *On Genesis* in which he unfolds the

---

40 Frederick, 'From Whale's Road to Water under the Earth', p. 28.
41 As with line 34b, Krapp and Dobbie suggest that a half-line was also dropped from line 39b (*The Exeter Book*, ed. by Krapp and Dobbie, p. 375).
42 Thanks to Gale Owen-Crocker for her help in translating this passage and others that eluded me.
43 See 'dwæscan', in Bosworth and Toller, *An Anglo-Saxon Dictionary*; and 'dwæscan', in Cameron, Amos, and Healey, *Dictionary of Old English*.

link between baptism and the Flood.⁴⁴ This commonplace link between the Flood and baptism — obviously, Bede's exegesis, which draws on a multiplicity of biblical and patristic sources, is not the first — may provide a way to clarify and unify the various elements of the riddle's description.⁴⁵ The riddle's language is not anomalous but clearly derives from a common cultural and theological understanding of baptism, which by definition takes away, destroys, the old life and brings in the new one in Christ,⁴⁶ the very symbiosis of destruction and redemption found in the riddle. Bede's argument is dense and complex, but his analysis establishes some fundamental ideas pertinent to the imagery in Riddle 84. In particular (and ultimately at great length), he observes of Genesis 6. 13–14 that

> Designet ergo arca ecclesiam, designet diluuium fontem baptismi quo abluitur, designet fluctus mundi temptantis quibus probatur, designet finem in quo coronatur.
>
> [The ark can represent the Church, the deluge represents the font of baptism in which one is washed, the surge is the temptations of this world which prove, it represents the end in which one is crowned.]⁴⁷

Daniel Anlezark notes that Bede is especially concerned to incorporate into his explication that 'The ark has room for all the world's nations and all stations of life. Bede's theology of salvation throughout *On Genesis* could generally be categorized as inclusive'.⁴⁸ We can see this sense of inclusiveness in the first third of Riddle 84 as the text continually asserts the utility of water to many kinds of creatures, emphasizing baptism's enfolding maternal embrace, even as the literal Flood embodies violent destruction.

The same sort of maternal embrace appears as well in the *Metrical Epilogue* to the Old English *Pastoral Care*:

> Ðis is nu se wæterscipe   ðe us wereda god
> to frofre gehet   foldbuendum.
> He cwæð ðæt he wolde   ðæt on worulde forð
> of ðæm innoðum   a libbendu

---

44 Bede, *Libri quatuor in principium Genesis*, ed. by Jones; a meticulous analysis of this trope appears in Daniel Anlezark's comprehensive study, *Water and Fire*. Thank you to all on the ISSEME list serve who answered my query about flowing water, especially Jay Gates, who pointed me to Bede's commentary.
45 See John J. Gallagher's contribution to this volume, '"Streams of Wholesome Learning": The Waters of Genesis in Early Anglo-Saxon Exegesis'.
46 See, for instance, Romans 6. 4, 'For we are buried together with him by baptism into death; that as Christ is risen from the dead by the glory of the Father, so we also may walk in newness of life', and John 3. 5, 'Truly, truly, I say to you, unless one is born of water and the Spirit he cannot enter into the kingdom of God'.
47 Bede, *Libri quatuor in principium Genesis*, ed. by Jones, II. 1127–29, quoted in Anlezark, *Water and Fire*, p. 53.
48 Anlezark, *Water and Fire*, p. 56.

```
wætru fleowen,    ðe wel on hine
gelifden under lyfte.    Is hit lytel tweo
ðæt ðæs wæterscipes    welsprynge is
on hefonrice,    ðæt is halig gæst. (ll. 1–8)
```

> [Now this is the water-source that the God of hosts promised as a solace for us earth dwellers. He said that he wanted ever-living waters to flow forth in the world from the very hearts of those under the sky who fully trusted in him. There is little doubt that this water-source's well-spring is in the kingdom of heaven, which is the holy spirit.][49]

In lines 1 and 7 of this passage, the word *wæterscipe* can be translated as 'body of water, piece of water',[50] but broken into its lexical parts, the term *scip* also simply denotes a ship. Here the water and the vessel become one, the means of salvation and salvation itself.

After these lines suggestive of baptism and salvation, the text of Riddle 84 breaks down almost completely because of the damage to its folios. Following line 44, 'Biþ stanum bestreþed, stormum' (It is covered over with stones, storms), no complete statements can be determined, only fragments of disconnected images — 'timbred weall' (timbered wall) (l. 45b), 'hrif wundigen' (a wounded belly) (l. 52b) — and incomplete action: 'hrusan hrineð' (lays hold of the earth) (l. 47a), '....tenge' (presses toward) (l. 48b). Only the final lines imply a conventional ending for the riddle, asking to 'reveal the word-hoard for the warriors', something — the hoard itself? — 'protected, opened with words', and reiterating the idea of the riddle creature's varied powers:

```
Hordword onhlid,    hæleþum ge[ ...
... ... ... ... ... ... ]wreoh,    wordum geopena,
hu mislic sy    mægen þara cy[ ... ]. (ll. 54–56)
```

> [Reveal the word-hoard for the warriors ... protected, opened with words, how various be the powers of the ....]

In their notes to these final lines of the riddle, Krapp and Dobbie observe that in line 55a Holthausen 'restores [*wisdom on*]*wreoh*, which agrees well with the sense of this passage', while in line 56b he restores *mægen þara cy*[*nna*], which 'is required by the sense'.[51] However, even without these conjectures, rather than open-ended, the riddle has a definitive conclusion: it has brought together a set of allusions — Father, Son, and even Holy Spirit (ll. 10–12) — that creates for water the transcendent powers of the Mother

---

49 *The Anglo-Saxon Minor Poems*, ed. by Dobbie, p. 111. See also Daniel Anlezark in this volume, 'Drawing Alfredian Waters: The Old English *Metrical Epilogue* to the *Pastoral Care*, Boethian Metre 20, and *Solomon and Saturn II*'.
50 See 'wæterscipe', in Bosworth and Toller, *An Anglo-Saxon Dictionary*.
51 *The Exeter Book*, ed. by Krapp and Dobbie, pp. 375–76.

Church in concert with the imagery of maternal protection. Bede explicates the waters of the Flood as temptations and tribulations of the world, so the closing lines of the riddle ask its audience to give a name to the paradoxically destructive and redemptive qualities of the riddle object. In fact, one might well suggest 'baptism' as a new solution to this text.

In its larger cultural context, then, Riddle 84 in many respects continues to demonstrate the Anglo-Saxon ambivalence towards water, employing language that acknowledges its power and force while simultaneously elevating that force from the earthly to the heavenly. While alluding to and drawing on the Latin tradition of Aldhelm, the Old English riddle recognizes the divine in the quotidian. Although scholars have sometimes questioned the Exeter Book's heterogeneity, asking how it includes so many texts that on the face of it might seem inappropriate in a volume apparently designed for a religious house, Riddle 84 may well provide a kind of archetype. It presents its audience with the effortless merger of the secular and sacred that characterizes so many of the earliest English literary texts, here attending to the ways in which Anglo-Saxon culture of necessity attends to natural forces with care, even as it recognizes and appreciates their terrible beauty as a reflection of the divine.

Works Cited

Manuscript

Exeter, Exeter Cathedral Library, MS 3501 (Codex Exoniensis)

Primary Sources

Aldhelm, *Aldhelm: The Poetic Works*, trans. by Michael Lapidge and James Rosier (Woodbridge: D. S. Brewer, 2009)
———, *The Riddles of Aldhelm*, ed. by James Hall Pitman (New Haven: Yale University Press, 1925; repr. Hamden, CT: Archon Books, 1970)
The Anglo-Saxon Minor Poems, ed. by Elliot Van Kirk Dobbie, Anglo-Saxon Poetic Records, 6 (New York: Columbia University Press, 1942)
Bede, *Libri quatuor in principium Genesis usque ad nativitatem Isaac et eiectionem Ismahelis adnotationum*, ed. by Charles W. Jones, vol. 1 of *Bedae Venerabilis Opera: Opera exegetica*, CCSL, 118A (Turnhout: Brepols, 1967)
The Exeter Anthology of Old English Poetry: An Edition of Exeter Dean and Chapter MS 3501, ed. by Bernard J. Muir, 2 vols, 2nd edn (Exeter: University of Exeter Press, 2000)
The Exeter Book, ed. by George Philip Krapp and Elliott Van Kirk Dobbie, Anglo-Saxon Poetic Records, 3 (New York: Columbia University Press, 1936)
The Holy Bible: Translated from the Latin Vulgate (Baltimore: John Murphy, 1914)

Klaeber's Beowulf, ed. by R. D. Fulk, Robert E. Bjork, and John D. Niles, Toronto Old English Studies, 4th edn (Toronto: University of Toronto Press, 2008)

The Old English Riddles of 'The Exeter Book', ed. by Craig Williamson (Chapel Hill: University of North Carolina Press, 1977)

Secondary Works

Anlezark, Daniel, *Water and Fire: The Myth of the Flood in Anglo-Saxon England* (Manchester: Manchester University Press, 2006)

Bitterli, Dieter, *Say What I Am Called: The Old English Riddles of the Exeter Book and the Anglo-Saxon Riddle Tradition* (Toronto: University of Toronto Press, 2009)

Bosworth, Joseph, and T. Northcote Toller, *An Anglo-Saxon Dictionary*, 2 vols: vol. I (Oxford: Clarendon Press, 1898), and Supplement, vol. II, ed. by T. Northcote Toller (Oxford: Oxford University Press, 1921), with *Enlarged Addenda and Corrigenda*, ed. by Alistair Campbell (Oxford: Oxford University Press, 1972)

Cameron, Angus, Ashley Crandell Amos, and Antonnette diPaolo Healey, eds, *Dictionary of Old English: A to H Online* (Toronto: Dictionary of Old English Project, 2016), <https://www.doe.utoronto.ca/pages/index.html> [accessed 15 February 2020]

Cavell, Megan, *Weaving Words and Binding Bodies: The Poetics of Human Experience in Old English Literature* (Toronto: University of Toronto Press, 2016)

Dale, Corinne, *The Natural World in the Exeter Book Riddles* (Cambridge: D. S. Brewer, 2017)

Frederick, Jill, 'From Whale's Road to Water under the Earth', in *Water and the Environment in the Anglo-Saxon World*, ed. by Maren Clegg Hyer and Della Hooke, Exeter Studies in Medieval Europe (Liverpool: Liverpool University Press, 2017), pp. 15–32

Fry, Donald K., 'Exeter Book Riddle Solutions', *Old English Newsletter*, 15.1 (1981), 22–33

Murphy, Patrick J., *Unriddling the Exeter Riddles* (University Park: Pennsylvania State University Press, 2011)

Neville, Jennifer, 'A Modest Proposal: Titles for the *Exeter Book Riddles*', *Medium Ævum*, 88 (2019), 116–23

―――, *Representations of the Natural World in Old English Poetry* (Cambridge: Cambridge University Press, 1999)

―――, *Truth Is Trickiest: Enigmatic Discourses in the Exeter Book Riddles* (forthcoming)

Niles, John D., *Old English Enigmatic Poems and the Play of the Text* (Turnhout: Brepols, 2006)

Williamson, Craig, trans., *A Feast of Creatures: Anglo-Saxon Riddle-Songs* (Philadelphia: University of Pennsylvania Press, 2011)

Index

Abingdon Abbey: 46, 51–52
abyss: 119, 136 n. 32, 137–38, 142, 168–70, 176 n. 48, 217–18, 253, 272 n. 22
Adomnán, saint: 64
 De locis sanctis: 157
 Vita S. Columbae: 64, 101
Ælfric of Eynsham, *Preface to Genesis*: 171
Aelred of Rievaulx, saint: 87
Æthelbald, King of Mercia: 45
Æthelberht, King of Kent: 61
Æthelred, ealdorman of Mercia: 53
Æthelthryth, saint: 41
Aidan, saint: 19, 36, 37, 38, 41, 65
Alban, saint: 61
Alcuin, of York, saint: 45, 263
Aldhelm of Malmesbury, saint: 87, 120, 179, 212
 Riddles: 270–71
 De Virginitate: 122–23, 219, 231
Alfred the Great, King of Wessex: 45, 72, 241, 249
Ambrose, of Milan, saint: 110, 113, 170, 173, 175, 182, 245
 De mysteriis: 172
 De Noe et arca: 171
Andersey Island: 51–52
Andreas, Old English poem: 191–207
Andromeda: 129–30
Anglesey, Isle of (*Ynys Môn*): 40
Anglo-Saxon Chronicle: 38, 94–96
Annals of Fulda: 96
Apollonius of Tyre, Old English: 92
Apostolic Tradition: 61
aqueduct: 242–43, 247–50
 Aqua Claudia: 248–49

Ark of the Covenant: 158, 159 n. 28
Athelney, Somerset: 45, 196
Augustine, Archbishop of Canterbury, saint: 46, 65, 248–49
Augustine, Bishop of Hippo, saint: 110–11, 138–39, 173, 182–83, 245
 De Genesi ad litteram: 183
Avalon: 42–43
Avebury Henge: 34
Avon, River: 34, 47

Bamburgh, Northumberland: 37, 74
baptism: 20–21, 35, 49, 59–67, 69–70, 156, 191, 201, 203–04, 277–78
baptisteries: 49, 62–63, 76, 148, 204
Bardney, Lincolnshire: 48
Basil of Caesarea, saint: 181–82, 185, 245
Bath, Somerset: 47, 213 n. 9
Bathley, Nottinghamshire: 98
Battle of Maldon, Old English poem: 21
beaches: 112, 197 n. 22
Beane, River: 75
Bede, Venerable, saint: 20, 24, 36, 59, 64, 66, 77, 87, 123, 131, 157, 170, 179, 263
 Commentary on Revelation: 150 n. 11
 Historia Ecclesiastica: 35, 37–38, 41, 48, 61–62, 65–66, 68–72, 92–94, 167, 204
 Judgement Day poem: 15
 Life of St Cuthbert: 37–38, 63, 74, 221
 De natura rerum: 178, 255
 On Ezra and Nehemiah: 136, 140

On Genesis: 140, 169 n. 8, 171, 175–76, 184, 205, 245–46, 277–78
Benedict Biscop: 112
Beowulf, Old English poem: 73, 88–92, 99–100, 191, 193, 198, 200, 223, 268, 275–76
Bermondsey, London: 46
Beuno, saint: 39
Bible
  Genesis: 61, 168–88, 192, 200–01, 205, 273
  Exodus: 61, 96, 119–20, 176, 200, 202
  Numbers: 61, 200
  Joshua: 160
  Job: 181
  Psalm 7: 252
  Psalm 42: 61, 148 n. 6
  Psalm 43: 118
  Psalm 46: 122
  Psalm 58: 220
  Psalm 66: 97
  Psalm 103: 128, 139–40
  Proverbs: 245, 254
  Wisdom: 183 n. 79
  Daniel: 228, 247
  Jonah: 129–30, 192
  Gospel of Matthew: 61, 128, 130, 132, 137, 139, 194
  Gospel of Mark: 61, 194
  Gospel of Luke: 61, 194
  Gospel of John: 61, 206, 244, 264
  Acts of the Apostles: 61
  I Corinthians: 122
  Ephesians: 122
  I Peter: 130, 170
  II Peter: 170
  Revelation: 61, 136–37, 149–54, 201
birds: 38, 40, 67 n. 30, 91, 115–16, 217, 224–25, 251–53, 260, 270–71
birdsong: 110, 115–16
Blythe, River: 75
boats *see* vessels
Boethius, *De consolatione philosophiae*: 256–58
  see also Metres of Boethius, Old English
bogs: 34, 222
  see also fenlands
bridges: 70, 76, 86, 95, 121, 254
Bridget, saint: 43
Brue, River: 42–43
burial: 39, 65, 70–77, 113, 203
Byrhtferth, of Ramsey: 120, 174 n. 23
  Vita Oswaldi: 120–23

Cadfan, saint: 39
Cadoc, saint: 40
Cædmon's Hymn, Old English poem: 193
Caelian Hill, Rome: 248–50
Càin Íarraith: 101
Caldey Island (*Ynys Bŷr*), Pembrokeshire: 40
Canterbury School commentaries: 167–79, 184–88
Canterbury, Kent: 46, 63, 200, 204 n. 46
Catterick, North Yorkshire: 35, 62, 69
Cedd, saint: 65
celeuma: 117–20
Cenwald, Bishop of Worcester: 264
Ceolfrith, Abbot of Monkwearmouth-Jarrow, saint: 111–14
Charlemagne, King of the Franks: 102, 158
charters: 20, 49, 68
Chedworth, Gloucestershire: 49
Chertsey Abbey: 46
Christ I, Old English poem: 272
Christ II, Old English poem: 194, 268–69
Christ: 117, 122, 131–41, 148–49, 152, 192–96, 203–04, 206–07, 231, 244, 246, 264, 270, 273, 278
  Ascension: 156
  Baptism: 20, 59–61, 74, 77, 157, 148

Crucifixion: 132–38, 130, 156, 198, 199 n. 28, 201
Resurrection: 130, 133, 136–37, 139, 142, 157
Transfiguration: 152
Ciarán of Saighir, saint: 87
clouds: 152, 156, 174 n. 29, 180–83, 185, 205, 234, 260
Coelred, King of Mercia: 45
Coln, River: 49
Columba, saint: 36, 63–65, 101
Comgall of Bangor, saint: 87
conduit *see* aqueduct
Coquet Island, Northumberland: 38
Cosmas Indicopleustas, *Topographia Christiana*: 174
creation: 133, 168–72, 174–77, 181, 184–88, 193, 205–07, 225, 272, 276–77
Cricklade, Wiltshire: 98
Cronica sive Antiquitates Glastoniensis Ecclesie see John of Glastonbury
Crowland, Lincolnshire: 41, 42, 211, 221, 225, 230–31
Cuthbert, Bishop of Lindisfarne, saint: 19, 23, 37, 38, 63–65, 68, 71, 74, 87, 225
Cuthwine, Anglian Bishop: 131
Cybi, saint: 40
Cynegisl, King of Wessex: 69

dams: 67, 243, 246–48
Daniel, Old English poem: 247
De mirabilibus sacrae scripturae: 177–78
Deben, River: 73, 75
Dee, River: 71
Deiniol, saint: 40
Deluge, biblical: 140, 142, 148, 168–71, 174–77, 179–80, 184–87, 201, 277–78
Deneheah, Abbot of Reculver: 53
deposits, votive: 76

Devil: 44, 133, 136–37, 139–42, 191, 255 n. 32
dew: 149, 181, 207, 211, 228, 234
Didache: 61
Dijle, River: 96
Doomsday *see* Last Judgement
Dorney, Buckinghamshire: 70
Dream of the Rood, The, Old English poem: 198
drinking: 67, 115, 197, 201, 206, 226, 228–29, 244–45
drowning: 85–86, 92–97, 102, 269
Dryhthelm of Melrose, saint: 87
Dubricius (Dyfrig), saint: 39
Dwynwen, saint: 40

Easby Cross: 157–59, 161
Easter: 65
Edgar, King of England: 71
Edred (Eadred), King of England: 45
Edward, Abbot of Crowland: 45
Edwin, King of Northumbria: 35, 62, 69
Ely, Cambridgeshire: 41
English Channel: 72
Epiphany *see* Christ, baptism
Evesham Abbey: 47, 53
Exodus, Old English poem: 120 n. 51, 174
Eynsham Minster: 46

Farne Islands, Northumberland: 37, 38, 40
fenlands: 41–42, 45, 54, 67, 196, 222–27, 235
Fens, Cambridgeshire: 18, 34, 211, 221–25, 233–34
Fionán, saint: 36
fish: 90, 128, 130, 134, 192, 260, 270–71
Flood, biblical *see* Deluge, biblical
flooding: 43, 46, 67, 75

Galilee, Sea of: 192
Genesis A, Old English poem: 175, 187
Germanus, saint: 93
Germigny-des-Prés, oratory: 157–59
Gifts of Men, The, Old English poem: 90
Gildas, saint: 40, 118–19
Glastonbury, Somerset: 35, 42–45, 52
Glen, River: 35, 62, 69
Gregory I, Pope: 69, 77, 140, 212
    Dialogi: 222, 245
    Moralia in Iob: 219, 227–28, 232
    Regula Pastoralis: 227, 245
    see also *Pastoral Care*, Old English
Gregory of Nyssa, saint: 181
Gregory of Tours, saint: 94
Guthlac, saint: 18, 34, 45, 211–12
    Felix, *Vita sancti Guthlaci*: 211–25, 227–35
    Old English *Life of Guthlac*: 212–25, 228–35
    Vercelli Homily 23: 213–15, 218
Guthrum, King: 20–1, 72

Hadrian, Abbot of St Augustine's: 167
hail: 217, 274
Harald (III) Hardrada, King of Norway: 95
harbour: 48, 220, 231–35, 254
Harold II, King of England: 95
Hartlepool, County Durham: 48
heaven: 61, 77, 134, 148–56, 161–62, 168–69, 173–76, 178, 180, 183, 186, 192, 195, 201, 205, 222, 226, 228, 242–43, 248, 256, 261, 279
Heliand, Old Saxon poem: 74
Helier, saint: 40
hell: 74, 128, 130, 135–42, 203, 213, 217–18, 231–32, 255, 268, 272
Hexameron: 171 n. 15, 174–75, 181–82, 185, 187
Humber, River: 20

Huna, saint: 42
hundred courts: 70–71
Husband's Message, The, Old English poem: 114
hydrological cycle: 13, 168, 177, 183–85

ice: 115–17, 173, 198, 205, 255, 274–75
Iona (*Ì Chaluim Chille*), Inner Hebrides: 36, 43, 45, 64, 221
Irish Sea: 72
Isidore, of Seville, saint: 140, 182
    Etymologies: 146, 149, 152, 184
    De natura rerum: 172–74, 183–84
Isle, River: 45
Itchen, River: 35, 51

Jaffa: 96, 127, 129, 136
Jarrow: 73, 112–13, 161, 178
Jerome, saint: 121
    Commentary on Daniel: 219
    Commentary on Jonah: 130–35, 137–38
    Vita sancti Pauli: 212, 221
Jersey, Channel Islands: 40
Jerusalem: 96, 186 n. 96, 251
    New (Heavenly) Jerusalem: 134, 142, 149–52, 154–55, 192, 201, 204
Jesus see Christ
Joffrid, Abbot of Crowland: 45
John of Glastonbury: 44
John the Archcantor: 112
John the Baptist, saint: 20, 59
Jonah, prophet: 127–41, 192
Joppa see Jaffa
Jordan, River: 20, 59, 77, 148, 157, 160, 182
Juliana, Old English poem: 191

Kempsey Minster: 47
Kenelm, saint: 49
Kennet, River: 34
Kevin of Glendalough, saint: 87
Kirkstead, Lincolnshire: 48

Lammana Priory: 40
Last Judgement: 15, 140–42, 151–52, 170, 252
Lastingham, Yorkshire: 48
Lateran Basilica, Rome: 63, 148
Lea, River: 20, 72
Lechlade, Gloucestershire: 98
Letter of Alexander to Aristotle, Old English: 91
Leviathan: 136 n. 32, 139–41
Lindisfarne (Holy Island), Northumberland: 36–38, 41, 45, 221
living waters *see* waters, living
Lleyn (*Llŷn*) Peninsula, Anglesey: 39
Loch Ness Monster: 23, 101

manuscripts
  Antwerp, Museum Plantin-Moretus, MS M. 17. 4: 127–28, 131–38
  Bern, Burgerbibliothek, Codex Bongarsianus 318 (Bern *Physiologus*): 149
  Cambridge, Corpus Christi College, MS 422: 250
  Exeter, Exeter Cathedral Library, MS 3501 (Codex Exoniensis, The Exeter Book): 114, 268, 272, 279
  London, British Library, Add. MS 49598 (Benedictional of St Æthelwold): 59–60, 77
  London, British Library, MS Cotton Claudius B.iv (Old English Hexateuch): 139
  London, British Library, MS Cotton Tiberius A.iii: 90
  London, British Library, MS Cotton Vitellius A.xv: 91
  London, British Library, MS Harley 2506: 128
  London, British Library, MS Harley 603 (Harley Psalter): 128, 138–41
  London, British Library, MS Harley 647: 128
  Oxford, Bodleian Library, MS Hatton 20 (Old English *Pastoral Care*): 241, 249, 264
  Paris, Bibliothèque nationale de France, MS lat. 14429: 149–50
  Rome, Biblioteca Apostolica Vaticana, Reg. lat. 12 (Bury Psalter): 97–98
  Stuttgart, Württembergische Landesbibliothek, Bibl. fol. 23 (Stuttgart Psalter): 135
  Utrecht, Universiteitsbibliothek, MS Bibl. Rhenotaiectinae I Nr. 32 (Utrecht Psalter): 128, 138–41
Marden, Wiltshire: 34
markets: 51–53, 67–68, 70
marshlands *see* fenlands
Martyrology, Old English: 178 n. 56, 184
Maxims II, Old English poem: 223
Mediterranean Sea: 43, 63, 250–51
Mellitus, Bishop: 77
Melrose, Scottish Borders: 37, 41, 63
Metres of Boethius, Old English: 250, 256–58, 260–64
Milred, bishop of Worcester: 53
Miriam, prophet: 120
Moluag (Lugid), saint: 101
Moses, prophet: 119–20, 200, 202–04
music: 111–13, 115, 117–18, 121–22

nakedness: 93, 129, 137–38
Ness, River: 23, 101
Noah, prophet: 140–41, 175 n. 38, 185–86, 193 n. 8, 204, 207
North Sea: 68, 72–73, 112, 120, 223

Ock, River: 46
Orderic Vitalis: 215
Orosius, Paulus, *History*, Old English: 92

Oswald, Archbishop of York: 42, 120
Oswald, King of Northumbria, saint: 69

Paradise: 139, 152, 180, 183, 186
  waters of: 61, 146–49, 151–52, 154, 176, 180, 183, 245 n. 10
Pastoral Care, Old English: 226–27, 241–48, 278–79
Pater Noster Prose Dialogue, Old English: 252
Patrick, saint: 65, 87
Paulinus of Nola, saint: 117
Paulinus, Bishop, saint: 35, 62
Peada, King of Mercia: 61
pearls: 149–55, 161
Penda, King of Mercia: 93
Peterborough, Cambridgeshire: 42
pilgrimage: 40, 45 n. 44, 96, 113–14, 117, 119, 194–95
Plegmund, Archbishop of Canterbury: 18, 241
ports: 53, 195, 231–32
prognostics: 90
proreta: 122–23
Pseudo-Isidore, *De ordinae creaturarum*: 177–79
Puffin Island (*Ynys Lannog*), Anglesey: 40

Radbod, King of Frisia: 74
rafts *see* vessels
rain: 67, 93, 168–69, 173–74, 176, 179–88, 255, 252, 255, 274
rainbow: 140, 185
Ramsey Abbey: 42, 120–23
Red Sea: 96, 110, 119, 172, 182, 200, 202
Repton, Derbyshire: 34, 63
Reykholt, Iceland: 100
Rhône, River: 94
Riddles, Old English: 114, 184, 267–80
Rígsþula, eddic poem: 90
Rogation Days: 121

Romsley, Worcestershire: 49
Rothbury Cross: 156
Ruin, The, Old English poem: 198, 268
Rune Poem, The, Old English poem: 223–24

Saewulf: 96–97
sagas, Icelandic: 100–01
sailors: 110, 115–18, 120, 122–23, 129, 132, 137, 251
Salisbury Plain, Wiltshire: 33
San Vitale, Ravenna: 147
Sancton, East Yorkshire: 73
Sant'Apollinare in Classe, Ravenna: 152–53
sea-bed *see* abyss
sea-monsters: 88, 91, 101, 127–41, 224
  see also Leviathan
Seafarer, The, Old English poem: 114–17
Seaxburh, saint: 41
Sedulius, *Pascale carmen*: 131–32
Seiriol, saint: 40
Selsey Minster: 41
Severn, River: 47–48, 68 n. 39, 76–77
Sheppey, Isle of: 41, 45
ships *see* vessels
shipwreck: 92, 110, 118, 231
Sigeberht, King of Essex: 61
Sigeburga, Abbess of Thanet: 53
Silbury Hill, Wiltshire: 33
singing *see* music
Skellig Michael (Great Skellig, *Sceilig Mhichíl*), Ireland: 36
Snape, Suffolk: 68
Snorri Sturluson: 100
snow: 254–56, 274–75
Solomon and Saturn I, Old English poem: 251–52
Solomon and Saturn II, Old English poem: 91, 250–56, 258–64
Spong Hill, Norfolk: 73
springs: 15, 24, 33–35, 49–51, 61, 69, 75, 119, 243–45, 247–48, 261, 279
  see also wells

St Frideswide's Minster: 46
St Michael's Mount: 40
Stamford Bridge, Battle of: 95
Stephen of Ripon, *Vita Wilfridi*: 118–20
Stour, River: 46
Stratford, Warwickshire: 51
Sulpicius Severus, *Vita Martini*: 222
Sutton Hoo, Norfolk: 68, 70, 73
Swale, River: 35, 62, 69, 77, 161
swimming: 86–102, 253, 270
Swithhelm, King of Essex: 61

Tancred, saint: 42
Tees, River: 73
Teviot, River: 63
Thames, River: 20, 46, 51, 53, 69–70, 72, 77, 95
Thanet, Isle of: 41, 94
Theodore of Tarsus, Archbishop of Canterbury, saint: 168, 179
Theodulf of Orléans: 158
Theophilus of Antioch: 181
Thorney Abbey: 41
Tone, River: 45
Torthred, saint: 42
Tova, saint: 42
Trent, River: 20, 35, 62, 69, 77
Tudwal, saint: 40
Twyning, Gloucestershire: 47
Tyne, River: 68

Vergil, *Aeneid*: 212–13, 216–18
vessels, water-borne: 38, 64, 67–68, 71, 73, 89–92, 96–97, 101, 109, 112–14, 117–23, 127–29, 132–34, 137–39, 142, 192–96, 211, 231
 Church as ship: 114, 117–18
Vita Antonii: 212, 221

Vita Ceolfridi see Ceolfrith
Vita sancti Guthlaci see Guthlac, saint

Wanderer, The, Old English poem: 114, 198
washing: 67, 87
water, holy: 35, 49, 77, 229, 232–33
waters, living: 61, 65 n. 19, 192, 206–07, 226, 241–48, 250, 279
watersheds: 72–74
Wear, River: 73, 111–14
Wearmouth: 73, 112–13
Wells, Somerset: 49–50
wells: 18, 35, 49–50, 65, 75, 206, 245, 26–62, 279
 see also springs
Werburgh, saint: 41
Whale, The, Old English poem: 191
Wilfrid, Bishop of Ripon, saint: 87, 93, 118–20
Winchester, Hampshire: 35, 49, 51, 63.
 Old Minster: 35, 49
Winwæd, Battle of the: 93
wisdom: 211, 219, 225–27, 234, 243, 246, 248, 252, 261
Witham, River: 24 n. 34, 48, 76
Withington, Gloucestershire: 47, 50
Worcester, Worcestershire: 46, 53
Wulf and Eadwacer, Old English poem: 114–15
Wulfram of Sens, saint: 74

Yeavering, Northumberland: 62, 69–70
Yeo, River: 45
York: 62, 75

Studies in the Early Middle Ages

All volumes in this series are evaluated by an Editorial Board, strictly on academic grounds, based on reports prepared by referees who have been commissioned by virtue of their specialism in the appropriate field. The Board ensures that the screening is done independently and without conflicts of interest. The definitive texts supplied by authors are also subject to review by the Board before being approved for publication. Further, the volumes are copyedited to conform to the publisher's stylebook and to the best international academic standards in the field.

Titles in Series

Cultures in Contact: Scandinavian Settlement in England in the Ninth and Tenth Centuries, ed. by Dawn M. Hadley and Julian D. Richards (2000)

On Barbarian Identity: Critical Approaches to Ethnicity in the Early Middle Ages, ed. by Andrew Gillett (2002)

Matthew Townend, *Language and History in Viking Age England: Linguistic Relations between Speakers of Old Norse and Old English* (2002)

Contact, Continuity, and Collapse: The Norse Colonization of the North Atlantic, ed. by James H. Barrett (2003)

Court Culture in the Early Middle Ages: The Proceedings of the First Alcuin Conference, ed. by Catherine Cubitt (2003)

Political Assemblies in the Earlier Middle Ages, ed. by P. S. Barnwell and Marco Mostert (2003)

Wulfstan, Archbishop of York: The Proceedings of the Second Alcuin Conference, ed. by Matthew Townend (2004)

Borders, Barriers, and Ethnogenesis: Frontiers in Late Antiquity and the Middle Ages, ed. by Florin Curta (2006)

John D. Niles, *Old English Enigmatic Poems and the Play of the Texts* (2006)

Teaching and Learning in Northern Europe, 1000–1200, ed. by Sally N. Vaughn and Jay Rubenstein (2006)

Narrative and History in the Early Medieval West, ed. by Elizabeth M. Tyler and Ross Balzaretti (2006)

People and Space in the Middle Ages, 300–1300, ed. by Wendy Davies, Guy Halsall, and Andrew Reynolds (2006)

John D. Niles, *Old English Heroic Poems and the Social Life of Texts* (2007)

The Crisis of the Oikoumene: The Three Chapters and the Failed Quest for Unity in the Sixth-Century Mediterranean, ed. by Celia Chazelle and Catherine Cubitt (2007)

Text, Image, Interpretation: Studies in Anglo-Saxon Literature and its Insular Context in Honour of Éamonn Ó Carragáin, ed. by Alastair Minnis and Jane Roberts (2007)

The Old English Homily: Precedent, Practice, and Appropriation, ed. by Aaron J. Kleist (2007)

James T. Palmer, *Anglo-Saxons in a Frankish World, 690–900* (2009)

Challenging the Boundaries of Medieval History: The Legacy of Timothy Reuter, ed. by Patricia Skinner (2009)

Peter Verbist, *Duelling with the Past: Medieval Authors and the Problem of the Christian Era, c. 990–1135* (2010)

Reading the Anglo-Saxon Chronicle: Language, Literature, History, ed. by Alice Jorgensen (2010)

England and the Continent in the Tenth Century: Studies in Honour of Wilhelm Levison (1876–1947), ed. by David Rollason, Conrad Leyser, and Hannah Williams (2010)

Early Medieval Northumbria: Kingdoms and Communities, AD 450–1100, ed. by David Petts and Sam Turner (2011)

Conceptualizing Multilingualism in Medieval England, c. 800–c. 1250, ed. by Elizabeth M. Tyler (2011)

Neglected Barbarians, ed. by Florin Curta (2011)

The Genesis of Books: Studies in the Scribal Culture of Medieval England in Honour of A. N Doane, ed. by Matthew T. Hussey and John D. Niles (2012)

Giselle de Nie, *Poetics of Wonder: Testimonies of the New Christian Miracles in the Late Antique Latin World* (2012)

Lilla Kopár, *Gods and Settlers: The Iconography of Norse Mythology in Anglo-Scandinavian Sculpture* (2012)

R. W. Burgess and Michael Kulikowski, *Mosaics of Time: The Latin Chronicle Traditions from the First Century BC to the Sixth Century AD, vol. I: A Historical Introduction to the Chronicle Genre from its Origins to the High Middle Ages* (2013)

Sacred Sites and Holy Places: Exploring the Sacralization of Landscape through Space and Time, ed. by Sæbjørg Walaker Nordeide and Stefan Brink (2013)

Christine Maddern, *Raising the Dead: Early Medieval Name Stones in Northumbria* (2013)

Landscapes of Defence in Early Medieval Europe, ed. by John Baker, Stuart Brookes, and Andrew Reynolds (2013)

Sara M. Pons-Sanz, *The Lexical Effects of Anglo-Scandinavian Linguistic Contact on Old English* (2013)

Society and Culture in Medieval Rouen, 911–1300, ed. by Leonie V. Hicks and Elma Brenner (2013)

Shane McLeod, *The Beginning of Scandinavian Settlement in England: The Viking 'Great Army' and Early Settlers, c. 865–900* (2014)

England and Rome in the Early Middle Ages: Pilgrimage, Art, and Politics, ed. by Francesca Tinti (2014)

Luigi Andrea Berto, *In Search of the First Venetians: Prosopography of Early Medieval Venice* (2014)

Clare Pilsworth, *Healthcare in Early Medieval Northern Italy: More to Life than Leeches* (2014)

Textus Roffensis: Law, Language, and Libraries in Early Medieval England, ed. by Bruce O'Brien and Barbara Bombi (2015)

Churches and Social Power in Early Medieval Europe: Integrating Archaeological and Historical Approaches, ed. by José C. Sánchez-Pardo and Michael G. Shapland (2015)

Niamh Wycherley, *The Cult of Relics in Early Medieval Ireland* (2015)

Ross Balzaretti, *The Lands of Saint Ambrose: Monks and Society in Early Medieval Milan* (2019)

Michael D. J. Bintley, *Settlements and Strongholds in Early Medieval England: Texts, Landscapes, and Material Culture* (2020)

Cities, Saints, and Communities in Early Medieval Europe: Essays in Honour of Alan Thacker, ed. by Scott DeGregorio and Paul Kershaw (2020)